Fractal and Fractional in Geomaterials

Fractal and Fractional in Geomaterials

Editors

Yifei Sun
Cheng Chen
Meisam Goudarzy

MDPI • Basel • Beijing • Wuhan • Barcelona • Belgrade • Manchester • Tokyo • Cluj • Tianjin

Editors
Yifei Sun
Ruhr-Universität Bochum
Germany

Cheng Chen
Wuhan University of Technology
China

Meisam Goudarzy
Ruhr-Universität Bochum
Germany

Editorial Office
MDPI
St. Alban-Anlage 66
4052 Basel, Switzerland

This is a reprint of articles from the Special Issue published online in the open access journal *Fractal and Fractional* (ISSN 2504-3110) (available at: https://www.mdpi.com/journal/fractalfract/special_issues/FFG).

For citation purposes, cite each article independently as indicated on the article page online and as indicated below:

LastName, A.A.; LastName, B.B.; LastName, C.C. Article Title. *Journal Name* **Year**, *Volume Number*, Page Range.

ISBN 978-3-0365-7054-9 (Hbk)
ISBN 978-3-0365-7055-6 (PDF)

Contents

About the Editors

Yifei Sun

Yifei Sun is a Professor of Civil Engineering at the College of Civil and Transportation Engineering. His current research focuses on the development of advanced numerical and analytical methodologies for geomaterials. He has received research funding from several national and international agencies, e.g., the NSFC Fund for Excellent Young Scientists (overseas), the Alexander von Humboldt Fellowship, and the Ulam Program.

Cheng Chen

Cheng Chen is an Associate Professor of Civil Engineering at the School of Civil Engineering and Architecture, Wuhan University of Technology. His current research focuses on the micromechanics of granular materials and railway engineering. He has received several projects, including the NSFC funding for young scientists.

Meisam Goudarzy

Meisam Goudarzy is a senior researcher at the Faculty of Civil and Environmental Engineering, Ruhr-Universität Bochum, Germany. His research focuses on the mechanical behavior of soils under monotonic, cyclic and dynamic loadings, from micro to macro scales. He has received several prestigious projects from the DFG.

fractal and fractional

MDPI

Editorial

Editorial for Special Issue "Fractal and Fractional in Geomaterials"

Yifei Sun [1,*], Cheng Chen [2] and Meisam Goudarzy [3]

1 Key Laboratory of Ministry of Education for Geomechanics and Embankment Engineering, Hohai University, Nanjing 210024, China
2 School of Civil Engineering and Architecture, Wuhan University of Technology, Wuhan 430062, China
3 Faculty of Civil and Environmental Engineering, Ruhr-Universität, 44801 Bochum, Germany
* Correspondence: yifei.sun@hhu.edu.cn

Citation: Sun, Y.; Chen, C.; Goudarzy, M. Editorial for Special Issue "Fractal and Fractional in Geomaterials". *Fractal Fract.* **2023**, *7*, 55. https://doi.org/10.3390/fractalfract7010055

Received: 23 December 2022
Accepted: 29 December 2022
Published: 1 January 2023

Geomaterials, such as clay, sand, rockfill and ballast, etc., in the field are usually exposed to complex physical or mechanical conditions, where anomalous behaviors, e.g., state-dependent non-associated flow, non-Fickian diffusion, or non-Darcy seepage, usually take place. In recent years, fractal laws and fractional mechanics have been developed as robust tools for solving such complex or anomalous behavior of different materials.

This Special Issue contains 16 published papers. In [1], a set of nuclear magnetic resonance tests are carried out on different granular materials, including calcareous and quartz sands, where it is found that the grain size significantly affected their pore size distributions (PSDs). As the grain size increases, the heterogeneity and fractal dimension of PSD increase remarkably.

A simplified unified model of the relative permeability coefficient of unsaturated soil with fractal dimension is developed by Tao et al. in [2]. A strong correlation of the model parameters with the fractal dimension of soil is suggested. Through model comparison, it is found that the predicted results are consistent with the measured values.

In [3], a simplified model for predicting the relationship between the saturated permeability coefficient and air-entry value is established, by combining the Tao-Kong model and the fractal model of the soil–water characteristic curve.

To consider the solute transport in highly heterogeneous media, a nonlinear-coupled 3D fractional hydro-mechanical model accounting for anomalous diffusion and advection–dispersion is proposed in [4]. A finite difference approximation is then proposed to solve the problem in 1D domains, where two scenarios were examined.

An elastoplastic interface model using the fractional plastic flow rule is developed by Xu et al. in [5], where the dependence of the cyclic mobilisation of the soil–structure interface on its material state and fabric is captured.

In [6], the fractal theory and discrete element method are introduced to quantify the fractal dimension of a particle size distribution and understand the scale effect in soil–rock mixtures.

The effect of the fractal distribution of particle size on the critical state characteristics of calcareous sand is investigated through laboratory tests and theoretical analysis by Shen et al. in [7]. The critical state lines of calcareous sand in the q–p' plane are unique, regardless of the fractal dimension, whereas those in the e–$(p')^{\alpha}$ plane rotate anticlockwise as the fractal dimension increases. Modified constitutive relations modified with the fractal dimension are also provided.

In [8], a series of nuclear magnetic resonance tests on frozen soils are carried out to study the effect of freezing on soil's micro-pore structure and fractal characteristics. A larger pressure during freezing–thawing or a higher freezing temperature can result in a lower fractal dimension of the soil's structure.

A series of physical model tests are conducted by Yuan et al. [9] to study the effect of groundwater depth on pile–soil mechanical properties cyclic loads. Combined with fractal theory, the relationships between the pile top displacement, cyclic stiffness, and cyclic steps are evaluated.

In [10], the effect of non-plastic fines and stress anisotropy on the dynamic shear modulus of sand with different contents of non-plastic fines are investigated by using macro- and micro-laboratory tests. A unified expression for the shear modulus of binary mixtures is proposed.

In [11], a fractal relation between the hydraulic conductivity and fractal dimension is derived analytically, based on the capillary model of porous soil. It is found that an increased fractal dimension will increase the connectivity, increase the hydraulic gradient, and reduce the hydraulic conductivity.

The permeability and pore structure of the concrete–rock interfacial transition zone are studied using multi-scale experiments, by Yue et al. [12], including NMR, SEM-EDS, and XRD. A fractal permeability model is proposed.

In [13], to study the effects of the relative density and grading on the particle breakage of granular materials, a series of monotonic drained triaxial tests are performed on granular materials with different initial gradings and relative densities. Due to particle breakage, the grading curves of granular materials after triaxial tests can be simulated by a power-law function with a fractal dimension.

A series of discrete element simulation of triaxial tests on breakable particles within a flexible membrane are carried out by Chen et al. [14]. A strong correlation between the particle breakage ratio and fractal dimension is found.

In [15], a series of SEM tests are carried out on overconsolidated soil to study the effects of consolidation pressure and overconsolidation ratio, etc., on the micro-pore structure and fractal characteristics of soil. The fractal dimension of soil can well represent the complex characteristics of its microstructure.

The authors develop a fractal-entropy model to predict the hydraulic conductivity of granular soils [16], where the pore size distribution (PSD) is discretized based on fractal entropy, while the effective diameter of soil is computed using the grading entropy theory.

Acknowledgments: The Guest Editors would like to express their appreciation to all of the authors of this Special Issue for their contribution. The kind help provided by the external reviewers as well as the Editorial Office is also greatly appreciated.

Conflicts of Interest: The author declares no conflict of interest.

References

1. He, S.-H.; Ding, Z.; Hu, H.-B.; Gao, M. Effect of Grain Size on Microscopic Pore Structure and Fractal Characteristics of Carbonate-Based Sand and Silicate-Based Sand. *Fractal Fract.* **2021**, *5*, 152. [CrossRef]
2. Tao, G.; Wang, Q.; Chen, Q.; Nimbalkar, S.; Peng, Y.; Dong, H. Simple Graphical Prediction of Relative Permeability of Unsaturated Soils under Deformations. *Fractal Fract.* **2021**, *5*, 153. [CrossRef]
3. Tao, G.; Wu, Z.; Li, W.; Li, Y.; Dong, H. Simplified Relation Model of Soil Saturation Permeability Coefficient and Air-Entry Value and Its Application. *Fractal Fract.* **2021**, *5*, 180. [CrossRef]
4. Salomoni, V.; De Marchi, N. Numerical Solutions of Space-Fractional Advection-Diffusion-Reaction Equations. *Fractal Fract.* **2022**, *6*, 21. [CrossRef]
5. Xu, J.; Shen, Y.; Sun, Y. Cyclic Mobilisation of Soil-Structure Interface in the Framework of Fractional Plasticity. *Fractal Fract.* **2022**, *6*, 76. [CrossRef]
6. Fu, X.; Ding, H.; Sheng, Q.; Zhang, Z.; Yin, D.; Chen, F. Fractal Analysis of Particle Distribution and Scale Effect in a Soil-Rock Mixture. *Fractal Fract.* **2022**, *6*, 120. [CrossRef]
7. Shen, X.; Shen, Y.; Xu, J.; Liu, H. Influence of the Fractal Distribution of Particle Size on the Critical State Characteristics of Calcareous Sand. *Fractal Fract.* **2022**, *6*, 165. [CrossRef]
8. Kong, B.; He, S.-H.; Tao, Y.; Xia, J. Pore Structure and Fractal Characteristics of Frozen-Thawed Soft Soil. *Fractal Fract.* **2022**, *6*, 183. [CrossRef]
9. Yuan, B.; Li, Z.; Chen, W.; Zhao, J.; Lv, J.; Song, J.; Cao, X. Influence of Groundwater Depth on Pile-Soil Mechanical Properties and Fractal Characteristics under Cyclic Loading. *Fractal Fract.* **2022**, *6*, 198. [CrossRef]
10. Goudarzy, M.; Sarkar, D. Macro-and Micromechanical Assessment of the Influence of Non-Plastic Fines and Stress Anisotropy on the Dynamic Shear Modulus of Binary Mixtures. *Fractal Fract.* **2022**, *6*, 205. [CrossRef]

11. Zhao, X.; Yang, B.; Yuan, S.; Shen, Z.; Feng, D. Seepage-Fractal Model of Embankment Soil and Its Application. *Fractal Fract.* **2022**, *6*, 277. [CrossRef]
12. Yue, J.; Sheng, J.; Wang, H.; Hu, Y.; Zhang, K.; Luo, Y.; Zhou, Q.; Zhan, M. Investigation on Pore Structure and Permeability of Concrete-Rock Interfacial Transition Zones Based on Fractal Theory. *Fractal Fract.* **2022**, *6*, 329. [CrossRef]
13. Yang, G.; Chen, Z.; Sun, Y.; Jiang, Y. Effects of Relative Density and Grading on the Particle Breakage and Fractal Dimension of Granular Materials. *Fractal Fract.* **2022**, *6*, 347. [CrossRef]
14. Chen, C.; Zhang, X.; Sun, Y.; Zhang, L.; Rui, R.; Wang, Z. Discrete Element Modelling of Fractal Behavior of Particle Size Distribution and Breakage of Ballast under Monotonic Loading. *Fractal Fract.* **2022**, *6*, 382. [CrossRef]
15. Kong, B.; Dai, C.-X.; Hu, H.; Xia, J.; He, S.-H. The Fractal Characteristics of Soft Soil under Cyclic Loading Based on SEM. *Fractal Fract.* **2022**, *6*, 423. [CrossRef]
16. Zhang, G.; Wang, H.; Israr, J.; Ma, W.; Yang, Y.; Ren, K. A Fractal Entropy-Based Effective Particle Model Used to Deduce Hydraulic Conductivity of Granular Soils. *Fractal Fract.* **2022**, *6*, 474. [CrossRef]

fractal and fractional

Article

Numerical Solutions of Space-Fractional Advection–Diffusion–Reaction Equations

Valentina Anna Lia Salomoni [1,2,*] and Nico De Marchi [1]

1 Department of Management and Engineering, University of Padua, Stradella S. Nicola 3, 36100 Vicenza, Italy; nico.demarchi@unipd.it
2 Department of Civil, Environmental and Architectural Engineering, University of Padua, Via F. Marzolo 9, 35131 Padua, Italy
* Correspondence: valentina.salomoni@unipd.it; Tel.: +39-049-827-5590

Abstract: Background: solute transport in highly heterogeneous media and even neutron diffusion in nuclear environments are among the numerous applications of fractional differential equations (FDEs), being demonstrated by field experiments that solute concentration profiles exhibit anomalous non-Fickian growth rates and so-called "heavy tails". Methods: a nonlinear-coupled 3D fractional hydro-mechanical model accounting for anomalous diffusion (FD) and advection–dispersion (FAD) for solute flux is described, accounting for a Riesz derivative treated through the Grünwald–Letnikow definition. Results: a long-tailed solute contaminant distribution is displayed due to the variation of flow velocity in both time and distance. Conclusions: a finite difference approximation is proposed to solve the problem in 1D domains, and subsequently, two scenarios are considered for numerical computations.

Keywords: fractional derivative; fractional diffusion equation; fractional advection–dispersion equation; solute transport; porous media

Citation: Salomoni, V.A.L.; De Marchi, N. Numerical Solutions of Space-Fractional Advection–Diffusion–Reaction Equations. *Fractal Fract.* **2022**, *6*, 21. https://doi.org/10.3390/fractalfract6010021

Academic Editor: Wojciech Sumelka

Received: 2 December 2021
Accepted: 29 December 2021
Published: 31 December 2021

1. Introduction

Fractional and tempered fractional [1] differential equations (FDEs) have proved to be strong tools in the modelling of many physical phenomena, including acoustics and thermal systems and rheology and modelling of materials, leading to significant developments of analytical and numerical methods for solving fractional ordinary and partial differential equations in recent times. They comprise, e.g., Laplace–Fourier transform techniques and Green function approach [2], Lie symmetries theory and group analysis [3–6], Adomian decomposition [7,8], and homotopy perturbation methods [9], as well as finite element [10,11] and finite volume schemes [12,13], finite difference methods [14], and spectral ones [15–18].

Among the numerous applications in the various fields of science and engineering, we recall solute transport in highly heterogeneous media and even neutron diffusion in nuclear environments [19]. In fact, field experiments demonstrate that solute concentration profiles exhibit anomalous non-Fickian growth rates and so-called "heavy tails", i.e., effects which cannot be predicted via the standard mass transport equation [20–23] but via fractional-order differential equations that may be viewed as long-time and long-space limits of a continuous time random walk (CTRW) [24]. Correspondingly, the Fickian advection–dispersion equation (ADE), unable to reflect the long tail dispersion process, is converted into a fractional one (FADE). The FADE approach appears to have the potential to predict non-Fickian dispersion processes, but its wide application is delayed by the difficulty to both obtain analytical solutions, especially when reaction terms are incorporated [25,26] or kernel non-singularities occur [27], and increased computational costs. Again, when considering water flow in low-permeability porous media, a nonlinear relationship between water flux and hydraulic gradient should be considered, thereby indicating a non-Darcian flow. Hence, for developing a more appropriate description able to include a memory effect, the existing relationships can be modified in a fractional fashion [24,28].

In case of time-fractional neutron diffusion models with delayed neutrons [19,29,30], non-local effects are established and sub-diffusive phenomena caught, coming from the heterogeneity of nuclear reactors. Anyway, analytical solutions for this problem are generally not available and only a few efficient numerical techniques have been developed in the literature to approximate the solution of even a 2D fractional [10,31] or non-fractional [32] diffusion models, even in the presence of a reaction term [26,33]. Alternatively, non-local problems accounting for long-range interactions can be treated via non-local integral models [34] and even combined time-fractional and space-nonlocal strategies [35].

Starting from an already-available 3D hydro-mechanical model in finite strains based on the modified mixture theory [36], an upgraded fractional version is proposed here, accounting for a fractional constitutive equation for both pore fluid and solute transport. Correspondingly, the model is suitable for performing anomalous diffusion (FD) and advection–dispersion–reaction (FADR) analyses; however, more importantly, the present formulation innovatively includes the contribution of deformation non-linearities for the solid skeleton, as well as a strain-dependent permeability tensor. Additionally, a modified finite difference algorithm is proposed to solve the problem in 1D domains, adopting the fractional-centered derivative scheme combined with the β-method, with the advective term treated via a Lax–Wendroff scheme. When accounting for advection–dispersion phenomena, including overpressure effects, the numerical algorithm was enriched by the split operator method [24,25]. Hence, the advection step was solved via an explicit second-order midpoint method, together with the β-method for the fractional diffusion–dispersion step and the trapezoidal rule for the fluid excess step. The algorithm additionally shifts into a backward implicit one in case of long-term analyses. Literature results are used to validate the model, and are subsequently applied for simulating radionuclide transport processes within cemented materials.

2. Materials and Methods

Fractional Constitutive Equations for Pore Fluid and Solute Transport

Darcy's law is used to describe the flow of fluids in porous media within the local fluid content continuity equation in isothermal and fully saturated conditions, i.e., [36].

$$\nabla \cdot \mathbf{v} - \frac{1}{\rho_w} \nabla \cdot \left(\rho_w \frac{\mathbf{k}}{\mu} \cdot (\nabla p - \rho_w \mathbf{g}) \right) = 0. \tag{1}$$

The porous medium is treated as inhomogeneous thanks to the adoption of fractional-order density gradients so to recover, at least at the phenomenological level, nonlocal (long-range) diffusion effects [37–39]. Time/space-fractional derivatives are proposed to describe anomalous dispersion, induced by fluid motion stacked around porous media inhomogeneities, e.g., fractures. This phenomenon is reflected as non-Gaussian particle dispersion with heavy tails [40].

Correspondingly, Equation (1) becomes

$$\rho_w \nabla \cdot \mathbf{v} - \nabla \cdot \left(\rho_w \frac{\mathbf{k}}{\mu} \cdot \left(\nabla^{\alpha-1} p - \rho_w \mathbf{g} \right) \right) = 0, \tag{2}$$

choosing the Grünwald definition for fractional derivative [25] which appears convenient for numerical solutions [41], so that the value of a fractional differential operator acting on the generic function $\Psi(x, t)$ is an infinite series, i.e.,

$$\frac{\partial^\alpha \Psi(x,t)}{\partial x^\alpha} = \lim_{N \to \infty} \frac{1}{h^\alpha \Gamma(-\alpha)} \sum_{j=0}^{N-1} \frac{\Gamma(j-\alpha)}{\Gamma(j+1)} \Psi(x + \frac{\alpha}{2}h - jh, t), \tag{3}$$

in which $h = \Delta x = x/N$, N is a positive integer.

It is to be noted that the fractor, α, physically reflects the heterogeneity of the soil medium in which solute is transported, and, for isotropic media, $\alpha = 2$, i.e., the more heterogeneous the medium, the smaller than the integer constant of 2. Its decrease reproduces an increase in the resistance of the medium to solute dispersion and transport, leading to a delay in the phenomenon.

The combined solute transport within the non-linear deformable matrix-fluid system can be written as [24]

$$(1-\phi)J\dot{C}(\rho_s K_d - 1) + JC\nabla\cdot\mathbf{v} = -\nabla\cdot\left[\phi\rho_w C\left(\frac{\mathbf{k}}{\mu}\cdot\left(\nabla^{\alpha-1}p - \rho_w\mathbf{g}\right)\right)\right] + \frac{\phi\mathbf{D}}{J}\cdot\nabla^{\alpha}C, \tag{4}$$

by having extended the approaches [42,43] to 3D domains and $J = \det(\mathbf{F}) > 0$, with $\mathbf{F}(X, t)$ being the deformation gradient of the solid skeleton.

Equation (4) neglects the effect of sorption (being the medium in saturated conditions), so that the concentration of solute in the solid phase can be assumed to be linearly dependent on the concentration of the solute in the pore fluid via the contaminant partitioning coefficient.

The system of FADE (2) (4) becomes partly uncoupled due to the independence of the transport equation for pore fluid on the solute concentration, so that the pressure values can be treated as known data within Equation (4). Additionally, having already studied the contribution of anomalous water diffusion elsewhere [24], it is assumed to associate the fractional feature to the contaminant only. Hence, by introducing Equation (2) into Equation (4) and neglecting spatial non-linearities for fluid properties, the 1D non-linear saturated model is governed by

$$(1-\phi)J(\rho_s K_d - 1)\dot{C} + \left(\phi\rho_w\frac{k}{\mu}\frac{\partial p}{\partial x}\right)\frac{\partial C}{\partial x} + (J + \phi\rho_w)\frac{k}{\mu}\frac{\partial^2 p}{\partial x^2}C - \frac{\phi D}{J}\cdot\nabla^{\alpha}C = 0. \tag{5}$$

Similar observations can be carried out for the jacobian, which introduces an additional coupling with the mechanical part; in fact, recalling that $\vartheta = \ln J$, the jacobian results directly related to pore pressure through [24]

$$\vartheta \approx \frac{p - p_0 - \sigma_0}{\kappa + \frac{4}{3}G}, \tag{6}$$

so even itself can be treated as a known term (Expression (6) is valid under the assumptions of negligible self-weight and isothermal conditions).

Equation (5) involves effectively three processes of solute transport: the second term denotes advection caused by fluid flow; the third term reflects the transfer of solute resulting from pore fluid excess; and the fractional-order derivative term represents the contribution of dispersion in a heterogeneous medium. After rearrangement, we obtain

$$\frac{\partial\overline{C}}{\partial t} + \overline{u}\frac{\partial\overline{C}}{\partial x} + \overline{YC} - \overline{K}\cdot\nabla^{\alpha}\overline{C} = 0, \tag{7}$$

with

$$\overline{u} = \frac{\phi\rho_w k}{(1-\phi)J(\rho_s K_d - 1)\mu}\frac{\partial p}{\partial x}, \ \overline{Y} = \frac{\overline{Q}_f + (J + \phi\rho_w)\frac{k}{\mu}\frac{\partial^2 p}{\partial x^2}}{(1-\phi)J(\rho_s K_d - 1)}, \ \overline{K} = \frac{\phi D}{J^2(1-\phi)(\rho_s K_d - 1)}. \tag{8}$$

It is to be noticed that Equation (7) has general validity and, considering expression (8), shows variable coefficients, leading to a nonlinear structure; for simplicity, in the following we assume linear-type pore pressure distributions, permeability weakly dependent on deformation, together with small volume changes. Such an assumption has helped to validate the model against available results, as reported below.

Solving the FADE (7) for the transport rate $\overline{C}$ [44,45], including solute transport by pore fluid excess $\overline{YC}$, requires the adoption of a proper numerical strategy; as stated previously, it is here assumed to refer to the Grünwald definition of the fractional derivative, Equation (3), particularly referring to the shifted Grünwald–Letnikow formula [41]. Correspondingly, the final iterative system now becomes

$$(\mathbf{I}+\beta\mathbf{A}-\beta\mathbf{B})C^{t+1}=[\mathbf{I}-(1-\beta)\mathbf{A}+(1-\beta)\mathbf{B}]C^{t}+(\mathbf{F}^{t}+\beta\Delta\mathbf{F})\Delta t, \tag{9}$$

with

$$\mathbf{B}=\frac{\overline{K}}{2}\frac{\Delta t}{\Delta x^{\alpha}}\left[(1+\gamma)\mathbf{L}+(1-\gamma)\mathbf{L}^{T}\right] \quad \mathbf{L}=\begin{bmatrix} p_1 & p_0 & 0 & \cdots & 0\\ p_2 & p_1 & p_0 & \cdots & 0\\ p_3 & p_2 & p_1 & \cdots & 0\\ \vdots & \vdots & \vdots & \ddots & \vdots\\ p_N & p_{N-1} & p_{N-2} & \cdots & p_0\\ 0 & p_N & p_{N-1} & \cdots & p_1 \end{bmatrix} \tag{10}$$
$$p_N=-\frac{\Gamma(N-\alpha)}{\Gamma(-\alpha)\Gamma(N+1)}$$

where γ is a parameter accounting for the distribution probability of the relative weight of solute particle, and $\mathbf{A}$ is a tridiagonal matrix related to the discretization of the advective term via the Lax–Wendroff scheme [46].

$$\mathbf{A}=\begin{bmatrix} v^2 & -\frac{v(v-1)}{2} & 0 & \cdots & 0\\ -\frac{v(v+1)}{2} & v^2 & -\frac{v(v-1)}{2} & \cdots & 0\\ 0 & -\frac{v(v+1)}{2} & v^2 & \cdots & 0\\ \vdots & \vdots & \vdots & \ddots & -\frac{v(v-1)}{2}\\ 0 & 0 & 0 & -\frac{v(v+1)}{2} & v^2 \end{bmatrix}, \quad v=V\frac{\Delta t}{\Delta x}, \tag{11}$$

where V is the average fluid velocity along x direction, substantially equivalent to the first term in Equation (8).

When accounting for pore fluid excess, the scheme proposed by [47] was followed, based on the split-operator method. In particular, it was assumed to solve the advection step via an explicit second-order midpoint method, the fractional diffusion/dispersion step via the aforementioned β-method, and the fluid excess step via the trapezoidal rule.

3. Results

Once validated, the fractional model against the numerical example of [48], taken as benchmark [24], suggests that a 1D situation of an advection–dispersion–reaction was first considered [47], driven by Equation (7), with the additional term $E(\overline{C}_s-\overline{C})$ reflecting the transfer of solute from an active source (generally included when modelling overland solute transport)

$$\frac{\partial\overline{C}}{\partial t}+\overline{u}\frac{\partial\overline{C}}{\partial x}+\overline{Y C}-\overline{K}\cdot\nabla^{\alpha}\overline{C}-E(\overline{C}_s-\overline{C})=0, \tag{12}$$

being

$$\begin{gathered}\overline{C}_s=C_0e^{-\overline{\mu}t}\\ \overline{C}_s(x,0)=C_0(x)\end{gathered} \tag{13}$$

As assumed by the data reported in Table 1, together with an initial condition $C_0 = 0.5$, Figure 1 depicts the spatial–temporal distribution of concentration in the case of a standard ADR (non-fractional) solution, whereas Figure 2 evidences the fractional effect, producing a greater and slightly delayed peak with decreasing fractor alpha (maximum 10%). This seems to confirm (as experienced in [47]) that the ADR model generally predicts a faster flushing process of solute as compared to the FADR one, considering that the fractional approach allows to account for the material heterogeneity.

Table 1. Material parameters for FADR analysis.

Parameter	Values	Unit [-]
a	−10	m
b	10	m
$\overline{K}$	10^{-4}	m^2/s
v	0.5	m/s
E	0.02	s^{-1}
$\overline{\mu}$	0.0001	s^{-1}
N	200	-
T	100	-
t_{end}	1000	s

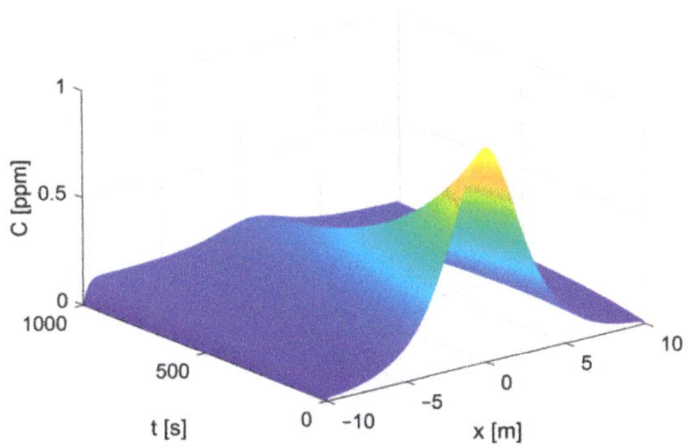

Figure 1. Concentration distribution (standard ADR model).

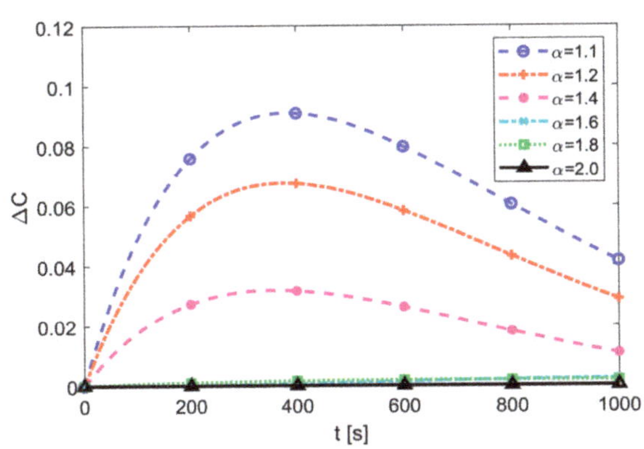

Figure 2. Time history of relative concentration variation at a reference point.

To additionally test the contribution of a fractional model, a log-term low-waste radionuclides transport process within cemented materials is considered. The transport rate through engineered barriers (compacted bentonite clay and cement) is mainly diffusion-controlled; however, due to the strong sorption behavior of the materials, the diffusion rate of the radionuclides is much slower than in pure water. Particularly, the case reported in [49] is reconstructed via a fractional approach, with material data of Table 2 and C(x,0) = 0. The backward-time implicit scheme is used with variable time steps, such that $t_{n+1} = 1.0125 \cdot \Delta t_n$.

Table 2. Material parameters for FD radionuclides transport.

Parameter	Values	Unit [-]
a	0	mm
b	150	mm
$\overline{K}$	10^{-4}	m^2/s
N	300	-
t_0	10^{-5}	y
t_{end}	993	y

The classical non-fractional solution is newly obtained in terms of the distribution of spatial–temporal radionuclides concentration (Figure 3). Interestingly, change in the transport mechanism can be noticed once fractors are lower than 1.6 and particularly when lower than 1.2 (Figure 4). A slower diffusion process is found again, suggesting that possibly thinner cement containers could be required for ensuring the same safety level and degraded zone as the standard AD model. The heterogeneous cement matrix may, in fact, form hierarchical storage-release zones of nuclides, leading to a delayed transport process of the contaminant (Figure 5).

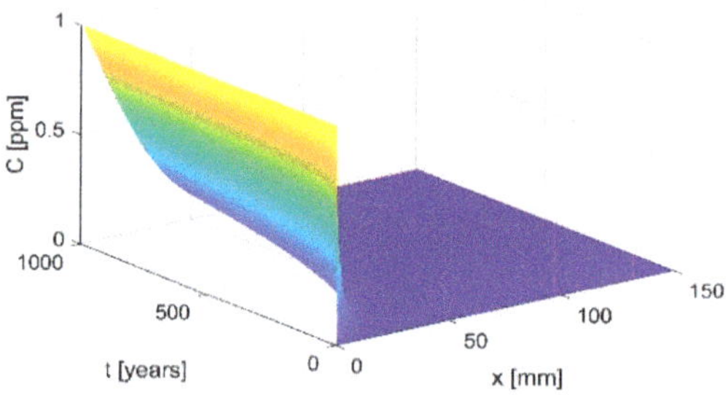

Figure 3. Radionuclide distribution (standard AD model).

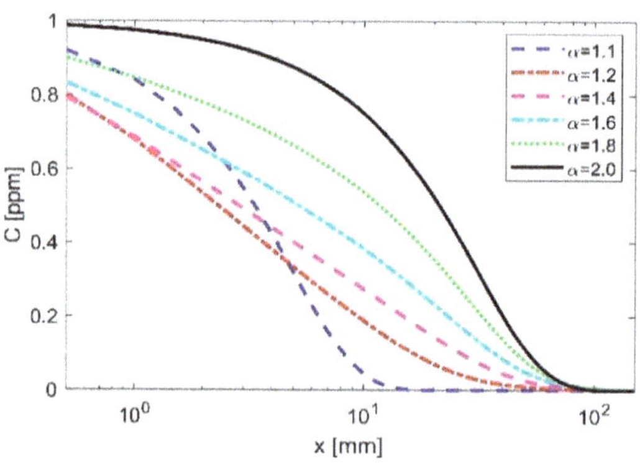

Figure 4. Spatial distribution of low-level radionuclides at a reference time-step.

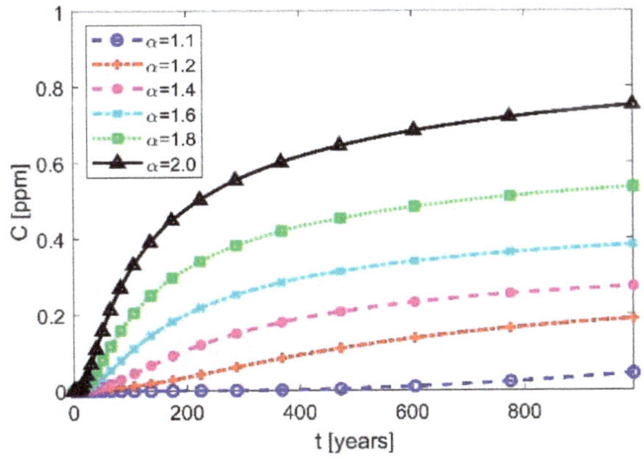

Figure 5. Time history of radionuclides at a reference node.

4. Discussion

A nonlinear coupled 3D fractional hydro-mechanical model accounting for anomalous diffusion (FD) and advection–dispersion (FAD) for solute flux is presented, accounting for a Riesz derivative treated through the Grünwald–Letnikow definition. The formulation innovatively includes the contribution of deformation non-linearities for the solid skeleton, as well as a strain-dependent permeability tensor. A finite difference approximation is built on purpose for modeling 1D domains and adopting the fractional-centered derivative scheme combined with the β-method, with the advective term treated via the Lax–Wendroff scheme. In particular, the algorithm shifts into a backward-time implicit one with variable time steps in case of long-term contaminants (radionuclides) transport. Under the assumptions of linear-type pore pressure distributions, permeability which is weakly dependent on deformation, together with small volume changes, means that the effects of the fractional dispersion term are analyzed.

A delayed release of solute pollutants or a long-tailed dispersion process caused by volume heterogeneities of the porous media was successfully described, evidencing a general gradual and slower solute contaminant transport. Additionally, when accounting for radionuclides transport within cemented materials and decreasing fractors, the model evidences a sort of solute-clog, with higher solute gradients and increasing solute levels. Correspondingly, it appears to be particularly suitable for reconstructing phenomena in which the kinetics of diffusion are slower than the kinetics of reaction and layers of very low solubility form and do not dissolve on observable time scales.

Author Contributions: Conceptualization, methodology, writing—original draft preparation and writing—review and editing, V.A.L.S.; software, data curation and investigation, N.D.M. All authors have read and agreed to the published version of the manuscript.

Funding: Financial support from the Italian Ministry of Education, University and Research (MUR) in the framework of the Project PRIN2017 #2017HFPKZY is gratefully acknowledged.

Institutional Review Board Statement: Not applicable.

Informed Consent Statement: Not applicable.

Data Availability Statement: The data (MatLab code lines) used in this study are available on request from the corresponding author. The data are not publicly available due to privacy reasons.

Conflicts of Interest: The authors declare no conflict of interest. The funder had no role in the design of the study; in the collection, analyses, or interpretation of data; in the writing of the manuscript, or in the decision to publish the results.

Nomenclature

v	solid velocity vector
$\rho_{w,s}$	Fluid/solid density
g	Gravity acceleration vector
k	Saturated permeability tensor
μ	Dynamic fluid viscosity
p	pore pressure
α	fractional differential order
$\Gamma(\cdot)$	gamma function
C	solute/contaminant concentration
ϕ	porosity
k_d	contaminant partitioning coefficient
D	hydrodynamic dispersion tensor
J	jacobian
ϑ	logarithmic strain
p_0, σ_0	initial values of pore pressure and axial stress
κ, G	bulk and shear modulus of porous material
E	transfer term
$\overline{C}_s$	solute concentration in the mixing soil zone

References

1. Sultana, F.; Singh, D.; Pandey, R.K.; Zeidan, D. Numerical schemes for a class of tempered fractional integro-differential equations. *Appl. Num. Math.* **2020**, *157*, 110–134. [CrossRef]
2. Miller, K.S.; Ross, B. *An Introduction to the Fractional Calculus and Fractional Differential Equations*; Wiley-Interscience: New York, NY, USA, 1993.
3. Prakash, P.; Sahadevan, R. Lie symmetry analysis and exact solution of certain fractional ordinary differential equations. *Nonlinear Dynam.* **2017**, *89*, 305–319. [CrossRef]
4. Jannelli, A.; Ruggieri, M.; Speciale, M.P. Numerical solutions of space fractional advection–diffusion equation, with nonlinear source term. *Appl. Numer. Math.* **2020**, *155*, 93–102. [CrossRef]
5. Bira, B.; Mandal, H.; Zeidan, D. Exact solution of the time fractional variant Boussinesq-Burgers equations. *Appl. Math.* **2021**, *66*, 437–449. [CrossRef]
6. Mandal, H.; Bira, B.; Zeidan, D. Power Series Solution of Time-Fractional Majda-Biello System Using Lie Group Analysis. In Proceedings of the International Conference on Fractional Differtiation and its Applications (ICFDA), Amman, Jordan, 16–18 July 2018. [CrossRef]
7. Daftardar-Geji, V.; Jafari, H. Adomian decomposition: A tool for solving a system of fractional differential equations. *J. Math. Anal. Appl.* **2005**, *301*, 508–518. [CrossRef]
8. Zeidan, D.; Chau, C.K.; Lu, T.T. On the characteristic Adomian decomposition method for the Riemann problem. *Math. Meth. Appl. Sci.* **2021**, *44*, 8097–8112. [CrossRef]
9. Momani, S.; Odibat, Z. Homotopy perturbation method for nonlinear partial differential equations of fractional order. *Phys. Lett. A* **2007**, *365*, 345–350. [CrossRef]
10. Zhang, N.; Deng, W.; Wu, Y. Finite difference/element method for a two-dimensional modified fractional diffusion equation. *Adv. Appl. Math. Mech.* **2012**, *4*, 496–518. [CrossRef]
11. Liu, Y.; Yu, Z.D.; Li, H.; Liu, F.W.; Wang, J.F. Time two-mesh algorithm combined with finite element method for time fractional water wave model. *Int. J. Heat Mass Transfer* **2018**, *120*, 1132–1145. [CrossRef]
12. Hejazi, H.; Moroney, T.; Liu, F. Stability and convergence of a finite volume method for the space fractional advection–dispersion equation. *J. Comput. Appl. Math.* **2014**, 684–697. [CrossRef]
13. Li, F.; Fu, H.; Liu, J. An efficient quadratic finite volume method for variable coefficient Riesz space-fractional diffusion equations. *Math. Methods Appl. Sci.* **2021**, *44*, 2934–2951. [CrossRef]
14. Zhou, Y.; Suzuki, J.L.; Zhang, C.; Zayernouri, M. Implicit-explicit time integration of nonlinear fractional differential equations. *Appl. Numer. Math.* **2020**, *156*, 555–583. [CrossRef]
15. Doha, E.H.; Bhrawy, A.H.; Baleanu, D.; Ezz-Eldien, S.S. Numerical schemes with high spatial accuracy for a variable-order anomalous subdiffusion equation. *Appl. Math. Comput.* **2013**, *219*, 8042–8056.
16. Zhao, T.; Mao, Z.; Karniadakis, G.E. Multi-domain spectral collocation method for variable-order nonlinear fractional differential equations. *Comput. Methods Appl. Mech. Engrg.* **2019**, *348*, 377–395. [CrossRef]
17. Zaky, M.A.; Hendy, A.S.; Macias-Diaz, J.E. Semi-implicit Galerkin–Legendre spectral schemes for nonlinear time-space fractional diffusion–reaction equations with smooth and nonsmooth solutions. *J. Sci. Comput.* **2020**, *82*, 13. [CrossRef]
18. Dwivedi, K.D.; Das, S.; Baleanu, D. Numerical solution of nonlinear space–time fractional-order advection–reaction–diffusion equation. *J. Comput. Nonlinear Dynam.* **2020**, *15*, 061005. [CrossRef]

19. Roul, P.; Rohil, V.; Espinosa-Paredes, G.; Obaidurrahman, K. Numerical simulation of two-dimensional fractional neutron diffusion model describing dynamical behaviour of sodium-cooled fast reactor. *Ann. Nuclear Energy* **2021**, *166*, 108709. [CrossRef]
20. Schumer, R.; Benson, D.A.; Meerschaert, M.M.; Baeumer, B. Fractal mobile/immobile transport. *Water Res. Res.* **2003**, *39*, 1296.
21. Zhang, Y.; Benson, D.A.; Reeves, D.M. Time and space non-localities underlying fractional-derivative models: Distinction and literature review of field applications. *Adv. Wat. Res.* **2009**, *32*, 561–581. [CrossRef]
22. Fomin, S.; Chugunov, V.; Hashida, T. The effect of non-Fickian diffusion for modelling the anomalous diffusion of contaminant from fracture into porous rock matrix with bordering alteration zone. *Transp. Por. Media* **2010**, *81*, 187–205. [CrossRef]
23. Fomin, S.A.; Chugunov, V.A.; Hashida, T. Non-Fickian mass transport in fractured porous media. *Adv. Wat. Res.* **2011**, *34*, 205–214. [CrossRef]
24. Salomoni, V.A.; De Marchi, N. A three-dimensional finite strain model of solute transport in saturated porous media with a fractional approach. *J. Eng. Sci.* 2021; *under review*.
25. Deng, Z.Q.; Singh, V.P.; Bengtsson, L. Numerical solution of fractional advection-dispersion equation. *J. Hydr. Engrg.* **2004**, *130*, 422–431. [CrossRef]
26. Kumar, S.; Zeidan, D. An efficient Mittag-Leffler kernel approach for time-fractional advection-reaction-diffusion equation. *Appl. Num. Math.* **2021**, *170*, 190–207. [CrossRef]
27. Yavuz, M.; Ozdemir, N. An Integral Transform Solution for Fractional Advection-Diffusion Problem. In Proceedings of the International Conference on Mathematical Studies and Applications, Karaman, Turkey, 4–6 October 2018; pp. 442–446.
28. Zhou, H.W.; Yang, S.; Zhang, S.Q. Modeling non-Darcian flow and solute transport in porous media with the Caputo-Fabrizio derivative. *Appl. Math. Mod.* **2009**, *68*, 603–615. [CrossRef]
29. Roul, P.; Rohil, V.; Espinosa-Paredes, G.; Prasad Goura, V.M.K.; Gedam, R.S.; Obaidurrahman, K. Design and analysis of a numerical method for fractional neutron diffusion equation with delayed neutrons. *Appl. Numer. Math.* **2020**, *157*, 634–653. [CrossRef]
30. Roul, P.; Rohil, V.; Espinosa-Paredes, G.; Obaidurrahman, K. An efficient numerical method for fractional neutron diffusion equation in the presence of different types of reactivities. *Ann. Nucl. Energy* **2021**, *152*, 108038. [CrossRef]
31. Kuila, S.; Raja Sekhar, T.; Zeidan, D. On the Riemann Problem Simulation for the Drift-Flux Equations of Two-Phase Flows. *Int. J. Com. Meth.* **2016**, *13*, 1650009. [CrossRef]
32. Zeidan, D.; Romenski, E.; Slaouti, A.; Toro, E.F. Numerical study of wave propagation in compressible two-phase flow. *Int. J. Num. Meth. Fluids* **2007**, *54*, 393–417. [CrossRef]
33. Erdogan, U.; Sari, M.; Kocak, H. Efficient numerical treatment of nonlinearities in the advection–diffusion–reaction equations. *Int. J. Num. Meth. Heat Fluid Flow* **2019**, *29*, 132–145. [CrossRef]
34. Pinnola, F.P.; Vaccaro, M.S.; Barretta, R.; Marotti de Sciarra, F. Random vibrations of stress-driven nonlocal beams with external damping. *Meccanica* **2021**, *56*, 1329–1344. [CrossRef]
35. Barretta, R.; Marotti de Sciarra, F.; Pinnola, F.P.; Vaccaro, M.S. On the nonlocal bending problem with fractional hereditariness. *Meccanica* **2021**. [CrossRef]
36. Salomoni, V.A. A mathematical framework for modelling 3D coupled THM phenomena within saturated porous media undergoing finite strains. *Comp. Part B Eng.* **2018**, *146*, 42–48. [CrossRef]
37. He, J.H. Approximate analytical solution for seepage flow with fractional derivatives in porous media. *Comput. Meth. Appl. Mech. Eng.* **1998**, *167*, 57–68. [CrossRef]
38. Schumer, R.; Benson, D.A.; Meerschaert, M.M.; Wheatcraft, S.W. Eulerian derivation of the fractional advection–dispersion equation. *J. Contam. Hydrol.* **2001**, *48*, 69–88. [CrossRef]
39. Ochoa-Tapia, J.A.; Valdes-Parada, F.J.; Alvarez-Ramirez, J. A fractional-order Darcy's law. *Phys. A Stat. Mech. Appl.* **2007**, *374*, 1–14. [CrossRef]
40. Cushman, J.H.; Moroni, M. Statistical mechanics with three-dimensional particle tracking velocimetry experiments in the study of anomalous dispersion I: Theory. *Phys. Fluids* **2001**, *13*, 75–80. [CrossRef]
41. Oldham, K.B.; Spanier, J. *The Fractional Calculus*; Academic Press: New York, NY, USA, 1974.
42. Zhang, H.J.; Jeng, D.S.; Barry, D.A.; Seymour, B.R.; Li, L. Solute transport in nearly saturated porous media under landfill clay liners: A finite deformation approach. *J. Hydrol.* **2013**, *479*, 189–199. [CrossRef]
43. Peters, G.P.; Smith, D.W. Solute transport through a deforming porous medium. *Int. J. Num. An. Meth. Geomech.* **2002**, *26*, 683–717. [CrossRef]
44. Chaves, A.S. A fractional diffusion equation to describe Levy flights. *Phys. Lett. A* **1998**, *239*, 13–16. [CrossRef]
45. Benson, D.A.; Wheatcraft, S.W.; Meerschaert, M.M. The fractional-order governing equation of Levy motion. *Water Resour. Res.* **2000**, *36*, 1413–1423. [CrossRef]
46. Owolabi, K.M.; Atangana, A. *Numerical Methods for Fractional Differentiation*; Springer: Singapore, 2019.
47. Deng, Z.Q.; de Lima, J.L.M.P.; de Lima, M.I.P.; Singh, V.P. A fractional dispersion model for overland solute transport. *Water Resour. Res.* **2006**, *42*, W03416. [CrossRef]
48. Çelik, C.; Duman, M. Crank–Nicolson method for the fractional diffusion equation with the Riesz fractional derivative. *J. Comp. Phys.* **2012**, *231*, 1743–1750. [CrossRef]
49. Salomoni, V.A.; Majorana, C.E. Parametric analysis of diffusion of activated sources in disposal forms. *J. Haz. Mat.* **2004**, *A113*, 45–56.

fractal and fractional

MDPI

Article

Simplified Relation Model of Soil Saturation Permeability Coefficient and Air-Entry Value and Its Application

Gaoliang Tao [1,2], Zhijia Wu [1], Wentao Li [1,*], Yi Li [1] and Heming Dong [3]

1 School of Civil Engineering, Architecture and Environment, Hubei University of Technology, Wuhan 430068, China; tgl1979@126.com (G.T.); wzj2301214319@163.com (Z.W.); yi.li10566@outlook.com (Y.L.)
2 School of Intelligent Construction, Wuchang University of Technology, Wuhan 430223, China
3 Shiyan City Transportation Investment Co., Ltd., Shiyan 442000, China; dhm.wayne@hotmail.com
* Correspondence: wli20201027@hbut.edu.cn

Abstract: Based on the Tao and Kong (TK) model and the fractal model of the soil–water characteristic curve, a simplified model of the relationship between the saturated permeability coefficient and the air-entry value is established in this study: $k_s = k_0\psi_a{}^{-2}$. It is shown that the saturated permeability coefficient of soil is determined by its maximum pore size. In order to facilitate the mutual prediction of saturation permeability coefficient and air-entry value, based on the data of five types of soil in the UNSODA database, the comprehensive proportionality constant k_0 of the five types of soil were obtained: sand $k_0 = 0.03051$; clay $k_0 = 0.001878$; loam $k_0 = 0.001426$; sandy loam $k_0 = 0.009301$; and silty clay loam $k_0 = 0.0007055$. Based on the obtained comprehensive proportionality constant k_0 and the relationship model between saturated permeability coefficient and air intake value, the air-entry value of five kinds of soils in the existing literature and the SoilVision database were calculated. Comparing the calculated air-entry value with the measured one, the results showed that the model simplifies the traditional air-entry value prediction method to some extent and can effectively predict the air-entry value of different types of soil. On the whole, the model better predicts the air-entry value for sandy, clay, and silty clay loam than loam and sandy loam.

Keywords: model; saturated permeability coefficient; air-entry value; maximum pore; comprehensive proportionality constant

Citation: Tao, G.; Wu, Z.; Li, W.; Li, Y.; Dong, H. Simplified Relation Model of Soil Saturation Permeability Coefficient and Air-Entry Value and Its Application. *Fractal Fract.* **2021**, *5*, 180. https://doi.org/10.3390/fractalfract5040180

Academic Editor: Norbert Herencsar

Received: 1 September 2021
Accepted: 18 October 2021
Published: 23 October 2021

1. Introduction

Soil in nature is a kind of three-phase material, so it is of great significance to study the related characteristics of soil in depth for engineering construction, environmental protection, and energy mining. As a way to describe soil characteristics, a soil–water characteristic curve (SWCC) contains an enormous amount of basic soil information. The soil–water characteristic curve characterizes the relationship between water content (including mass water content w, volume water content θ_w, saturation Sr, and water content e_w) and suction in the soil (mainly referring to the matrix suction, which reflects the influence and effect of the boundary surface). The soil–water characteristic curve can reflect the water-holding capacity of the soil and the size and distribution of pores in the soil. Research on the shear strength, consolidation theory, and permeability theory of unsaturated soil involves its application by measuring the balance state of soil samples. The soil–water characteristic curve can be drawn with the water content and corresponding suction. The SWCC is usually divided into the boundary effect zone, the transition zone, and the residual zone according to the degree of saturation of soil. Soils change from the saturated state to the unsaturated state corresponding to the change from the boundary effect zone to the transition zone. Usually, these two zones are differentiated by their air-entry values. Therefore, as an important parameter of the SWCC, the air-entry value has a significant effect on soil shear strength, deformation, permeability, and even the interaction mechanism between water, air, and soil particles within the soil. The air-entry value is a critical point on the soil–water characteristic curve, which means that the largest pores in the soil can hardly resist the applied suction and cause water loss. At this time, the applied matrix suction is the air-entry value.

Many researchers in the world have made outstanding contributions to the study of air-entry value. Bouwer [1] conducted field measurements of soil air-entry values with a cylindrical lid permeameter equipped with a riser and a vacuum gauge, and they established a step function to explore hydraulic conductivity. Fredlund and Xing [2] proposed a fitting formula for the measured SWCC and obtained the air-entry value of the soil by plotting and solving the fitted SWCC. Fallow and Elrick [3] externally added a tension bottle to the permeameter and measured the air-entry value of the soil on site, thereby providing an improved description of the hydraulic properties of soils. Nemati et al. [4] proposed a method for measuring the air-entry value in a rapid tension chamber. Different potting substrates were in contact with unsaturated bodies located outside the tension chamber, and the potential in the cabin was detected to calculate the air-entry values of peat substrates. Sakaki et al. [5] found that there was a close relationship between the diameter of the characteristic particle and the air-entry value. The results showed that the average pore size had a dominant effect on the air-entry value. Based on this, a simple air-entry value prediction was proposed. Chui et al. [6] analyzed two geological cases in Singapore, and they used the ordinary leather method and the simple Kriging method (SK) to estimate the spatial variability of the air-entry value of the soil. Ni et al. [7] used the instantaneous profile method to quantify the influence of plant roots on the air-entry value, and they found that the presence of roots significantly increased the air-entry value of silt. Under mixed planting with rotted roots, the vegetation soil showed a significant decrease in the air-entry value. Slowik et al. [8] proposed an air-entry value identification method through regarding the air-entry value as an indicator of the early cracking of materials. Soltani et al. [9] investigated the relationship between air-entry value and adaptive parameters by establishing an explicit equation method based on the air-entry value and residual state suction. Wijaya et al. [10] obtained the shrinkage curves of multiple soil samples, and their results showed that the minimum void ratio of the shrinkage curve has a significant effect on the air-entry value.

At present, the air-entry value is usually determined by using the SWCC. However, the measurement of the SWCC requires experimental instruments. The SWCC can be measured by a variety of methods and materials, including a pressure plate test, a vacuum dryer, the cooling mirror dew point method, filter paper, the unsaturation test, and an unsaturated triaxial apparatus [11–18]. The method used to test the soil–water characteristic curve usually provides a series of discrete data points between the matrix suction and water content. However, to use the soil–water characteristic curve to study the phenomena of the permeability, stress, and deformation of unsaturated soil, one needs complete curves and continuous mathematical formulas. At the same time, direct test measurement is often time-consuming and labor-intensive, with harsh conditions, and the measured test data are insufficient. Therefore, it is necessary and meaningful to use the soil–water characteristic curve model for data fitting and prediction. These limitations are mainly manifested in expensive equipment and operation difficulty. In the absence of experimental equipment, it is difficult for people to measure the air-entry value of soil, which can be more quickly and effectively obtained via indirect prediction. Therefore, the authors of this study combined the SWCC fractal model with the existing permeability model and further established a model of constant of proportionality for the permeability coefficient. This model was found to simplify the traditional air-entry value prediction method and overcome the limitations of practical measurement methods. The applicability and effectiveness of the model were verified by using the relevant data of existing soil.

2. Establishment of Prediction Model

The prediction model of the saturated permeability coefficient can be obtained from the Tao and Kong (TK) model [19], namely Formula (1):

$$k_s = k_c \int_{\theta_r}^{\theta_s} \frac{dx}{\psi^2(x)} \tag{1}$$

where k_s is the saturated permeability coefficient, θ_s is the saturated volumetric water content, θ_r is the residual volumetric water content, and ψ is the matrix suction.

In Formula (1):

$$k_c = \frac{\gamma {T_S}^2 \cos^2 \alpha}{2 p_i \mu} \tag{2}$$

where γ is the unit weight of water, μ is the absolute viscosity of water, T_s is the table surface tension, α is the contact angle, and p_i is the ratio of the actual length of the i-th pore channel to the length of the soil sample. For the same type of soil, k_c is a constant.

According to the fractal model of the soil–water characteristic curve proposed in the literature [20], the volumetric water content can be expressed as follows:

$$\begin{cases} \theta = \frac{e_0}{1+e_0} - 1 + \left(\frac{\psi_a}{\psi}\right)^{3-D} & \psi \geq \psi_a \\ \theta = \theta_s = \frac{e_0}{1+e_0} & \psi < \psi_a \end{cases} \tag{3}$$

where e_0 is the void ratio, ψ_a is the air-entry value; D is the fractal dimension, and θ_s is the volumetric water content.

Differentiate Formula (3):

$$dx = d\theta = (D-3){\psi_a}^{3-D}\psi^{D-4}d\psi \tag{4}$$

Substitute Formula (4) into Formula (1):

$$k_s = k_c(D-3)\int_{\psi_d}^{\psi_a} {\psi_a}^{3-D}\psi^{D-6}d\psi \tag{5}$$

After calculation and simple transformation, Formula (6) can be obtained:

$$k_s = k_c \frac{D-3}{D-5} {\psi_a}^{-2} \left[1 - \left(\frac{\psi_d}{\psi_a}\right)^{D-5}\right] \tag{6}$$

where $\psi_a \ll \psi_d$, $D-5 < -2$.

Formula (6) can be simplified to Formula (7):

$$k_s = \frac{D-3}{D-5} k_c {\psi_a}^{-2} \tag{7}$$

It can be seen from Formula (7) that the saturated permeability coefficient of the soil and the square of the air-entry value have an inverse proportional relationship. According to the literature [19,21], the clay content of natural soil has a direct relationship with the fractal dimension D as follows.

The clay content is within a certain range for the same type of soil. The corresponding fractal dimension does not vary much, and compared to the square of the air-entry value, the fractal dimension has a negligible effect on permeability [19,21]. In order to simplify the calculation, the authors of this article set the part containing the fractal dimension D and the permeability coefficient proportional constant k_c as a constant. This constant is named the comprehensive proportionality constant k_0. Then, Formula (7) is transformed into:

$$k_s = k_0 {\psi_a}^{-2} \tag{8}$$

According to the law of the capillary, the air-entry value is inversely proportional to the maximum pore diameter of the soil, so Formula (8) shows that the saturated permeability coefficient is directly proportional to the maximum pore diameter. This indicates that the saturated permeability coefficient of the soil is essentially controlled by the maximum pore diameter.

3. Determine Constant of Proportionality k_0 for the Model

3.1. Source of Experimental Data

The main sources of the database collection in this study were the soil-related data in the UNSODA database [22]. The five soil types in this database are sand, clay, loam, sandy loam, and silty clay loam. The database only contains data on the SWCC, without the air-entry value.

The authors of a previous study [23] obtained the relationship between the air-entry value and the basic parameters in the VG model, which is presented as follows:

$$\alpha = \left(\frac{0.76}{\psi_a}\right)^{1.26} \tag{9}$$

where α is the fitting parameter of the model and ψ_a is the air-entry value of the soil.

In this paper, the VG model was firstly used to fit the volumetric water content and suction of the soil in the UNSODA database to obtain the standard SWCC and the corresponding fitting parameters. Then, the corresponding air-entry value was calculated by Formula (9).

The VG model equation is given by Formula (10):

$$\theta_w = \theta_r + \frac{\theta_s - \theta_r}{\left[1 + (\alpha\psi)^n\right]^m} \tag{10}$$

where θ_w is the volumetric water content; θ_r is the residual water content; θ_s is the saturated water content; ψ is the matrix suction; and α, m, and n are the fitting parameters of the model, usually $m = 1 - 1/n$.

The data obtained from the database are summarized in Table 1, and Figure 1 shows the results of statistical data processing and box diagram plotting.

Table 1. Basic statistical data obtained from the UNSODA database.

Type of Soil	Sample	Mean Value		Standard Deviation	
		AEV (kPa)	k_s (cm/s)	*AEV* (kPa)	k_s (cm/s)
Sand	50	2.23	0.024	1.41	0.051
Clay	47	2.51	0.0053	5.21	0.0092
Loam	38	2.94	0.0007	2.39	0.0008
Sandy loam	43	4.44	0.0009	2.99	0.0011
Silty clay loam	40	6.12	0.000024	1.79	0.000011

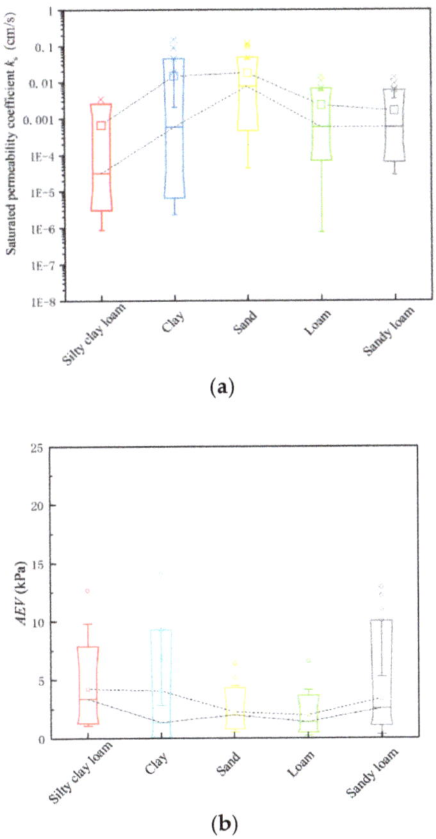

Figure 1. (**a**) Saturated permeability coefficient for various soils; (**b**) air-entry values for various soils.

3.2. Determination of the Comprehensive Proportionality Constant

Formula (8) was used to fit the data to obtain an image with ${\psi_a}^{-2}$ as the horizontal axis and k_s as the vertical axis. The collected soil data were plotted into a scatter plot and then linearly fitted. The linear function with the highest correlation coefficient was selected as the fitting result. The slope of the linear function was taken as the comprehensive constant of proportionality. The fitting result is shown in Figure 2, and the comprehensive constants of proportionality of the five soils are shown in Table 2.

Table 2. The comprehensive proportional constants of proportionality of the five soils obtained by fitting.

Type of Soil	Comprehensive Proportionality Constant k_0	Fitted Correlation Coefficient R^2
Sand	0.03051	0.9158
Clay	0.001878	0.8797
Loam	0.001426	0.8373
Sandy loam	0.009301	0.9067
Silty clay loam	0.0007055	0.8267

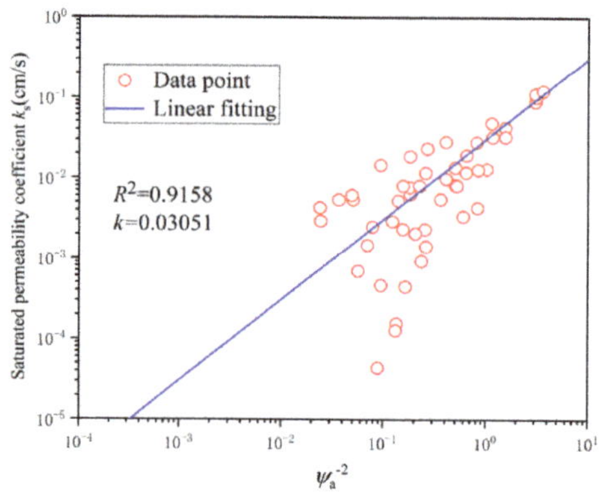

(a)

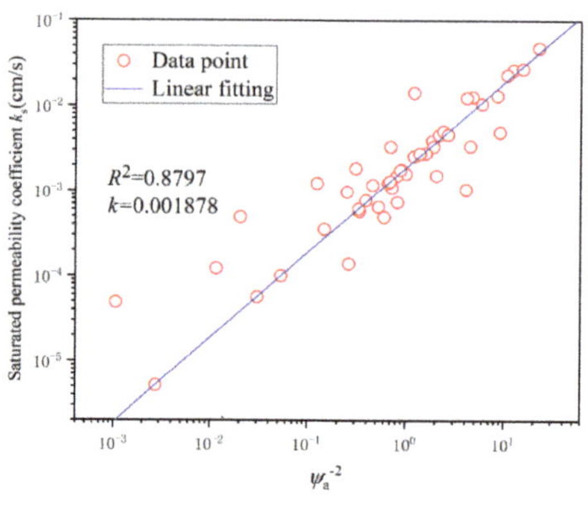

(b)

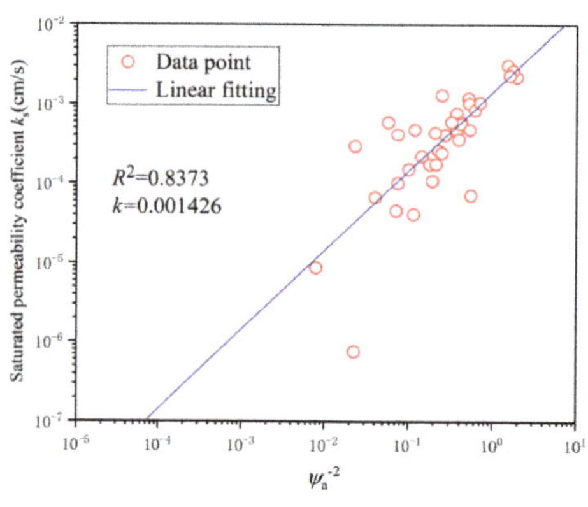

(c)

Figure 2. *Cont.*

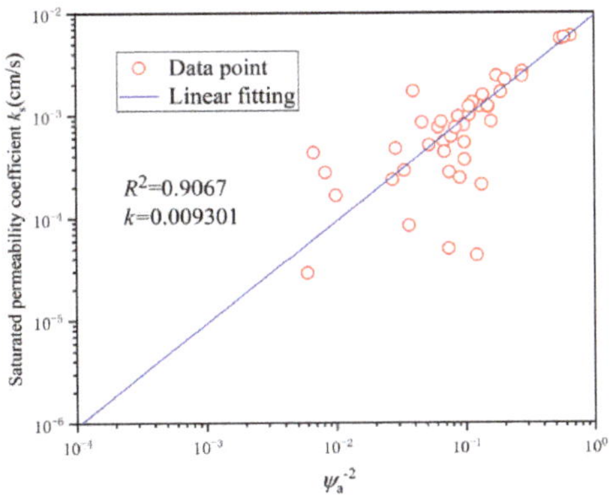

(d)

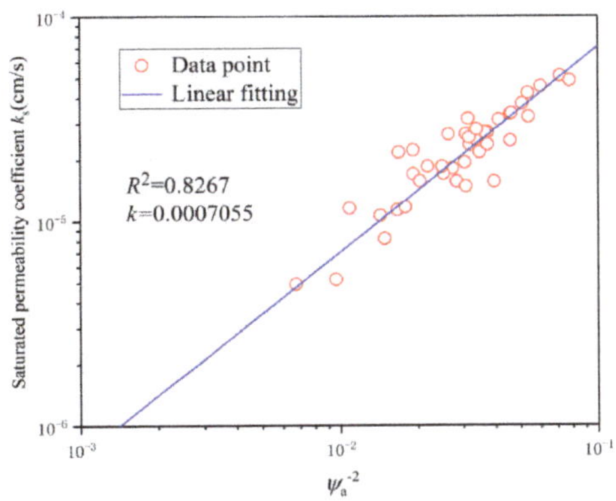

(e)

Figure 2. Comprehensive constant of proportionality of (**a**) sand, (**b**) clay, (**c**) loam, (**d**) sandy loam, and (**e**) silty clay loam.

In the process of solving the comprehensive proportionality constant, the correlation between sand and sandy loam was found to be over 0.9 when fitting the relevant data and within 0.8–0.9 among the other three soils.

4. Model Verification and Discussion

4.1. Data Verification

In order to verify the accuracy of the obtained comprehensive proportionality constant and the rationality of this method, the authors of this paper used a simplified relationship model between the saturated permeability coefficient and air-entry value of soil and the comprehensive proportionality constant k_0 determined above to calculate the corresponding air-entry value through the saturated permeability coefficient; then, they conducted a comparative analysis with the measured values. The validated soil data came from the clay and SoilVision database in [24] for sand, loam, sandy loam, and silty clay loam.

(i). Experimental data of Hunan clay

Part of the verification data came from the literature [24]. In this paper, the SWC-150 Fredlund SWCC instrument produced by the American Soilmoisture Company was used for the pressure plate test, and SWCC data of Hunan red clay with different initial dry densities were obtained. Falling head permeability tests were performed to obtain the saturated permeability coefficients of Hunan red clay with different initial dry densities.

The soil–water characteristic curve obtained by fitting the measured scatter diagram of the soil–water characteristic curve with the VG model is shown in Figure 3. The corresponding air-entry value was obtained using the tangent line graphing method. Specifically, the tangent line was plotted through the inflection point on the curve. There was an intersection point between the tangent line and the initial volumetric water content (vertical axis). The abscissa of the intersection point was taken as the air-entry value. We obtained the inflection point by taking the second-order derivation of $\lg(\psi)$ and setting the result of the second-order derivation equal to zero. The coordinates of the inflection point are shown in Figure 3. For different initial dry densities of 1.3, 1.35, 1.4, 1.45, 1.5, and 1.6 g/cm^3, the air-entry values were 2.65, 4.33, 6.70, 8.25, 9.77, and 13.42 kPa, respectively.

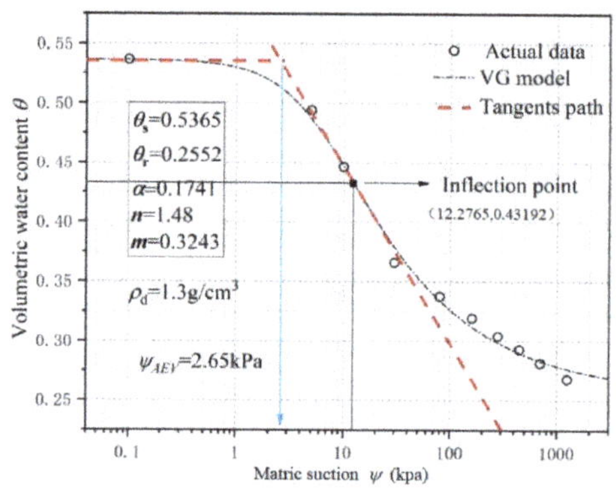

(a)

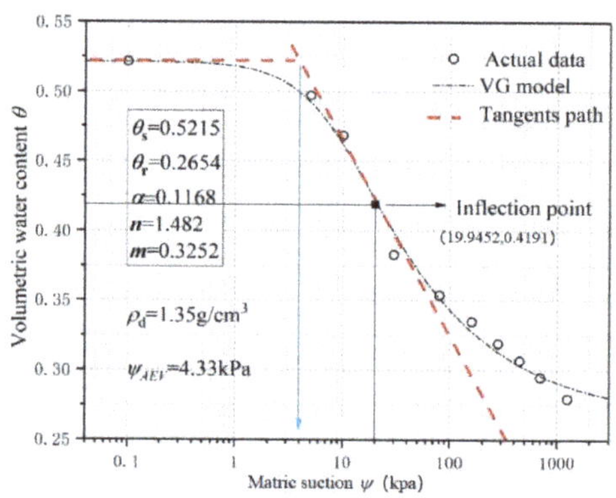

(b)

Figure 3. *Cont.*

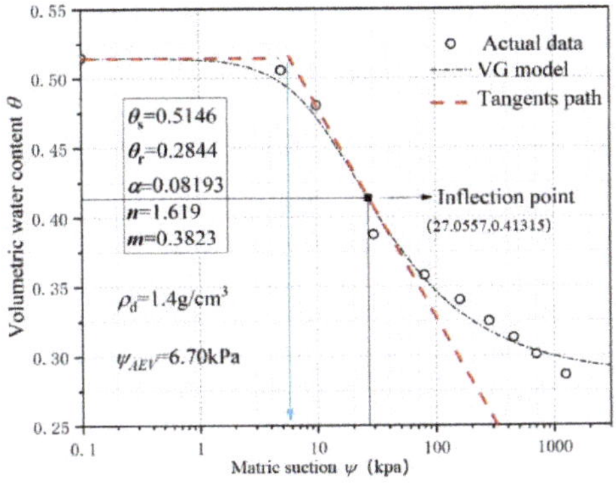

(c)

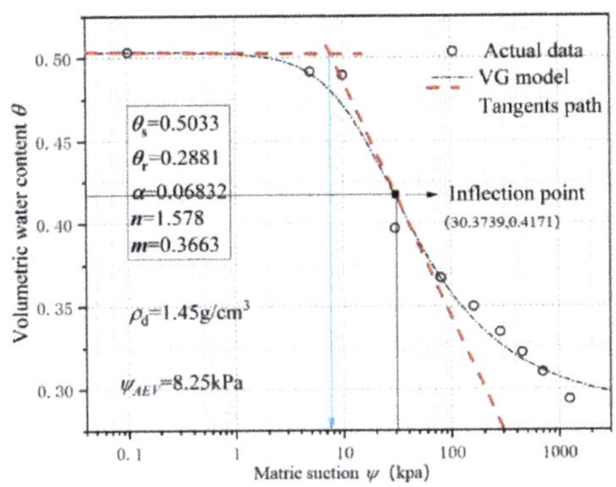

(d)

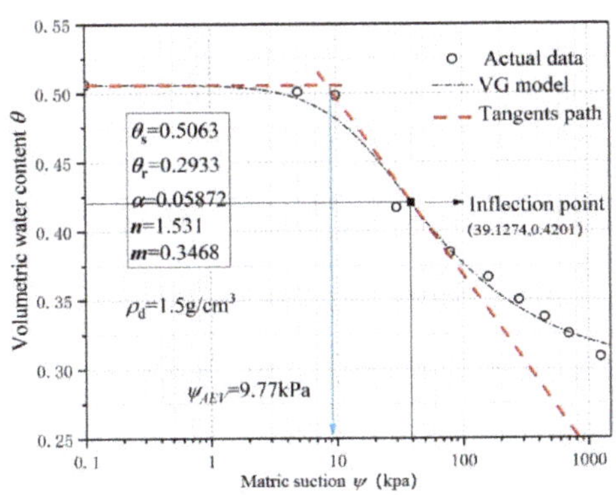

(e)

Figure 3. *Cont.*

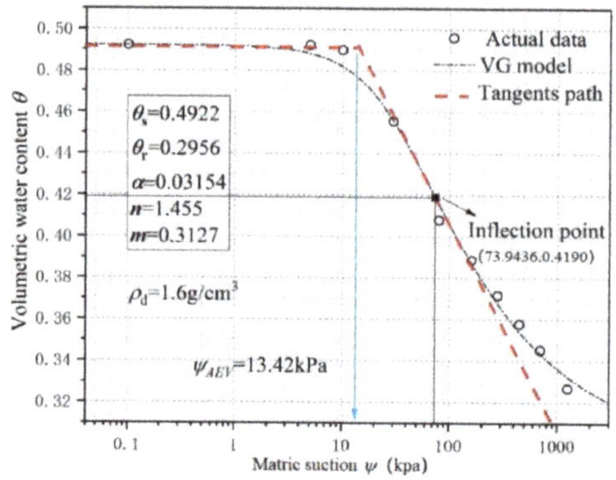

(f)

Figure 3. Soil–water characteristic curve and air-entry value of Hunan clay with different initial dry densities: (**a**) 1.3 g/cm^3, (**b**) 1.35 g/cm^3, (**c**) 1.4 g/cm^3, (**d**) 1.45 g/cm^3, (**e**) 1.5 g/cm^3, and (**f**) 1.6 g/cm^3.

(ii). SoilVision Database

The soil data in the SoilVision database include the saturated permeability coefficient, the standard soil–water characteristic curve, and the air-entry value of the soil [25]. The authors of this paper selected the relevant data of four soils in the database: sand, loam, sandy loam, and silt clay loam. Multiple samples were selected for each type of soil, and these samples were numbered, as shown in Tables 3–7.

Table 3. Basic sand information.

The Serial Number of the Figure	The Serial Number of the Database	Saturated Permeability Coefficient k_s (cm/s)	AEV in the Database (kPa)	Predicted AEV (kPa)
1	4283	0.0133	1.36	1.51
2	4346	0.020083	1.56	1.23
3	4417	0.010944	1.64	1.66
4	4461	0.0089444	1.64	1.84
5	4986	0.021167	0.84	1.19
6	3906	0.013417	2.28	1.50
7	5136	0.00875	2.9	1.86
8	4247	0.0040278	2.34	2.74
9	4333	0.00825	1.14	1.91
10	4498	0.0048889	2.81	2.48

Table 4. Basic clay information.

The Serial Number of the Figure	Dry Density (g/cm^3)	Saturated Permeability Coefficient k_s (cm/s)	AEV Obtained by Drawing (kPa)	Predicted AEV (kPa)
1	1.3	7.72×10^{-4}	2.65	1.56
2	1.35	4.15×10^{-4}	4.33	2.13
3	1.4	2.49×10^{-4}	6.7	2.75
4	1.45	1.73×10^{-4}	8.25	3.29
5	1.5	4.78×10^{-5}	9.77	6.27
6	1.6	9.92×10^{-6}	13.42	13.76

Table 5. Basic loam information.

The Serial Number of the Figure	The Serial Number of the Database	Saturated Permeability Coefficient k_s (cm/s)	AEV in the Database (kPa)	Predicted AEV (kPa)
1	1451	8.3333×10^{-5}	8.94	4.14
2	4407	4.860×10^{-3}	0.56	0.54
3	4274	2.7778×10^{-5}	10.49	7.16
4	4273	5.5556×10^{-5}	4.72	5.07
5	4401	2.7778×10^{-5}	3.65	7.16
6	4790	1.8333×10^{-5}	6.2	2.79
7	1441	2.2222×10^{-5}	5.37	8.01
8	1426	5.5556×10^{-6}	14.39	16.02
9	1402	4.1667×10^{-5}	8.62	5.85
10	4409	4.8889×10^{-3}	2.81	2.48

Table 6. Basic sandy loam information.

The Serial Number of the Figure	The Serial Number of the Database	Saturated Permeability Coefficient k_s (cm/s)	AEV in the Database (kPa)	Predicted AEV (kPa)
1	4991	2.50×10^{-4}	2.59	6.10
2	5099	5.56×10^{-5}	8.33	12.94
3	3955	3.53×10^{-4}	3.57	5.13
4	3199	8.33×10^{-5}	6.38	10.56
5	5238	8.19×10^{-4}	1.12	3.37
6	5184	9.72×10^{-4}	3.1	3.09
7	4384	2.92×10^{-2}	0.48	0.56
8	4429	2.39×10^{-4}	3.02	6.24
9	4284	3.61×10^{-4}	2.63	5.08
10	5238	8.19×10^{-4}	1.12	3.37

Table 7. Basic silty clay loam information.

The Serial Number of the Figure	The Serial Number of the Database	Saturated Permeability Coefficient k_s (cm/s)	AEV in the Database (kPa)	Predicted AEV (kPa)
1	4919	7.95×10^{-3}	0.34	0.30
2	4930	6.29×10^{-3}	0.56	0.33
3	4850	2.00×10^{-3}	0.31	0.59
4	4867	5.14×10^{-4}	1.08	1.17
5	4923	7.08×10^{-3}	0.34	0.32
6	4906	4.20×10^{-3}	0.37	0.41
7	4916	5.04×10^{-3}	0.35	0.37
8	4845	8.29×10^{-3}	0.29	0.29
9	4779	2.78×10^{-5}	8.67	5.04
10	4910	4.02×10^{-3}	0.65	0.42
11	4908	1.27×10^{-2}	0.37	0.24
12	4913	9.59×10^{-3}	0.25	0.27
13	4912	2.47×10^{-2}	0.35	0.17

4.2. Model Verification

Based on the values of k_0 determined in Section 3, (sand $k_0 = 0.03051$; clay $k_0 = 0.001878$; loam $k_0 = 0.001426$; sandy loam $k_0 = 0.009301$; and silty clay loam $k_0 = 0.0007055$), the saturated permeability coefficients of multiple soil samples were used in Formula (8) to obtain predicted soil air-entry values. A comparison of the measured and predicted results of the four types of soils is shown in Figure 4. The specific calculated and measured values are shown in Tables 3–7.

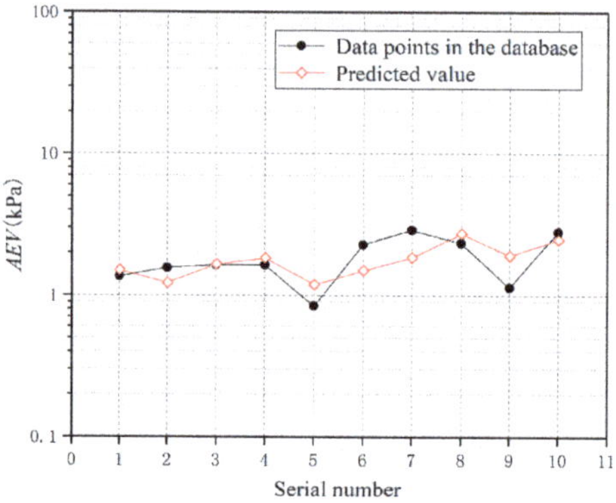

(a)

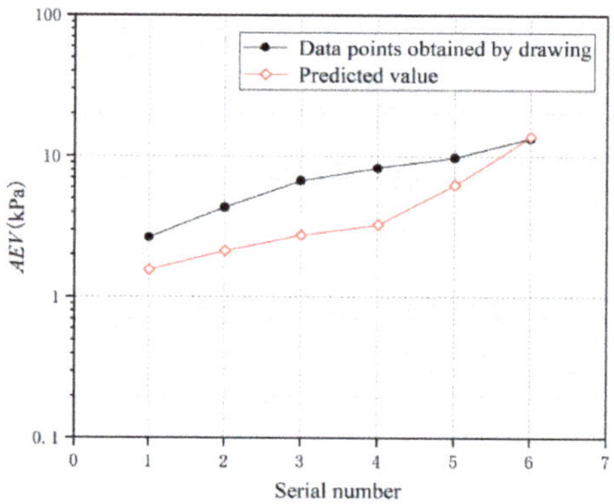

(b)

Figure 4. *Cont.*

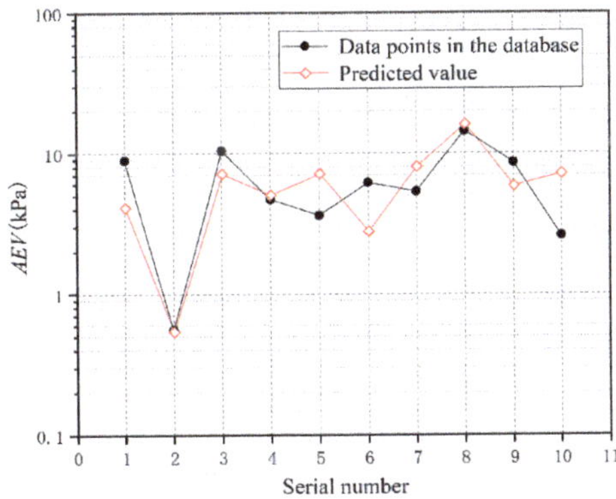

(c)

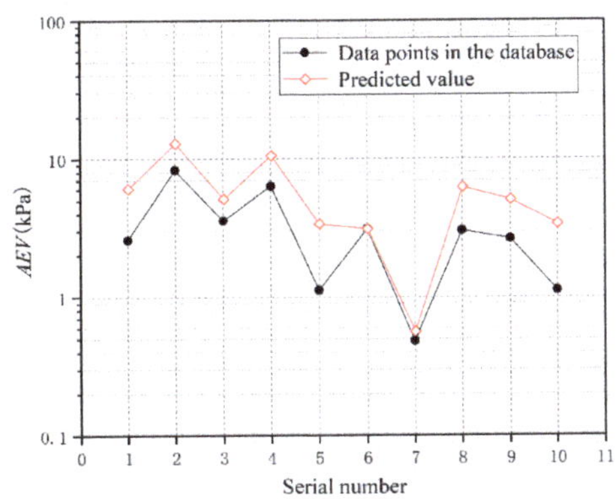

(d)

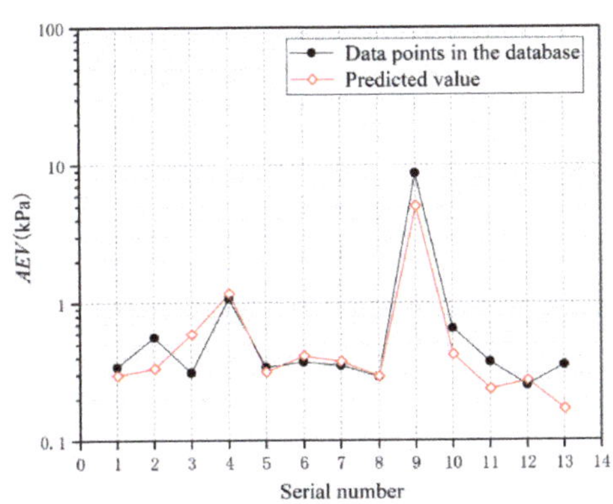

(e)

Figure 4. Measured and predicted values of air-entry values of various soils: (**a**) sand, (**b**) clay, (**c**) loam, (**d**) sandy loam, and (**e**) silty clay loam.

It can be seen from Figure 4 that the model could better predict the air-entry value of sand. Specifically, there were five measured data points that were close to the predicted points. Hunan clay was used as a verified clay soil sample, and its predicted values under different initial dry densities were close to the measured values. When the dry densities were 1.5 and 1.6 g/cm^3, the measured values were in good agreement with the predicted values. When the dry density was less than 1.5 g/cm^3, the predicted value was lower than the measured value. Overall, the predicted value was close to the measured value. The model predicted the air-entry value worse for loam and sandy loam than clay and sand. The predictions for loam and sandy loam soils were overall lower than those for clay and sandy soils. Of the selected data points, only three and two in the loam and sandy loam soils, respectively, were close to the measured values. The predicted and measured values of the silt clay loam soil were in the same order of magnitude, and there were six data points that were in good agreement with the predicted values.

4.3. Discussion

The above mentioned theoretical calculations and related experimental verification results showed that the prediction model is applicable to predict air-entry values to a certain extent. The prediction model was based on fractal theory and the TK model, and a new prediction model was obtained through the combination of the two theories. Taking sand as an example, the fractal dimension D of sand was found to be roughly in the range of 2.317~2.493 [25]. The reciprocal $\psi_a{}^{-2}$ of the square of the air-entry value of sand was found to vary in the range of 0.1–1. The change of fractal dimension relative to the square of the air-entry value was found to have a negligible effect on the permeability coefficient. According to capillary theory, the reciprocal of the square of the air-entry value is the square of the maximum pore size, indicating that the saturated permeability coefficient is determined by the maximum pore size.

The introduction of the comprehensive proportionality constant k_0 of the permeability coefficient significantly reduces the calculation work of the air-entry value. In this study, the k_0 of different soils could be approximated as a constant, which facilitated the mutual prediction between the air-entry value and the saturated permeability coefficient. That is to say, once the type of soil was determined, the saturated permeability coefficient could be used to predict the air-entry value, and vice versa.

For the determination of the comprehensive proportionality constant, the selection of data could have certain limitations. The fitting correlation of some soil data is relatively high. It is not easy to select the right amount of data and obtain fitting results with high correlations. For common soil types, the verification in this study generally presented good results, which could facilitate the establishment of this simplified model. For other soil types, the applicability of this simplified model needs to further verification. In this study, we collected a total of 218 sets of data, which was not enough to verify all soil types. More experimental data need to be considered in future research.

5. Conclusions

In this study, a simplified relationship model between the saturated permeability coefficient and the air-entry value for five types of soil was investigated, and a large amount of experimental data were collected to determine the model proportionality constant k_0. The air intake values of different types of soils were separately predicted by using the simplified relationship model of the saturation permeability coefficient of the soil, the air intake value proposed in this paper, and the determined proportionality constant k_0. A comparative analysis was carried out with the available experimental data. The main conclusions are as follows:

(1) The relationship model between the saturated permeability coefficient and air-entry value was determined to be $k_s = k_0\psi_a{}^{-2}$, where k_0 is the comprehensive proportional constant.

(2) The comprehensive constant of proportionality k_0 of the five soils was determined as follows: sand k_0 = 0.03051; clay k_0 = 0.001878; loam k_0 = 0.001426; sandy loam k_0 = 0.009301; and silty clay loam k_0 = 0.0007055.

(3) The model better predicted the air-entry value for sand, clay, and silty clay loam compared to loam and sandy loam.

(4) The model in this study can be used to make mutual predictions of the saturation permeability coefficient and air-entry value. Limited data on soil types were used to validate the model, so more experimental data need to be considered in future research.

Author Contributions: G.T. and Z.W. mostly contributed to the design of the manuscript. Y.L. carried out data collection and processing. W.L. and H.D. were involved in the statistical analysis. W.L. revised the paper. All authors have read and agreed to the published version of the manuscript.

Funding: The work in this paper was supported by grants from Natural Science Foundation of China (No.51978249) and the Innovation Group Project of Hubei Science and Technology Department (No. 2020CFA046), China.

Institutional Review Board Statement: Not applicable.

Informed Consent Statement: Not applicable.

Data Availability Statement: All data, models, and codes generated or used in the research process were obtained through experiments and calculation formulas in the text.

Conflicts of Interest: The authors declare no conflict of interest.

Notation

The following symbols are used in this paper:

k_s the saturated permeability coefficient
θ_s the saturated volumetric water content
θ_r the residual volumetric water content
ψ the matrix suction
γ the unit weight of water
μ the absolute viscosity of water
T_s the table surface tension
α the contact angle
p_i the ratio of the actual length of the i-th pore channel to the length of the soil sample
k_c a constant for the same type of soil
e_0 the void ratio
ψ_a the air-entry value
D the fractal dimension

References

1. Bouwer, H. Rapid field measurement of air entry value and hydraulic conductivity of soil as significant parameters in flow system analysis. *Water Resour. Res.* **1966**, *2*, 729–738. [CrossRef]
2. Fredlund, D.G.; Xing, A. Equations for the soil-water characteristic curve. *Can. Geotech. J.* **1994**, *31*, 521–532. [CrossRef]
3. Fallow, D.; Elrick, D. Field measurement of air-entry and water-entry soil water pressure heads. *Soil Sci. Soc. Am. J.* **1996**, *60*, 1036–1039. [CrossRef]
4. Nemati, M.; Caron, J.; Banton, O.; Tardif, P. Determining air entry value in peat substrates. *Soil Sci. Soc. Am. J.* **2002**, *66*, 367–373. [CrossRef]
5. Sakaki, T.; Komatsu, M.; Takahashi, M. Rules-of-Thumb for Predicting Air-Entry Value of Disturbed Sands from Particle Size. *Soil Sci. Soc. Am. J.* **2014**, *78*, 454–464. [CrossRef]
6. Ip, C.Y.S.; Rahardjo, H.; Satyanaga, A. Spatial variations of air-entry value for residual soils in Singapore. *Catena* **2019**, *174*, 259–268. [CrossRef]
7. Ni, J.; Leung, A.K.; Ng, C.W. Unsaturated hydraulic properties of vegetated soil under single and mixed planting conditions. *Géotechnique* **2019**, *69*, 554–559. [CrossRef]
8. Slowik, V.; Schmidt, M.; Fritzsch, R. Capillary pressure in fresh cement-based materials and identification of the air entry value. *Cem. Concr. Compos.* **2008**, *30*, 557–565. [CrossRef]

9. Soltani, A.; Azimi, M.; Deng, A.; Taheri, A. A simplified method for determination of the soil–water characteristic curve variables. *Int. J. Geotech. Eng.* **2019**, *13*, 316–325. [CrossRef]
10. Wijaya, M.; Leong, E.C.; Rahardjo, H. Effect of shrinkage on air-entry value of soils. *Soils Found.* **2015**, *55*, 166–180. [CrossRef]
11. Khanzode, R.; Vanapalli, S.; Fredlund, D. Measurement of soil-water characteristic curves for fine-grained soils using a small-scale centrifuge. *Can. Geotech. J.* **2002**, *39*, 1209–1217. [CrossRef]
12. Li, J.; Lu, Z.; Guo, L.; Zhang, L.M. Experimental study on soil-water characteristic curve for silty clay with desiccation cracks. *Eng. Geol.* **2017**, *218*, 70–76. [CrossRef]
13. Lu, N.; Wayllace, A.; Carrera, J.; Likos, W.J. Constant flow method for concurrently measuring soil-water characteristic curve and hydraulic conductivity function. *Geotech. Test. J.* **2006**, *29*, 230–241. [CrossRef]
14. Manahiloh, K.N.; Meehan, C.L. Determining the soil water characteristic curve and interfacial contact angle from microstructural analysis of X-ray CT images. *J. Geotech. Geoenviron. Eng.* **2017**, *143*, 04017034. [CrossRef]
15. Nishiumura, T.; Koseki, J.; Fredlund, D.G.; Rahardjo, H. Microporous membrane technology for measurement of soil-water characteristic curve. *Geotech. Test. J.* **2012**, *35*, 201–208. [CrossRef]
16. Peranić, J.; Arbanas, Ž.; Cuomo, S.; Maček, M. Soil-water characteristic curve of residual soil from a flysch rock mass. *Geofluids* **2018**, *2018*, 6297819. [CrossRef]
17. Sahin, H.; Gu, F.; Lytton, R.L. Development of soil-water characteristic curve for flexible base materials using the methylene blue test. *J. Mater. Civ. Eng.* **2015**, *27*, 04014175. [CrossRef]
18. Wang, X.; Benson, C.H. Leak-free pressure plate extractor for measuring the soil water characteristic curve. *Geotech. Test. J.* **2004**, *27*, 163–172. [CrossRef]
19. Tao, G.; Kong, L. A model for determining the permeability coefficient of saturated and unsaturated soils based on micro pore channel and its application. *J. Hydraul. Eng.* **2017**, *48*, 702–709. (In Chinese) [CrossRef]
20. Tao, G.; Kong, L.; Xiao, H.; Ma, Q.; Zhu, Z. Fractal characteristics and fitting analysis of soil–water characteristic curves. *Rock Soil Mech.* **2014**, *35*, 2443–2447. (In Chinese) [CrossRef]
21. Li, D.; Zhang, T. Fractal features of particle size distribution of soils in China. *Soil Environ. Sci.* **2000**, *9*, 263–265. (In Chinese) [CrossRef]
22. Nemes, A.; Schaap, M.; Leij, F. *The UNSODA Unsaturated Soil Hydraulic Database Version 2.0*; US Salinity Laboratory: Riverside, CA, USA, 1999.
23. Tinjum, J.M.; Benson, C.H.; Blotz, L.R. Soil-water characteristic curves for compacted clays. *J. Geotech. Geoenviron. Eng.* **1997**, *123*, 1060–1069. [CrossRef]
24. Zhu, X. Research on the Relationship between Fractal Characteristics of Soil Particle Size and Hydraulic Properties for Unsaturated Soils. Master's Thesis, Hubei University of Technology, Wuhan, China, 2019. (In Chinese).
25. Fredlund, M. *Soilvision 2.0, a Knowledge-Based Database System for Unsaturatedsaturated Soil Properties, Version 2.0*; Soilvision Systems Ltd.: Saskatoon, SK, Canada.

fractal and fractional

Article

A Fractal Entropy-Based Effective Particle Model Used to Deduce Hydraulic Conductivity of Granular Soils

Gang Zhang [1], Hongyu Wang [2], Jahanzaib Israr [1,3,*], Wenguo Ma [1], Youzhen Yang [1] and Keliang Ren [1]

1 Institute of Solid Mechanics, School of Physics and Elecronic-Electrical Engineering, Ningxia University, Yinchuan 750021, China
2 School of Civil and Hydraulics Engineering, Ningxia University, Yinchuan 750021, China
3 Department of Civil Engineering, University of Engineering and Technology Lahore, Punjab 54890, Pakistan
* Correspondence: jisrar@uet.edu.pk

Abstract: In this study, a rigorous mathematical approach used to compute an effective diameter based on particle size distribution (PSD) has been presented that can predict the hydraulic conductivity of granular soils with enhanced rigor. The PSD was discretized based on an abstract interval system of fractal entropy, while the effective diameter of soil was computed using the grading entropy theory. The comparisons between current entropy-based effective diameter (D_E) and those computed using existing procedures show that the current D_E can capture the particle size information of a given soil more accurately than others. Subsequently, the proposed D_E was successfully implicated into Kozeny–Carman's formula to deduce the saturated hydraulic conductivity of soils with enhanced accuracy. The proposed model was tested using current and previously published experimental data from literature. Not surprisingly, the results of the current model and those from previous experimental studies were found to be consistent, which can sufficiently verify the proposed entropy-based effective diameter model.

Keywords: effective particle diameter; fractal entropy; abstract interval; hydraulic coefficient

Citation: Zhang, G.; Wang, H.; Israr, J.; Ma, W.; Yang, Y.; Ren, K. A Fractal Entropy-Based Effective Particle Model Used to Deduce Hydraulic Conductivity of Granular Soils. *Fractal Fract.* **2022**, *6*, 474. https://doi.org/10.3390/fractalfract6090474

Academic Editor: Wojciech Sumelka

Received: 12 July 2022
Accepted: 26 August 2022
Published: 28 August 2022

1. Introduction

The particle size distribution (PSD) curve is an important soil characteristic that is widely used for preliminary estimates of numerous physical properties of granular soils, such as potentials of internal erosion, particle breakage, sedimentation, and saturated hydraulic conductivity, etc. [1–5]. For instance, Seelheim [6] and Hazen [7] pioneered the determination of saturated hydraulic conductivity based on the semi-logarithmic PSD curve. Ever since, several researchers have attempted to propose more rigorous and accurate correlations, such as the semi-empirical method of Kozeny–Carman (henceforth abbreviated as K-C), which combines soil properties, including the void ratio and specific particle sizes, to predict saturated hydraulic conductivity [8]. Nevertheless, the differences in the proportion and fraction of a PSD significantly affect the evaluation of sedimentary mixtures, such as loess, paleosol, river sediments, and glacial deposits [9]. For instance, the soil modulus extracted from a soil's PSD also influences its mechanical properties, including the shear strength and dynamic shear modulus [10]. Similarly, PSD is an important factor in assessing the potential of the internal erosion of soils [3,11–14]. Lately, Indraratna et al. [15] combined the PSD with the relative density considering the condition of particle packing and proposed the meaningful constriction size distribution (CSD)-based method. Israr and Zhang [16] adopted Loincz's model [17] to extract full grading information through entropy theory and proposed a fractal (grading) entropy-based method to promptly as well as accurately assess the internal erosion potential of granular soils.

Thus far, several end-member modeling algorithms have been proposed for decomposing and extracting valuable information from the PSD curves of soils [18]. For example, parametric curve-fitting, a statistical method such as the end-member modeling algorithm, has been used to identify the sub-populations of geological materials by decomposing its PSD [19]. Similarly, Chapuis and Saucer [20] proposed a modal decomposition method (MDM) to extract the sub-populations or modes from a soil's PSD curve to deduce its specific surface and to subsequently assess its internal erosion potential, while the PSD of sediments was fitted by the gradient descent (GD) method. Nonetheless, the above methods mainly account for a finer fraction of a non-uniform soil, whereas a coarser fraction is characterized through additional mathematical models with certain fitting parameters, such as the Gates–Gaudin–Schuhmann model, the Fuller model, the Gaudin–Meloy model (GMM), and the Fredlund unimodal (FUM) [21]. Meanwhile, a set of characteristic diameters determined from a PSD such as D10, D15, D60, and D85 and the coefficients of uniformity and curvature have been widely used in several geotechnical engineering applications (where the numeric value represents the percentage finer by mass). For example, D15 and D85 are widely adopted particle sizes in both filter design and the potential of internal instability assessment criteria [22–27].

Based on a simple and semi-automated sampling procedure, Hazen [8] proposed using D10 as a representative size of a PSD to capture its saturated hydraulic conductivity (k), which was later adopted for capturing the heterogeneity of a soil mixture through the coefficient of uniformity (Cu = D60/D10). Subsequently, several empirical formulae have been developed for estimating k based on Hazen's specific size D10 [1,28–30], D17 [31], D20 [32], D50 [33,34], and D75 [35]. However, in the widely accepted Kozeny–Carman (K-C) equation for k, an effective particle size is used instead of Hazen's D10 [8]. This effective particle size represents the entire PSD and is used for extracting material properties [19]; however, it is computed through complex procedures involving over-simplified assumptions [8,36–41]. A brief review of some of the most adopted methods is given in Appendix A.

Notably, the existing methods compute effective particle sizes based on different particle sizes and their relative weightage in terms of the percentage finer on the PSD curve. As an approximate and over-simplified approach, the PSD curves are plotted based on the results of sieve analysis, while only specific sieve sizes are used in delineating the distribution of particles. The accuracy of a PSD curve significantly depends on the interval chosen between various particle sizes and their corresponding percentage finer by either mass, number, or surface area. It is therefore recommended not to further simplify a given PSD curve for deducing the approximate effective particle size to indirectly represent the soil's pore size distributions and hence the hydraulic conductivity. Thus, the current study purports extracting PSD information more rigorously using the theory of grading entropy to compute the effective diameter for direct use in the K-C formula to capture saturated hydraulic conductivity with enhanced accuracy. The current proposition has been verified using an independent experimental dataset to demonstrate its enhanced rigor and the practical implication of this study has been demonstrated for utility to practitioners.

2. Effective Particle Diameter Based on Fractal Entropy

The effective particle size of a soil depends on the distribution of particle sizes and their relative proportion in a PSD curve, which is generally obtained from sieve analysis. For instance, the soil's particle sizes are plotted against the percentage finer by the mass of those particle sizes in a semi logarithmic coordinate system in order to obtain a PSD curve. Given that the group of discrete data can be analyzed through either a frequency distribution or cumulative frequency distribution diagram, wherein both number of discrete data intervals and their sizes are important, not surprisingly, the discrete intervals of a group of data will have a different impact on each dataset. For instance, a group of discrete data consisting of a single class interval (too wide) would compromise a lot of useful information, whereas the same dataset discretized into many small class intervals would yield a large amount of sparse data, which may not be conducive for subsequent analysis. Nevertheless, the entropy-based discretization method is an optimal solution [16]. In this paper, the grading entropy is used to discretize the PSD data to deduce a reasonable effective particle diameter that can accurately represent the grading information of soils.

2.1. Interval Class Discretization Based on Grading Entropy

Discretization based on the grading entropy principal was applied on the results of sieve analysis, whereby soil is divided into different class intervals (i.e., sieve opening sizes) and the size of a discretized interval class is not fixed. For instance, Lőrincz et al. [42] pointed out that the size of the subsequent interval class is twice the size of the former class interval. For instance, for a given series of sieve openings 0.063, 0.125, 0.25, 0.5, 1, 2, 4, 8, ... mm, the class size multiple is 2, whereas the suggested size of the elementary class interval class, also known as minimum soil particle size, remains $d_0 = 2^{-17}$ mm. This sequence of the interval class constitutes the primary statistical interval class system (referred to as the primary interval class), which is shown in the red brackets in Figure 1b. To accurately extract information from each interval class, a primary interval class can be sub-divided further into hypothetical secondary classes called an abstract interval system.

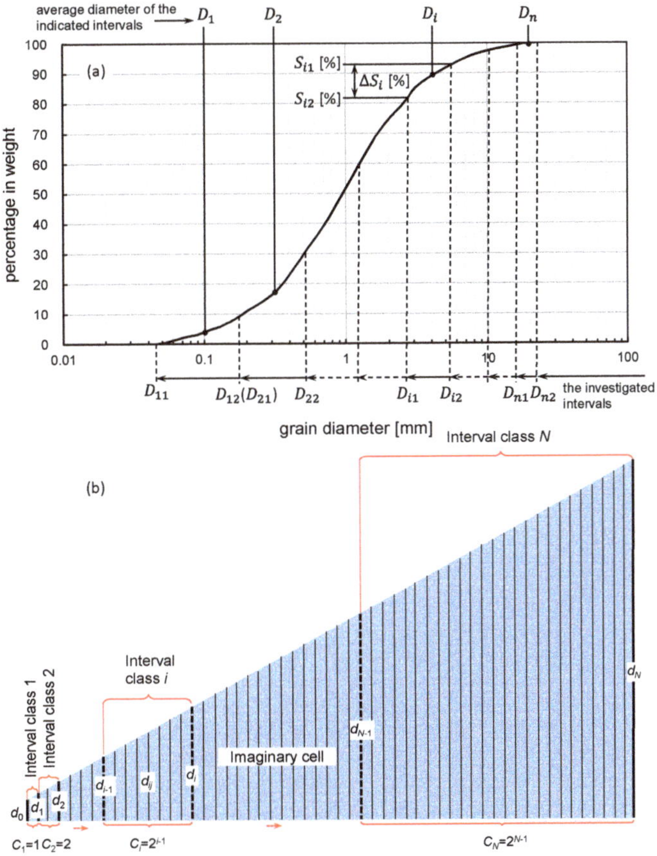

Figure 1. Discretization method of (**a**) Kozeny's effective diameter; and (**b**) abstract interval system based on grading entropy.

According to the abstract interval system (Figure 2), the PSD of a soil can be discretized into N different size interval classes, and the following formula can be obtained:

$$d_i = 2^{i-1} d_0,\ i = 1, 2, \ldots N \tag{1}$$

$$d_{ij} = d_i + (j-1)d_0,\ j = 1, 2, \ldots C_i \tag{2}$$

where d_i is the upper diameter size of the i interval class; d_{ij} is the upper diameter size of the j imaginary cell within the i interval class; C_i is the number of imaginary cells within the i interval class, and

$$C_i = \frac{d_{i+1} - d_i}{d_0} = 2^{i-1} \tag{3}$$

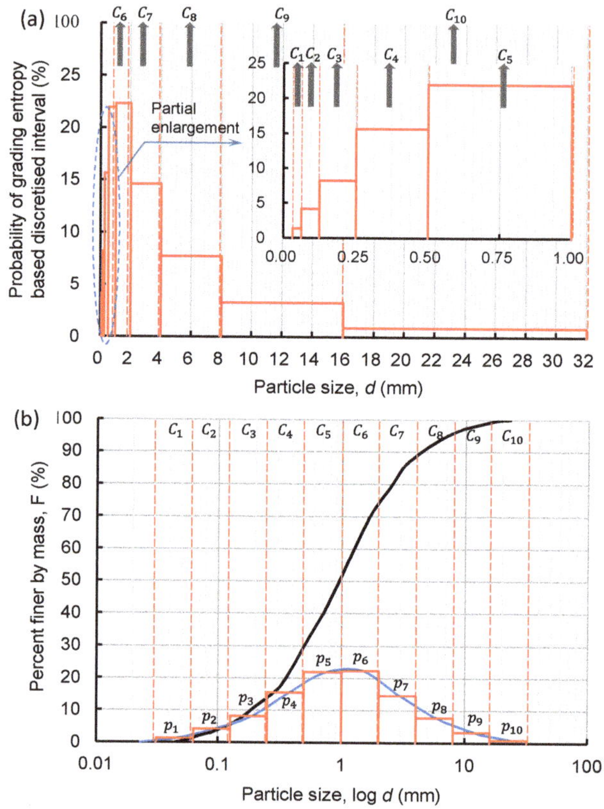

Figure 2. (**a**) Schematic diagram of grading entropy-based discretization method; and (**b**) PSD curve discretized by using entropy-based method in semi logarithmic coordinate system (where C_1–C_{10} and p_1–p_{10} represent abstract fractions and relevant percentages, respectively).

Figure 2a shows the abstract size fraction system of the PSD curve shown previously in Figure 1a, whereas Figure 2b demonstrates the application of the method of abstract size fractions in a semi logarithmic coordinate system. It can be seen from the comparison between Figures 1a and 2b that the information extracted by the grading entropy discretization method is more comprehensive.

2.2. Effective Particle Diameter Based on Grading Entropy

Given that a PSD curve can be discretized into several different size interval classes by using a series of data sequences (μ_{ij}, d_{ij}) within each interval class that can be substituted into Equation (A3), the effective particle diameter can be given by:

$$D_E = \frac{1}{\sum_{i=1}^{N} \sum_{j=1}^{C_i} \frac{\mu_{ij}}{d_{ij}}} = \frac{1}{\sum_{i=1}^{N} \sum_{j=1}^{C_i} \frac{1}{d_{ij}} \frac{p_i}{C_i}} \tag{4}$$

where D_E is the grading entropy-based effective particle diameter, μ_{ij} is the probability of the j imaginary cell within the i interval class, $\overline{d}_{ij}$ is the average diameter in the j imaginary cell within the i interval class, and

$$\overline{d}_{ij} = \frac{3}{\frac{1}{d_{i,j-1}} + \frac{2}{d_{i,j-1}+d_{i,j}} + \frac{1}{d_{i,j}}} \tag{5}$$

The above equation assumes that the soil particles are spherical, which does not take the shape factor into consideration. In this paper, the shape coefficient of Kovács [39] is adopted to account for the particle shape. For simplicity, the values of the shape coefficient of some regular geometries α_D are shown as follows: for sphere: $\alpha_D = 6$; for cube: $\alpha_D = 10.4$; for octahedron: $\alpha_D = 10.4$; for tetrahedron: $\alpha_D = 18$; for other complicated shapes, the values can be interpolated using these values.

3. Effective Particle Diameters for Different Soils

Considering the large variability of the non-uniformity coefficient and characteristic particle size distribution (PSD) parameters of non-uniform soils, the effective particle size of both widely graded PSDs and gap-graded PSDs have been analyzed in the following section:

3.1. Widely Graded Soils

As Figure 3 shows, a total of ten widely graded PSDs from Israr and Zhang [16] were selected for analysis, and consist of various proportions of clay, silt, sand, and gravel. The conventional grading parameters, such as D10, D50, and Cu, were used for the analysis in this paper.

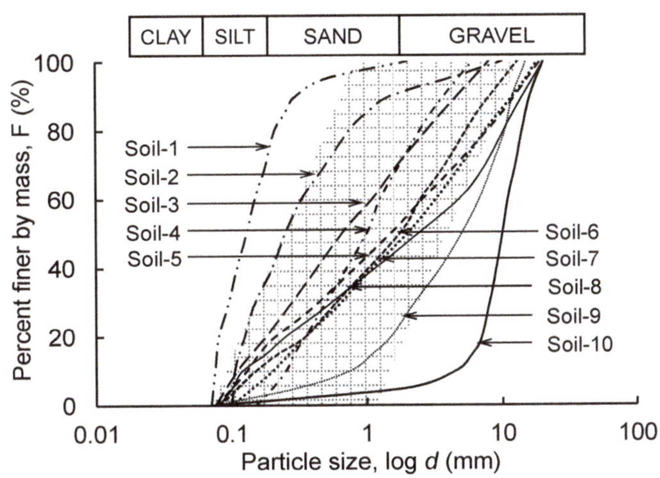

Figure 3. PSD curves of ten widely graded soils (adopted with permissions from Israr and Zhang [16], 2021).

Figure 4 presents the relative deviation data for 10 soils shown previously in Figure 3, whereby two characteristic particle sizes D10 (Hazen's effective particle size) and D50 (mean particle size on PSD curve) were plotted against their uniformity coefficients.

Apparently, there is no obvious correlation observed between either of the characteristic particle sizes and the uniformity coefficient. However, with the increase in Cu values of soils, a larger relative deviation is apparent in D50 values than D10; thus, the latter seems to be a relatively more reasonable option to represent uniform soils bearing low uniformity coefficients. Nevertheless, the relative deviation increases proportionally with the width of the PSD curve, thus indicating that neither of these sizes would effectively represent the statistical distribution of a large number of particles in a wider PSD curve. It is noteworthy that the particles finer than D10 and those coarser than D50 do not significantly influence the flow through porous media [43]. Thus, there may exist a corresponding particle size between D10 and D50 that can effectively represent and characterize both non-uniform PSD curves and can be chosen as an optimum effective particle diameter.

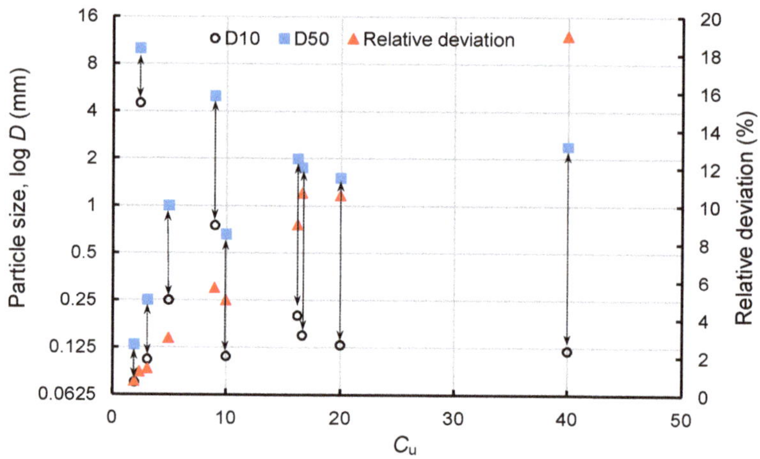

Figure 4. Relative deviation data of two characteristic particle sizes D10 (Hazen's effective particle size) and D50 (mean particle size on PSD curve) versus uniformity coefficients for 10 soils shown previously in Figure 3.

Figure 5 shows effective particle diameters computed from different existing methods, including the currently proposed approach in Equation (4), and plotted against the uniformity coefficients of soils shown previously in Figure 3. Notably, the range between D10 and D50 is marked with a blue area for the reader's convenience. The column height is the grading entropy-based effective particle diameter, while other effective particle diameters are marked with different markers. As shown, the grading entropy-based effective particle diameters are all located in the middle of the blue area between D10 and D50. A total of three effective particle diameters computed from the method of Carrier [8] were plotted out of the blue region, whereas the rest were plotted inside but closer to the D10 size. This shows that Carrier's method exhibits a large variability and cannot reasonably represent the widely graded PSD curves. Similarly, all values calculated from Fedorenko's model were plotted in the lower half of the blue region but closer to D10. Notably, the difference between Fedorenko's effective particle diameter and D10 gradually increases with the increase in Cu, indicating that Fedorenko's effective particle diameter can only represent the PSD curve within a certain grading width. Nevertheless, Kozeny's effective particle diameter calculation method lacks a description of the discretization scheme and the size of the interval class. Therefore, the calculated effective particle diameters obtained through different size intervals vary significantly, while the size interval is not specified in the literature [39]. For brevity, Figure 5 also shows the Kozeny's effective particle diameter computed with different interval sizes, where interval sizes of 2 mm and 0.01 mm are marked with Kozeny-2 and Kozeny-0.01, respectively. The value of Kozeny's effective particle diameter with a large interval size (e.g., 2 mm) is too small, whereas that with a small interval size (e.g., 0.01 mm) is close to the value of the entropy-based effective particle diameter, with an exception (see soil with Cu = 2.4).

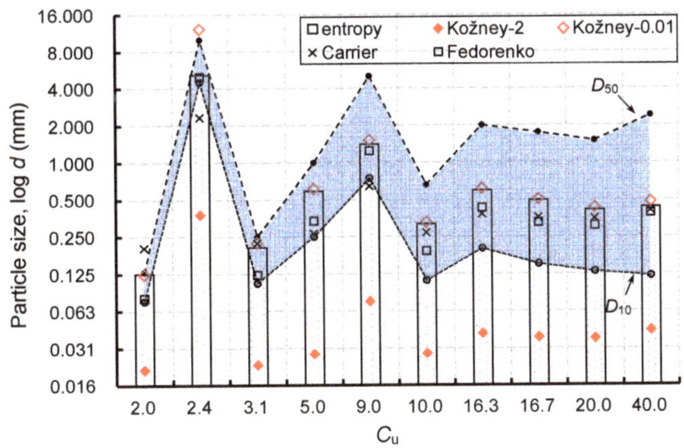

Figure 5. Calculated effective particle diameter of 10 PSDs using different effective particle diameter methods.

Figure 6 presents the values of Kozeny's effective particle diameters computed for the same 10 PSD curves shown previously in Figure 3 using different class interval sizes. As shown, the calculated values of the Kozeny's effective particle diameters decrease with the increase in the discretized interval size, thus indicating that the interval size has a great influence on the value of Kozeny's effective particle diameter. In general, the value of Kozeny's effective particle diameter has large variability when its interval size is large, which cannot represent the grading information of a PSD curve with reasonable confidence. Nonetheless, it can effectively represent the information of the PSD of soil when its interval size is small enough that it would consequently increase the computational costs by manifolds. Furthermore, there is the same problem of Carrier's effective particle diameter in the discretization of the interval class.

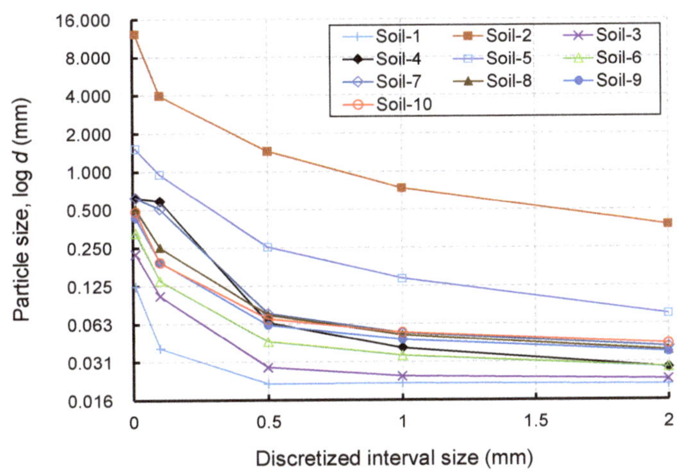

Figure 6. Values of Kozeny's effective particle diameter of 10 PSDs in different interval sizes.

As Figure 7 shows, both Kozeny's and Carrier's methods are greatly affected by the class interval size. For instance, with the increase in interval size, the value of Kozeny's effective diameter decreases gradually. This indicates that some grading information could not be extracted due to larger interval classes, thus resulting in reduced accuracy. Similarly,

Carrier's effective diameter initially increases and then decreases with the increase in interval size. Likewise, Feorenko's method obtains relatively consistent values of effective diameter sizes; however, this method is limited to relying on soil's Cu values, which do not apply to gap-graded soils. However, the values of grading entropy-based effective particle diameter are not affected by the interval size and soil's Cu values, owing to its abstract interval system, which can transverse all of the grading size ranges to comprehensively extract and consider grading information. Nevertheless, the values of entropy-based effective particle diameter are stably located in the region between D10 and D50, which is more reasonable than other methods.

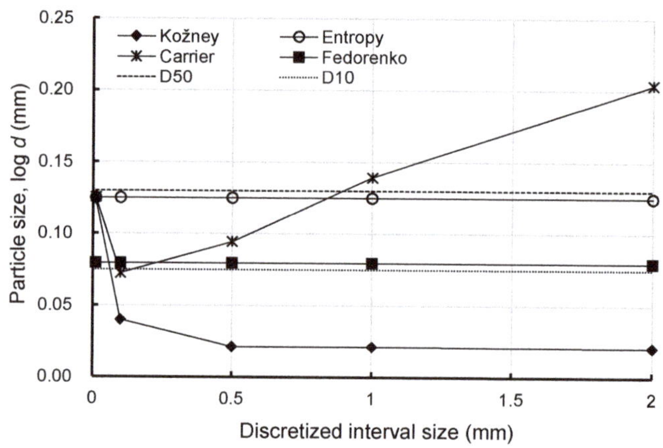

Figure 7. Values of different effective particle diameters computed from different methods for Soil-1 using different discretized interval sizes.

3.2. *Gap-Graded Soils*

Gap-graded soils exhibit a markedly higher potential of internal instability due to the absence of certain particle sizes in their PSD curves, which distinguishes them from other soils in terms of geo-mechanical and hydraulic properties [3,16,44]. It will induce large computational errors when some characteristic particle sizes and information, such as D10, D50, and Cu, are used to delineate hydraulic properties of gap-graded soils. For instance, two gap-graded PSDs (Gap30, Gap50) and one continuous PSD (Con100) were considered for analysis in this study (Figure 8). Notably, all three PSD curves have the same D50, but different grading shapes. It is obvious that there will be a large error when using D50 as a single characteristic diameter to deduce hydraulic conductivity or to assess the potential for seepage failure. In addition, continuous PSD Con100 and gap-graded PSD Gap50 have the same value of D10; therefore, should Hazen's approach be used to mimic the hydraulic conductivity of these two graded soils, the same value would be returned. This would not be the case when hydraulic conductivity is determined through a laboratory experiment.

In this study, six gap-graded soils from Andrianatrehina et al. [45] and Li [46] were selected for determining their effective particle diameter using existing and proposed models. Figure 9 shows that the median particle size (D50) of gap-graded soils increases exponentially with Cu; however, no obvious correlation was observed with the gap ratio Gr (= coarser particle size of the gap/finer particle size of the gap). In addition, the values of other effective particle diameter methods do not have any obvious regularity with both Cu and Gr values, which is consistent with the results of widely graded soils discussed in the previous section. Apparently, values of effective diameter from all four methods (i.e., Kozeny, Carrier, Fedorenko, and grading entropy) are plotted inside the region between D10 and D50 (see Figure 9a).

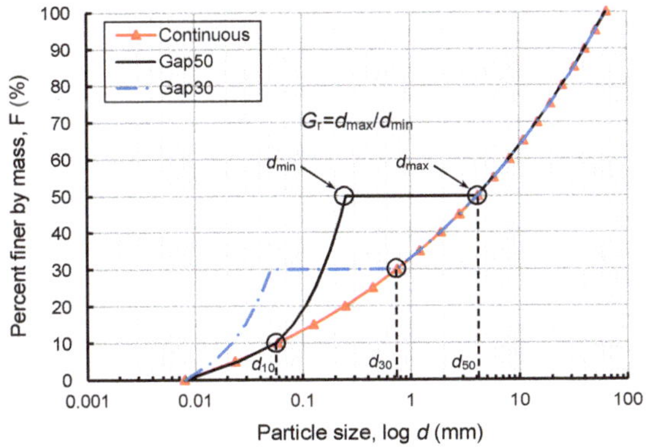

Figure 8. Two gap-graded and a continuous PSD curves for effective particle size computations using existing plus methods.

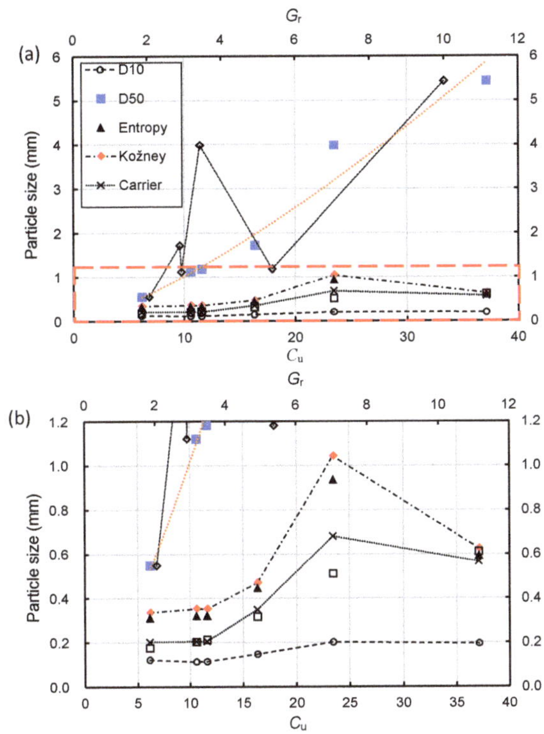

Figure 9. (**a**) Variations in gap ratio (G_r) and uniformity coefficient (C_u) versus soil particle sizes, and (**b**) enlarged view of highlighted inset of Figure 9a.

However, a closer look at this plot in Figure 9b shows that the values of both the Carrier and Fedorenko plot closer to D10, whereas that of Kozeny is higher than both D10 and D50. This may be attributed to the fact that the discretized interval sizes of Carrier and Kozeny are 2 mm and 0.01 mm, respectively. the larger the interval size, the lesser the grading information extracted, thus resulting in smaller values from Carrier's method and relatively higher values from Kozeny's method. Similarly, Fedorenko's method depends on Cu, which may be significantly affected by the width of the gap in gap-graded soils, thereby causing undesirably larger errors in computations. Not surprisingly, the results of the current method are closer to those from Kozeny's method due to an enhanced accuracy of up to a 0.01 mm interval size, which shows that the current method may also be used for gap-graded soils with enhanced confidence.

4. Implication of Proposed Model into K-C Formula

4.1. Modified K-C Formula Based on Grading Entropy Effective Diameter

The K-C formula is a semi empirical model for estimating the saturated hydraulic conductivity of granular soils, and is given by [8,47]:

$$k = (\frac{\gamma}{\mu})(\frac{1}{C_{K-C}})(\frac{1}{S_0^2})\frac{e^3}{1+e} \tag{6}$$

where γ is the unit weight of liquid; μ is the dynamic viscosity of liquid; C_{K-C} is Kozeny–Carman's empirical coefficient [8]; S_0 is the specific surface area per unit volume of soil particles; e is the void ratio. For uniform sphere particles, C_{K-C} = 4.8 ± 0.3, whereas, in other conditions, C_{K-C} is equal to 5. The calculation of the specific surface area S_0 is very important for the K-C formula. This study adopted the method of Chapuis and Aubertin [48], which assumes that the specific surface area per unit volume of approximately spherical and cubical soil particles [8,39] is:

$$S_0 = 6/D_E \tag{7}$$

Substituting Equation (7) into (6) yields:

$$k = \frac{1}{36}(\frac{\gamma}{\mu})(\frac{1}{C_{K-C}})\frac{D_E^2 e^3}{1+e} \tag{8}$$

Notably, the above formula does not take the particle shape factor into consideration. Therefore, when considering the influence of the shape factor on k, Equation (7) can take the following form:

$$S_0 = SF/D_E \tag{9}$$

where SF is the shape coefficient, where a different shape has a different SF value. The SF value is suggested as: spherical, SF= 6.0; rounded, SF = 6.1; worn, SF = 6.4; sharp, SF = 7.4; angular, SF = 7.7. Now, substituting Equation (9) into (8), the K-C formula modified based on grading entropy and the particle shape can be given by:

$$k = \left(\frac{\gamma}{\mu}\right)\left(\frac{1}{SF^2 C_{K-C}}\right)\frac{D_E^2 e^3}{1+e} \tag{10}$$

4.2. Determination of Saturated Hydraulic Conductivity and Comparisons

Figure 10 shows that six soil PSD curves were used in this study for determining their saturated hydraulic conductivity using the K-C formula with effective particle diameters from four different methods, including the proposed approach. Adopted from Choo et al. [49], the test parameters, including hydraulic conductivity results, could be used for verification. For completeness, ASTM D2434 was adopted to deduce the hydraulic conductivity of soils at room temperature to minimize losses in soil properties. The flow was introduced against the gravity under constant head conditions, while the head drop was temporally monitored through a series of manometers. The hydraulic conductivity was quantified as the slopes of flow velocity versus hydraulic gradient curves (i.e., Darcy's law) [49].

As Figure 11 shows, the calculated values of hydraulic conductivity using Kozeny's effective particle diameter have a relatively larger error for all six soils (above 50%), while some individual errors are even closer to 100%, as shown in Figure 11a. The errors between the measured value and predicted values using Carrier's method are also large, and the maximum error even exceeds 200%, although there are two errors within 20% to 50%, as shown in Figure 11b. These results may be because both Kozeny's and Carrier's effective diameter have a relatively higher error caused by their uncertain discretized interval sizes when used for calculating the hydraulic conductivity of soils. Likewise, the errors between predicted values from Fedorenko's method and the measured values remain between 20% and 50% (see Figure 11c). Fedorenko's effective particle diameter can be used to calculate the hydraulic conductivity for uniform soils; however, caution must be exercised when used for widely graded and gap-graded soils. It can be seen from Figure 11d that the errors of the K-C formula based on the effective particle diameter proposed in this study are much smaller (i.e., under 20%), such that the maximum error is 18.4%, whereas the minimum error is only 0.05%. This clearly establishes that the prediction accuracy of the proposed method is markedly higher than others.

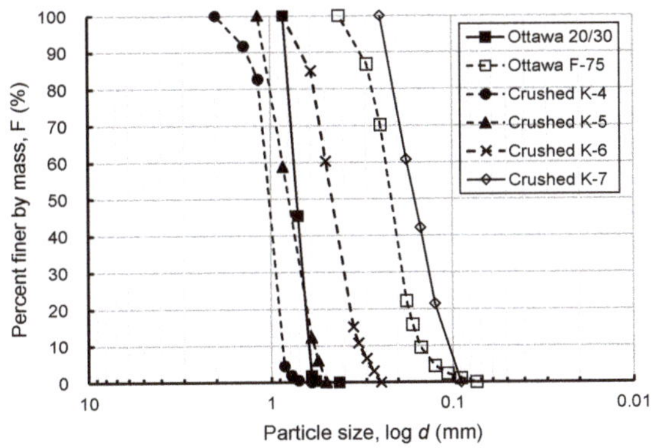

Figure 10. PSD curves adopted with permission from Choo et al. [49] used for permeability testing.

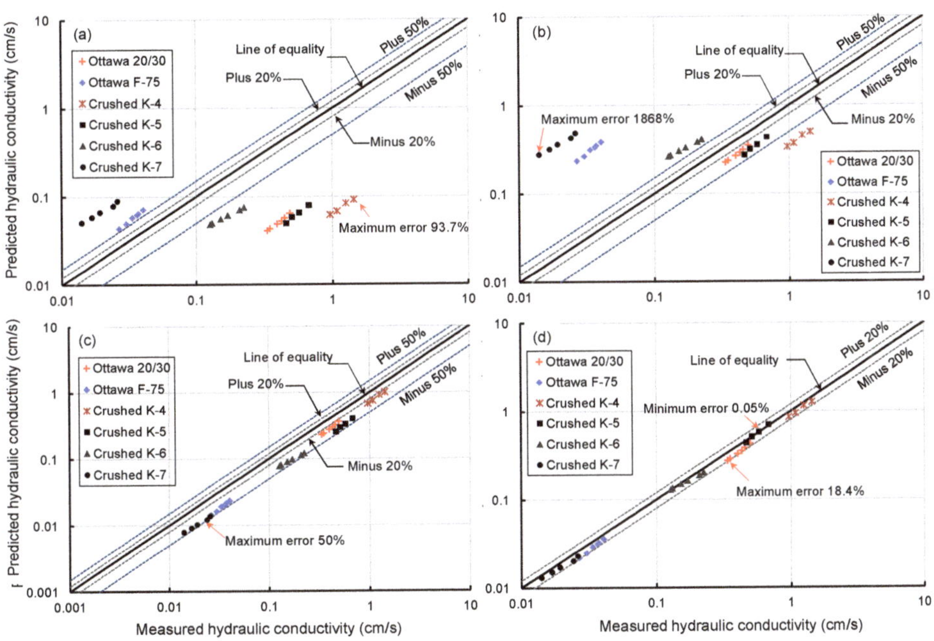

Figure 11. Comparison of the calculated values using four different effective particle diameter methods: (**a**) Kozeny's effective particle diameter method, (**b**) Carrier's method, (**c**) Fedorenko's method, and (**d**) current method proposed in this study.

For further verification, a larger experimental dataset of 30 laboratory results of hydraulic conductivity tests were adopted from Feng et al. [50]. The proposed entropy-based effective diameter was incorporated into the K-C formula and the hydraulic conductivity was estimated, which was then plotted against the experimental results, as shown in Figure 12. It is noteworthy that there are only three points plotted beyond the ±50% error line from the line of equality. For additional comparison, predictions from other methods that exceed a ±50% error have been summarized elsewhere by Feng et al. [50].

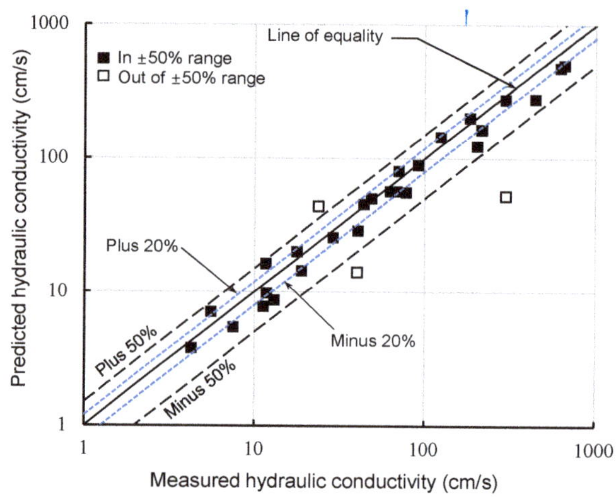

Figure 12. Verification of entropy-based effective particle diameter implications into K-C model through permeability test data from literature.

For brevity, out of 30 predictions, Hazen's [8] model results in 12 plotted beyond the ±50% error line, 9 for Shepherd's model, 7 for Kozeny–Carman's [8] model, 9 for Chapuis' model [1], and 6 for Feng et al.'s [50] model. Not surprisingly, the prediction accuracy of the hydraulic conductivity for the current model within the ±50% error range reaches 90%, which is higher than all five methods. Through further comparison, it was found that 17 out of the 30 calculated results of the current method are within the ± 20% error line, which means that the prediction accuracy within a 20% error range of this method exceeds 56%, which further shows that the current K-C formula has a higher prediction accuracy.

Given that the proposed effective particle diameter based on grading entropy could reasonably consider the grading information and particle shape of the soil, it could be conveniently integrated into existing K-C formula for estimating the hydraulic conductivity of a larger dataset of natural soils, thus showing a higher prediction accuracy than several well-accepted existing criteria tested in this study. While the authors still believe that the standard laboratory procedures remain the most reliable approaches, the proposed model may be conveniently adopted for only a prompt and preliminary measurement of the hydraulic conductivity of a soil.

5. Conclusions

Based on the discretization method of an abstract interval system of grading entropy, a novel approach for determining effective particle diameter was proposed. It was then compared with the existing effective diameter methods, such as Kozeny's, Carrier's, and Fedorenko's methods. In addition, the grading entropy-based effective particle diameter was integrated into Kozeny–Carman's formula for predicting the saturated hydraulic conductivity of soils. The specific findings from this study are as follows.

Although both Kozeny's and Carrier's effective particle diameters could be applied to non-uniform soils with an acceptable accuracy, they are sensitive to the discretized interval size, as the calculation error will become higher with the increase in interval size.

While effective particle sizes from Hazen (D10), Fedorenko (depending on C_u), and the median diameter (D50) are easy to compute, they cannot reasonably represent the grading information of widely and gap-graded soils, thus resulting in large errors. In contrast, the effective particle size proposed here is based on the grading entropy theory, which can more accurately represent the grading information of a soil's PSD than others.

The proposed entropy-based effective particle size was successfully implicated into the K-C model for determining the saturated hydraulic conductivity of granular soils. This proposed implication could be comprehensively demonstrated through the analysis of a large dataset from published studies, thereby showing above 90% and 56% accuracies with standard errors of up to 50% and 20%, respectively. Nevertheless, the proposed grading entropy-based effective particle size is expected to have broader implications in other geo-hydraulic problems, such as internal erosion, where the authors envisage extending this work further.

Author Contributions: Conceptualization, G.Z. and J.I.; methodology, G.Z.; software, J.I.; validation, J.I., W.M., Y.Y., and H.W.; formal analysis, G.Z.; investigation, J.I.; resources, H.W., K.R.; data curation, W.M.; writing—original draft preparation, J.I.; writing—review and editing, G.Z. and H.W.; visualization, W.M.; supervision, J.I.; project administration, G.Z.; funding acquisition, G.Z., W.M., Y.Y., K.R., and H.W. All authors have read and agreed to the published version of the manuscript.

Funding: This research was funded by National Natural Science Foundation of China (No. 41962016 and No. 51768059).

Institutional Review Board Statement: Not Applicable.

Informed Consent Statement: Not Applicable.

Data Availability Statement: All relevant supplementary material has already been reported in the main manuscript in the form of data in two tables and governing equations in both main manuscript and an Appendix A.

Acknowledgments: Financial supports from National Natural Science Foundation of China (No. 41962016 and No. 51768059) are appreciated. The corresponding author acknowledges financial support of Helan Mountain research scholarship program.

Conflicts of Interest: The authors declare no conflict of interest. The funders had no role in the design of the study; in the collection, analyses, or interpretation of data; in the writing of the manuscript, or in the decision to publish the results.

Appendix A. Review of Existing Methods of Effective Diameter Computations

A particle size distribution (PSD) curve is a group of data with continuous characteristics, wherein data are discretized into different groups. However, different discretization methods may yield different analysis results. Notably, the existing data discretization methods mainly include the equal width method [50], equal frequency method [51,52], entropy-based discretization method [53–55], clustering-based method [56], etc. Full et al. [57] pointed out that a more rigorous approach for analyzing a PSD curve could be through its discretization using interval classes. Nevertheless, discretization using variable size interval classes can make up for the defect of equal size discretization, such as sparse data. Not surprisingly, the existing approaches adopt characteristic methods based on discrete data with equal size intervals, which may not capture the grading information fully for the widely and gap-graded PSD curves.

Hazen [28] pioneered the empirical determination of saturated hydraulic conductivity based on the soil's particle size corresponding to a 10 percent finer by mass on the PSD curve (D10). Assuming that the flow through soil is analogous to the pipe flow, it was proposed that the finer fraction of a soil bears a close relationship with the pore sizes governing the flow, and hence hydraulic conductivity. Consequently, the particle size D10 was presented as being an approximate representative of soil's finer fraction and thus the hydraulic conductivity, while the contributing physical characteristics of soil was given by an empirical shape factor C, which varies with the soil type. For instance, it varies between

120 to 150 and 40 to 80 for well-graded coarse sand and very fine sand, respectively. Later, Fedorenko proposed an effective particle size based on Hazen's D10 and soil's uniformity coefficient, given by Equation (A1):

$$D_F = 1/2(D_{10} + D_{60})\sqrt{D_{10}/D_{60}} \tag{A1}$$

Kozeny [40] discretized the PSD curve into several uniform-sized intervals given by the number n, with an average diameter of the i-th interval given by D_i and the mass percentage of this interval given by ΔS_i. Assuming that all particles have same surface area and volume ratio:

$$\frac{N\pi {D_h}^2}{N\pi {D_h}^3/6} = \frac{\sum \frac{G_i}{\gamma_s}\frac{6}{D_i}}{\sum \frac{G_i}{\gamma_s}} = \frac{\sum \Delta S_i \frac{6}{D_i}}{\sum \Delta S_i} \tag{A2}$$

$$D_h = \frac{1}{\sum \frac{\Delta S_i}{D_i}} \tag{A3}$$

where D_h is the effective particle diameter, N is the number of spherical particles in the soil sample, γ_s is the unit weight of soil solids, G_i is the specific gravity of soil, and D_i is the mean particle size for adjacent intervals:

$$D_i = \frac{3}{\frac{1}{D_{i1}} + \frac{2}{D_{i1}+D_{i2}} + \frac{1}{D_{i2}}} \tag{A4}$$

Bear [37] proposed an effective particle diameter based on the harmonic mean value of select particle sizes from the soil's PSD curve:

$$D_{\text{eff}} = \sum m_i / \sum (m_i/D_i) \tag{A5}$$

Subsequently, Koltermann and Gorelick [38] observed that the harmonic mean value only bears a good relationship with the finer fraction of a PSD rather than the coarser fraction. For instance, it may not represent the PSD curves with less fine contents rationally. However, the geometric mean bears a stronger correlation with the PSD curves that have larger coarse contents. Similarly, Vienken and Dietrich [41] presented the following formula for the effective particle size:

$$d_e = \frac{0.1}{\frac{3\Delta g_m}{2d_m} + \sum_{i=2}^{i=n} \frac{\Delta g_i}{d_i}} \tag{A6}$$

where Δg_m is the weight of the finer fraction, Δg_i is the weight of the i-th class interval, d_m is the diameter of the last fraction, $\frac{1}{d_i} = \frac{1}{2} \times \left(\frac{1}{d_u} + \frac{1}{d_l}\right)$, d_u is the upper fraction limit, and d_l is the lower fraction limit.

More recently, Carrier [8] proposed an effective particle diameter for non-uniform spherical particles from soil's PSD as follows:

$$D_{\text{eff}} = 100\% / \sum (f_i / D_{\text{ave},i}) \tag{A7}$$

where f_i is the mass percentage (%) of particles between adjacent interval sizes; $D_{\text{ave},i} = D_{li}^{0.5} \cdot D_{si}^{0.5}$, D_{li}, and D_{si} is the larger and smaller fraction size in the i-th interval, and $D_{\text{ave},i}$ becomes the geometric average particle size of the i-th interval. Notably, Carrier's calculation procedure is like that presented by Kozney, with the only difference being in the choice of the size of calculation intervals. In essence, considering the PSD curve as log-linear in each interval size range, Carrier [8] corrected the calculation of the geometric average particle size of each interval as follows [46,48]:

$$D_{\text{ave},i} = D_{li}^{0.404} \cdot D_{si}^{0.595} \tag{A8}$$

References

1. Chapuis, R.P. Predicting the saturated hydraulic conductivity of sand and gravel using effective diameter and void ratio. *Can. Geotech. J.* **2004**, *41*, 787–795. [CrossRef]
2. Dyka, I.; Srokosz, P.E.; Bujko, M. Influence of grain size distribution on dynamic shear modulus of sands. *Open Eng.* **2017**, *7*, 317–329. [CrossRef]
3. Kenney, T.; Lau, D. Internal stability of granular filters. *Can. Geotech. J.* **1985**, *22*, 215–225. [CrossRef]
4. Tong, C.-X.; Burton, G.J.; Zhang, S.; Sheng, D. A simple particle-size distribution model for granular materials. *Can. Geotech. J.* **2018**, *55*, 246–257. [CrossRef]
5. Tsuda, A.; Henry, F.S.; Butler, J.P. Particle transport and deposition: Basic physics of particle kinetics. *Compr. Physiol.* **2013**, *3*, 1437–1471. [PubMed]
6. Seelheim, F. Methoden zur bestimmung der durchlassigkeit des bodens. *Z. Anal. Chem.* **1880**, *19*, 387–402. [CrossRef]
7. Hazen, A. *Some Physical Properties of Sands and Gravels, with Special Reference to Their Use in Filtration*; 24th Annual Report; Massachusetts State Board of Health: Boston, MA, USA, 1892; pp. 539–555.
8. Carrier, W.D. Goodbye, Hazen; Hello, Kozeny-Carman. *J. Geotech. Geoenviron. Eng.* **2003**, *129*, 1054–1056. [CrossRef]
9. Chen, H.; Lv, X.; Qiao, Y. Fitting four-modal sedimentary grain-size distribution with gradient descent method. In Proceedings of the 2011 International Conference on Remote Sensing, Environment and Transportation Engineering, Nanjing, China, 24–26 June 2011; IEEE: Piscataway, NJ, USA, 2011; pp. 5987–5990.
10. Wichtmann, T.; Triantafyllidis, T. Influence of the Grain-Size Distribution Curve of Quartz Sand on the Small Strain Shear Modulus Gmax. *J. Geotech. Geoenviron. Eng.* **2009**, *135*, 1404–1418. [CrossRef]
11. Burenkova, V. Assessment of suffusion in non-cohesive and graded soils. In Proceedings of the 1st International Conference "Geo-Filters", Filters in geotechnical and hydraulic engineering, Karlsruhe, Germany, 20–22 October 1992; pp. 357–360.
12. Chapuis, R.P. Similarity of internal stability criteria for granular soils. *Can. Geotech. J.* **1992**, *29*, 711–713. [CrossRef]
13. Kezdi, A. *Soil Physics: Selected Topics*; Elsevier Science: Amsterdam, The Netherlands, 1979.
14. Lőrincz, J. Relationship between grading entropy and dry bulk density of granular soils. *Period. Polytech. Civ. Eng.* **1990**, *34*, 255–263.
15. Indraratna, B.; Israr, J.; Rujikiatkamjorn, C. Geometrical method for evaluating the internal instability of granular filters based on constriction size distribution. *J. Geotech. Geoenviron. Eng.* **2015**, *141*, 4015045. [CrossRef]
16. Israr, J.; Zhang, G. Geometrical assessment of internal instability potential of granular soils based on grading entropy. *Acta Geotech.* **2021**, *16*, 1961–1970. [CrossRef]
17. Lőrincz, J.; Imre, E.; Fityus, S.; Trang, P.Q.; Tarnai, T.; Talata, I.; Singh, V.P. The grading entropy-based criteria for structural stability of granular materials and filters. *Entropy* **2015**, *17*, 2781–2811. [CrossRef]
18. Hateren, J.A.; Prins, M.A.; van Balen, R.T. On the genetically meaningful decomposition of grain-size distributions: A comparison of different end-member modelling algorithms. *Sediment. Geol.* **2018**, *375*, 49–71. [CrossRef]
19. Varga, G.; Újvári, G.; Kovács, J. Interpretation of sedimentary (sub) populations extracted from grain size distributions of Central European loess-paleosol series. *Quat. Int.* **2019**, *502*, 60–70. [CrossRef]
20. Chapuis, R.P.; Saucier, A. Assessing internal erosion with the modal decomposition method for grain size distribution curves. *Acta Geotech.* **2020**, *15*, 1595–1605. [CrossRef]
21. Fredlund, M.D.; Fredlund, D.; Wilson, G.W. An equation to represent grain-size distribution. *Can. Geotech. J.* **2000**, *37*, 817–827. [CrossRef]
22. Foster, M.; Fell, R. Assessing embankment dam filters that do not satisfy design criteria. *J. Geotech. Geoenviron. Eng.* **2001**, *127*, 398–407. [CrossRef]
23. Indraratna, B.; Raut, A.K.; Khabbaz, H. Constriction-based retention criterion for granular filter design. *J. Geotech. Geoenviron. Eng.* **2007**, *133*, 266–276. [CrossRef]
24. Khor, C.H.; Woo, H.K. Investigation of crushed rock filters for dam embankment. *J. Geotech. Eng.* **1989**, *115*, 399–412. [CrossRef]
25. Lafleur, J. Filter testing of broadly graded cohesionless tills. *Can. Geotech. J.* **1984**, *21*, 634–643. [CrossRef]
26. Sherard, J.; Dunnigan, L. Critical filters for impervious soils. *J. Geotech. Eng.* **1989**, *115*, 927–947. [CrossRef]
27. Terzaghi, K. Der grundgruch an stauwerken und seine verhutung (The failure of dams by piping and Its prevention). *Die Wasserkraft* **1922**, *17*, 445–449.
28. Trani, L.D.O.; Indraratna, B. The use of particle size distribution by surface area method in predicting the saturated hydraulic conductivity of graded granular soils. *Géotechnique* **2010**, *60*, 957–962. [CrossRef]
29. Beyer, W. On the determination of hydraulic conductivity of gravels and sands from grain-size distribution. *Wasserwirtsch. Wassertech.* **1964**, *14*, 165–169.
30. Harleman, D.R.F.; Melhorn, P.F.; Rumer, R.R. Dispersion-permeability correlation in porous media. *J. Hydraul. Div.* **1963**, *89*, 67–85. [CrossRef]
31. Terzaghi, C. Principles of soil mechanics: III—Determination of permeability of clay. *Eng. News Rec.* **1925**, *95*, 832–836.
32. Kozeny, J. Ueber kapillare Leitung des Wassers im Boden. *Sitzungsber Akad. Wiss.* **1927**, *136*, 271–306.
33. Vukovic, M.; Soro, A. *Determination of Hydraulic Conductivity of Porous Media from Grain-Size Composition*; Water Resources Publications: Littleton, CO, USA, 1992.
34. Koenders, M.A.; Williams, A.F. Flow equations of particle fluid mixtures. *Acta Mech.* **1992**, *92*, 91–116. [CrossRef]

35. Waters, T.J. A study of water infiltration through ASPHALT road surface materials. In Proceedings of the International Symposium on Subdrainage in Roadway Pavements and Subgrades, Granada, Spain, 11–13 November 1998; pp. 311–317.
36. Vardanega, P.J.; Waters, T.J. Analysis of Asphalt Concrete Permeability Data Using Representative Pore Size. *J. Mater. Civ. Eng.* **2011**, *23*, 169–176. [CrossRef]
37. Bear, J. *Dynamics of Fluids in Porous Media*; Dover Publications, Inc.: New York, NY, USA, 1972.
38. Koltermann, C.E.; Gorelick, S.M. Fractional packing model for hydraulic conductivity derived from sediment mixtures. *Water Resour. Res.* **1995**, *31*, 3283–3297. [CrossRef]
39. Kovács, G. *Seepage Hydraulics*; Akadénmiai Kiadó: Budapest, Hungary, 1981.
40. Kozeny, J. Das Wasser im Boden. Grundwasserbewegung. In *Hydraulik*; Springer: Vienna, Austria, 1953; pp. 380–445.
41. Vienken, T.; Dietrich, P. Field evaluation of methods for determining hydraulic conductivity from grain size data. *J. Hydrol.* **2011**, *400*, 58–71. [CrossRef]
42. Lőrincz, J.; Imre, E.; Gálos, M.; Trang, Q.P.; Rajkai, K.; Fityus, S.; Telekes, G. Grading entropy variation due to Soil crushing. *Int. J. Geomech.* **2005**, *5*, 311–319. [CrossRef]
43. Zunker, F. Fertilization and soil science. *J. Plant Nutr. A* **1932**, 25.
44. Israr, J. Internal Instability of Granular Filters under Cyclic Loading. Ph.D Thesis, University of Wollongong, Wollongong, Australia, 2016.
45. Andrianatrehina, L.; Souli, H.; Fry, J.-J.; Phan, Q.; Fleureau, J.-M. Internal stability of granular materials in triaxial tests. In Proceedings of the 6th International Conference on Scour and Erosion, France, Paris, 27–31 August 2012. ISSMGE.
46. Li, M. Seepage Induced Instability in Widely Graded Soils. Ph.D. Thesis, University of British Columbia, Vancouver, BC, Canada, 2008.
47. Das, B.M. *Advanced Soil Mechanics*; Tayor & Francis Group: New York, NY, USA, 2008.
48. Chapuis, R.P.; Aubertin, M. On the use of the Kozeny—Carman equation to predict the hydraulic conductivity of soils. *Can. Geotech. J.* **2003**, *40*, 616–627. [CrossRef]
49. Choo, H.; Kim, J.; Lee, W.; Lee, C. Relationship between hydraulic conductivity and formation factor of coarse-grained soils as a function of particle size. *J. Appl. Geophys.* **2016**, *127*, 91–101. [CrossRef]
50. Feng, S.; Vardanega, P.J.; Ibraim, E.; Widyatmoko, I.; Ojum, C.; O'Kelly, B.C.; Nogal, M. Permeability assessment of some granular mixtures. *Géotechnique* **2019**, *70*, 845–847. [CrossRef]
51. Saleh, A.; Puspita, K.; Sanjaya, A. Implementation of equal width interval discretization on smarter method for selecting computer laboratory assistant. In Proceedings of the 2018 6th International Conference on Cyber and IT Service Management (CITSM), Parapat, Indonesia, 7–9 August 2018; pp. 1–4.
52. Hacibeyoglu, M.; Ibrahim, M.H. EF_Unique: An improved version of unsupervised equal frequency discretization method. *Arab. J. Sci. Eng.* **2018**, *43*, 7695–7704. [CrossRef]
53. McDougall, J.R.; Imre, E.; Barreto, D.; Kelly, D. Volumetric consequences of particle loss by grading entropy. *Géotechnique* **2013**, *63*, 262–266. [CrossRef]
54. Imre, E.; Lőrincz, J.; Szendefy, J.; Trang, P.Q.; Nagy, L.; Singh, V.P.; Fityus, S. Case studies and benchmark examples for the use of grading entropy in geotechnics. *Entropy* **2012**, *14*, 1079–1102. [CrossRef]
55. Singh, V.P. *Entropy Theory in Hydraulic Engineering: An Introduction*; ASCE Press: Reston, VA, USA; ProQuest Ebook Central: Ann Arbor, MI, USA, 2014.
56. Henry, D.; Dymnicki, A.B.; Mohatt, N.; Allen, J.; Kelly, J.G. Clustering methods with qualitative data: A mixed-methods approach for prevention research with small samples. *Prev. Sci.* **2015**, *16*, 1007–1016. [CrossRef]
57. Full, W.E.; Ehrlich, R.; Kennedy, S.K. Optimal definition of class intervals for frequency tables. *Part. Sci. Technol.* **1983**, *1*, 281–293. [CrossRef]

fractal and fractional

MDPI

Article

Investigation on Pore Structure and Permeability of Concrete–Rock Interfacial Transition Zones Based on Fractal Theory

Juan Yue [1], Jinchang Sheng [1,*], Huimin Wang [1], Yunjin Hu [2,*], Kailai Zhang [1], Yulong Luo [1], Qing Zhou [1] and Meili Zhan [1]

1 College of Water Conservancy and Hydropower Engineering, Hohai University, Nanjing 210098, China; yuejuan@hhu.edu.cn (J.Y.); huimin.wang@hhu.edu.cn (H.W.); zhangkailai@hhu.edu.cn (K.Z.); lyl8766@hhu.edu.cn (Y.L.); 130402010004@hhu.edu.cn (Q.Z.); 191302010021@hhu.edu.cn (M.Z.)
2 School of Civil Engineering, Shaoxing University, Shaoxing 312000, China
* Correspondence: jinchang@hhu.edu.cn (J.S.); huyunjin@tsinghua.org.cn (Y.H.)

Abstract: The concrete–rock interfacial transition zone (ITZ) is generally considered the weak layer in hydraulic engineering, for it is more permeable than the intact concrete or rocks. The water permeability of the ITZ is a critical parameter concerned with structural safety and durability. However, the permeability and pore structure of the ITZ has not been investigated previously, and the mathematical model of ITZ permeability has not been established. This study performed multi-scale experiments on the concrete–rock ITZ with various rock types (limestone, granite, and sandstone). A series of quantitative and qualitative analysis techniques, including NMR, SEM-EDS, and XRD, characterize the ITZ pore structures. The controlled constant flow method was used to determine the permeability of the concrete, rock, and ITZ. The mathematical model of ITZ permeability was proposed using the fractal theory. The consistency between the experimental data and the proposed model indicates the reliability of this study. The results of the experiment show that ITZ permeability is between 4.08×10^{-18} m^2 and 5.74×10^{-18} m^2. The results of the experiment and the proposed model could determine ITZ permeability in hydraulic structure safety and durability analysis.

Keywords: concrete–rock ITZ; pore structure; permeability; fractal theory

Citation: Yue, J.; Sheng, J.; Wang, H.; Hu, Y.; Zhang, K.; Luo, Y.; Zhou, Q.; Zhan, M. Investigation on Pore Structure and Permeability of Concrete–Rock Interfacial Transition Zones Based on Fractal Theory. *Fractal Fract.* **2022**, *6*, 329. https://doi.org/10.3390/fractalfract6060329

Academic Editor: Zine El Abiddine Fellah

Received: 7 May 2022
Accepted: 9 June 2022
Published: 13 June 2022

1. Introduction

The concrete–rock interfacial transition zone (ITZ) widely exists in hydraulic engineering, such as arch dams, concrete gravity dams, and tunnels. The concrete–rock ITZ is characterized by high porosity, poor densification, calcium hydroxide (CH) enrichment, high permeability, and low shear strength, which is generally considered the weak layer and adversely impacts hydraulic structures. Thus, research on the properties of the ITZ has been a hot topic in hydropower engineering projects [1–3]. The durability of the concrete–rock structure depends on the penetration of water and corrosive ions into the matrix [4]. As the weakest layer of the whole system, the concrete–rock interface is considered one of the main flow channels for the erosion solution. The deterioration of the mechanical properties of the ITZ is vital to the stability of hydropower engineering projects [5].

The macroscopic physical properties of the ITZ are generally dependent on ITZ microstructures. Therefore, many researchers have conducted experiments to investigate the mineral composition and microstructures of the ITZ. Wang et al. [6] investigated the microstructure of the ITZ between the aggregate and cement slurry at different stages. The results showed that the porosity and micro-fracture length gradually increased while the average pore size decreased with the curing process. Several researchers [7–11] have studied the effect of the concrete mixture on the ITZ microstructures by adding silica fume, nano-silica, rice husk ash, kaolin, slag, fly ash, pozzolan, etc. It is believed that the addition of colloids can fill the accumulation pores of cement particles and reduce the content of calcium hydroxide (CH), increasing the ITZ compactness. The influences of aggregates (including light aggregates, heavy aggregates, different particle sizes and shapes) on the ITZ microstructural characteristics were discussed by Lin [12], Bentz et al. [13], and Lyu et al. [14], respectively. The results indicate that the ITZ between the light porous aggregate and cement slurry generates a denser ITZ microstructure.

Permeability is a crucial parameter closely related to the project's seepage characteristics and benefits in hydraulic engineering. As a potential leakage channel, the permeability of the ITZ has been investigated by many researchers. Skalny [15,16] first proposed to investigate the permeability characteristics of the ITZ by comparing the permeability test results of pure rock, cement-based and cement slurry–rock. Subsequently, many scholars have obtained the permeability of the ITZ by this method and proved that the permeability of the ITZ is far greater than mortar [17,18]. However, some scholars have assumed that although the permeability of the ITZ is large, it is not necessarily the permeability channel of the whole structure [15,16]. Since the experimental determination process is deemed time-consuming, and only some discrete data points can be obtained in the experimental operation, researchers attempted to use numerical simulation methods and analytical models to calculate the permeability of the ITZ. Snyder et al. [19] investigated the seepage mechanism of the ITZ around the spherical aggregate in mortar by the sphere model. Bentz et al. [13,20] and Zheng et al. [21,22] simplified the aggregates into spherical and polygonal agglomerates and developed a multiscale spherical stacking model to estimate the effects of the ITZ thickness, aggregate particle size distribution and aggregate type on the seepage characteristics of the ITZ. In order to obtain a more realistic three-dimensional model of the ITZ, Kim et al. [23] used μ-CT images to describe the cement slurry samples with different porosity by a low-order probability function. They reconstructed a three-dimensional ITZ microstructure framework with porosity gradients. Li et al. [24,25] constructed the cement slurry based on the Discrete Element Method (DEM), and the porosity and permeability of the cement matrix and the ITZ were obtained. Jia et al. [26,27] proposed an effective and convenient approach to obtain the intrinsic permeability of the tight porous media and constructed a 3D model, including the fracture penetrating the shale core to simulate the flow process in the fracture-matrix system realistically. Sun et al. [28] reconstructed the 3D microstructure image of the ITZ based on SEM imaging technology and obtained the total porosity, seepage porosity, and permeability of the ITZ using the permeability solving code.

Although the permeability properties and the microstructure characteristics of the ITZ have been consistently studied, the relationship between permeability and microstructure has not been established. Most existing permeability models are based on the intrinsic permeability calculation model of porous media. It was thought that the permeability of the ITZ is in direct proportion to the porosity, pore connectivity, and pore size [28–30]. Some models considered the effect of simplified ITZ microstructures on permeability [30]. However, the high complex and irregular geometry features have been ignored [31–34]. Many scholars have simulated cement and rock internal pore structures based on a fractal model; pore structure characteristics can be described by fractal parameters such as fractal dimensions of the pore surface, pore tortuosity, and fracture surface [35–40]. The studies provide a theoretical background to the proposed functional relationships between fractal dimensions and basic macro properties.

The objective of this study was to investigate the permeability of the concrete–rock ITZ and develop a mathematical model based on the fractal theory. Firstly, most of the previous microstructure research is focused on the ITZ between the aggregate and cement slurry. There is a lack of systematic research on the ITZ between the bedrock and concrete structure, especially concerning the permeability of the ITZ. Secondly, the influence of the ITZ on the leakage is still uncertain, and some scholars even believe that the ITZ may not be a seepage channel. Finally, the relationship between the permeability and microstructure of the ITZ has not been fully established. It was therefore necessary to introduce the micropore structure characteristics into the permeability model and quantify the effect of microcracks and pores on the seepage characteristics in the ITZ.

This study performed macro and micro experiments on the concrete–rock ITZ with different rock types (limestone, granite, and sandstone). A series of quantitative and qualitative analysis techniques, including NMR, SEM-EDS, and XRD, were used to characterize the ITZ pore structures and mineralogy. The controlled constant flow method was used to determine the permeability of the concrete, rock, and the ITZ, improving the reliability of the test results. Based on the experimental results and fractal theory, the ITZ permeability model was proposed to relate pore structure parameters and permeability. The proposed model could provide reasonable permeability parameters for hydraulic engineering seepage analysis.

2. Materials and Methods

2.1. Sample Preparation

The rocks used in the experiment (sandstone, limestone and granite) were taken from Huaqi Quarry in Nanjing, Jiangsu Province. Selected bedrocks were cut into 100 × 100 × 50 mm rectangular and 50 × 100 mm cylindrical specimens with an accuracy control of ±0.1 mm The compressive strength grade of concrete is C30, prepared by P.O. 42.5 ordinary silicate cement, standard sand (2.3~3.0 mm) and continuous graded gravel (particle size 3~15 mm). The mixing proportion of the concrete is shown in Table 1.

Table 1. Concrete grout mixture proportions.

Material	Water	Cement	Crush Stone	Sand	Admixtures
Concrete	0.36	1	2.824	1.392	0.005

The casting steps were as follows:

(1) The cleaned rock was placed into the molds and the raw materials were prepared strictly according to the C30 concrete mixing ratio. The weighed cement, sand, stones, admixtures, and water were poured into the concrete mixer and mixed for 60 s.

(2) After mixing was completed, the concrete was poured sequentially into the steel molds. Then, the molds were placed on a vibrating table to remove air bubbles from the samples. Vibrating time should not exceed 30 s to avoid the effect of water secretion.

(3) The composite specimen was dismantled within 24 h, put into a standard curing box for 28 days, and finally, a cube specimen of 100 × 100 × 100 mm was obtained.

(4) According to the sample size required for the penetration test, a coring machine was used to take the core from the cube specimen. Finally, a cylindrical example of 50 × 100 mm was obtained. Figure 1 shows the fabrication process of the cylindrical concrete–rock mixed specimen.

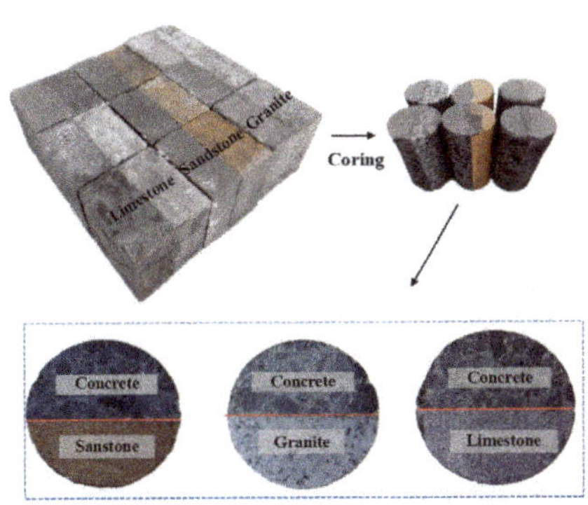

Figure 1. Preparation of concrete–rock mixed samples.

2.2. Permeability Test and Procedure

After sampling was completed, bedrock-concrete samples were selected and a laser controlled cutting machine was used to cut the samples into the size required by the macro and micro test instruments. As shown in Figure 2, the operation method for each test was as follows:

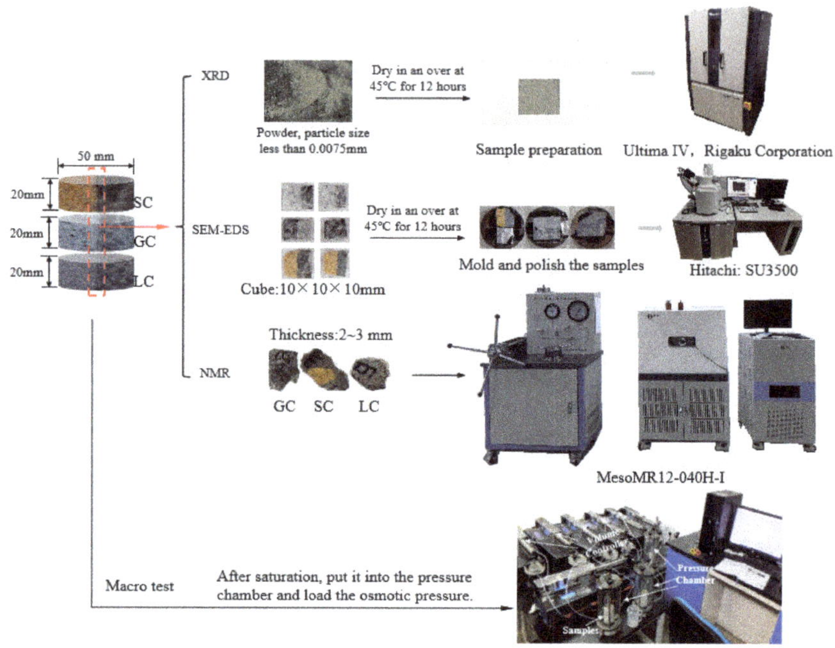

Figure 2. Macro and micro test schemes.

The size of the X-ray Diffractometer (XRD) test sample was powder with a particle size less than 0.0075 mm. The cement slurry of the ITZ was ground into powder and then placed in a drying oven at 45 °C for 12 h until it reached a constant weight. Then, it was placed on the observation lens in the Ultima IV observation cavity. The diffraction angle range was 5~80° and the scanning speed was 3°/s. Finally, the phase content of the ITZ was obtained.

- The samples for the scanning electron microscope-energy spectrum analysis (SEM-EDS) were a cube of 10 × 10 ×10 mm, and the sample contained both rock and concrete. Then, the samples were dried in a drying oven at 45 °C for 12 h to a constant weight. The samples were scanned using a SU3500-HITACHI SEM-EDS at an accelerating voltage of 3000 V. The image was focused, adjusted, and photographed at different magnifications. Under the observation condition of magnification of 500, a regular line was drawn in the adjacent area of rock and concrete. The starting point of the standard line was inside the rock, and the endpoint was inside the concrete slurry. The length of the entire element distribution observation line was within 800 μm.
- The samples for the Nuclear Magnetic Resonance (NMR) were a 2~3 mm block mortar of ITZ, and the samples were saturated with vacuum-saturation equipment (type: MesoMR12-040H-I). Subsequently, the saturated samples were wrapped in plastic film and placed inside the sample chamber. After finding the appropriate center frequency, the proper relaxation time and echo time were selected according to the sample and the pore size distribution curve was obtained. Finally, the porosity and pore size distribution curve of the samples were obtained by inversion.

- The samples for the macro test were cylindrical specimens of 50 × 20 mm. Firstly, the samples were saturated with vacuum-saturation equipment. Subsequently, the saturated samples were placed inside the pressure chamber of cement-based materials steady-state permeation test equipment. According to the project operation environment, the seepage pressure was set at 2 MPa until the test ended.

3. Results

3.1. Pore-Structure of Bedrocks and Concrete

The porosity of limestone, granite, sandstone, and concrete is 0.52~0.64%, 0.31~0.48%, 16.5~17.8%, and 4.27~4.95%, respectively. Figure 3 shows the pore size distribution characteristics of the bedrock and concrete matrix. As shown in Figure 2, the pore size distribution of limestone ranged from 0.000802 to 4.709 μm with a peak size of 0.05 μm. In comparison, the pore size ranged from 0.03212 to 23.246 μm and from 0.001216 to 40.510 μm for sandstone. Compared with limestone, the pore size distribution of granite and sandstone drifts to the large pore, and the peak sizes are about 1 μm and 3 μm, respectively. The pore size of concrete ranged from 0.000429 to 0.0341 μm; 90% of the pore diameters are concentrated within 0.03 μm.

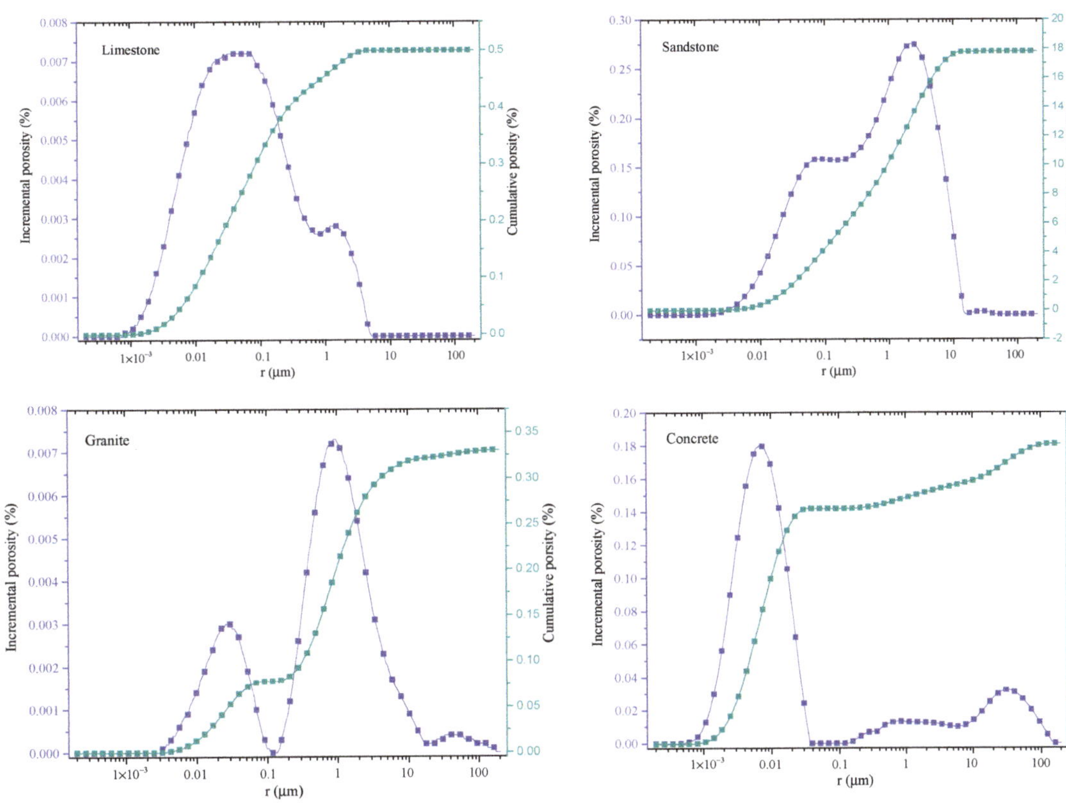

Figure 3. Comparison of pore size distribution of rock.

This study used Jullien's aperture classification method [41]. Pores were divided into gel pores (0.00~0.01 μm), medium capillary pores (0.01~0.05 μm), large capillary pores (0.05~10.00 μm), and macropores (>10 μm). Figure 4 gives the pore size distribution ratios of three types of rocks and concrete. Granite and sandstone have similar pore size distribution curves, mainly composed of medium and large capillary pores (more than 90%). While limestone contains more gel pores, it has a smaller average pore size.

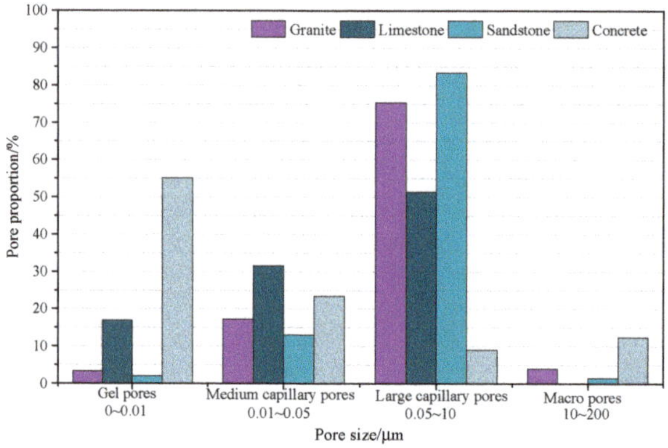

Figure 4. Mortar pore size proportion around rock and concrete.

Since sandstone and granite contain more medium and large pores, the cement slurry will penetrate into the pores near the bedrock interface and accumulate in the rock joint's surface area, forming a better interface structure. Limestone contains more gel pores than granite and sandstone, which leads to poor bonding of the concrete–rock interface. However, the infiltration of cement slurry is also affected by porosity. Therefore, the microstructure characteristics of different concrete–rock interface transition zones need further study.

3.2. Calcium Compounds of ITZ

The bonding behavior between the rock and concrete is determined by the density of the ITZ and the content of calcium silicate hydrate (CSH) [42]. In order to investigate whether different phase compositions on the ITZ are related to rock types, the difference in hydration products on the surface was monitored by XRD. The XRD patterns in the range of 5–80° 2θ form the granite–concrete, sandstone–concrete, limestone, and concrete surface powders in Figure 5 with the same intensity scales. The crystalline phases identified in the ITZ include some typical cement hydration products, such as calcium hydroxide (CH), calcium silicate hydrate (CSH), and ettringite (Aft). The main hydration products are CSH gel, which is in line with the finding from [43].

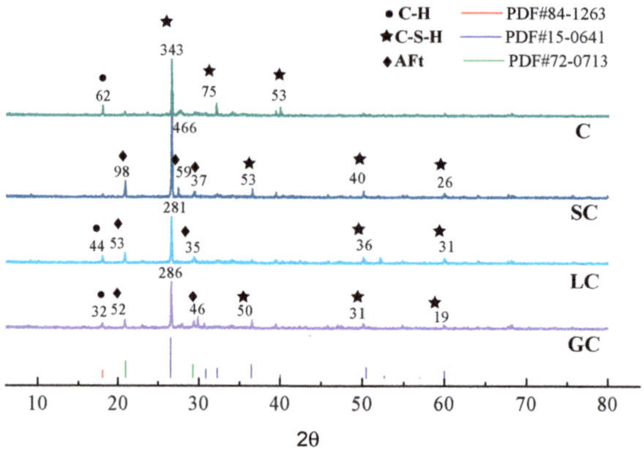

Figure 5. XRD patterns of concrete and ITZ.

Furthermore, the cement hydration products of concrete and the ITZ of rock and concrete are given in Figure 6. Previous studies have shown that the more CSH gel was formed, the denser ITZ was formed, and the more CH was formed, the higher the permeability of the ITZ [44,45]. XRD analysis shows that the content of CSH gel in the ITZ of concrete–rock is less than that of the cement slurry inside the concrete. This is because rock hinders the migration of water in the cement slurry, which makes water accumulate in the surface area of rock and forms a water film, providing a channel for the rapid migration of calcium ions, sulfate ions, and hydroxide ions. These ions eventually enrich at the rock interface and produce large amounts of calcium hydroxide and ettringite. The specimen around sandstone has the most hydrated products of CSH and has a lesser number of CH. In contrast, there is the least amount of CSH and the most amount of CH in the ITZ of limestone–concrete. The sandstone has the worst compactness from the pore structure characteristics of the three types of bedrock (the most significant porosity). The water absorption rate is much higher than that of limestone and granite, resulting in the lowest water–cement ratio in the sandstone–concrete interface transition zone. Compared with granite, limestone has the best compactness (minimum porosity), but macro pores dominate its pore size distribution. A larger pore can provide more contact surfaces and generate more CSH gel to fill micro-cracks and pores.

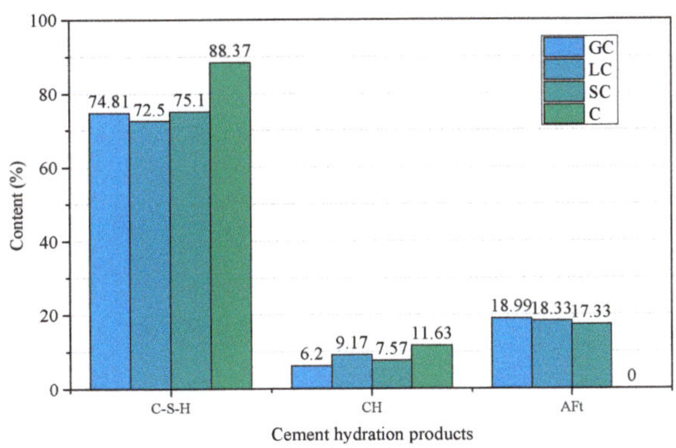

Figure 6. Content of Calcium compounds.

3.3. *Thickness of ITZ*

The content of CH in the ITZ was higher than the cement in the concrete, resulting in a region near the aggregate containing predominately fine particles and having a higher porosity [46]. Figure 7 shows the SEM images of concrete, limestone–concrete ITZ, granite–concrete ITZ and sandstone–concrete ITZ. In particular, three magnifications 100×, 500×, and 5000× were tested. As shown in Figure 7a, CSH gel, Aft and CH were formed on the concrete surface. The SEM images (Figure 7b) show no apparent boundary between limestone and concrete. However, at 5000× magnification, fractures can be easily observed, and the fracture width is measured at 1.08~1.35 μm. Compared with the limestone–concrete ITZ, the granite surface is layered and granular, strengthening the connection between rock and concrete. There is cement slurry immersion between the quartz and particle. We observed that a small amount of calcium hydroxide crystals were filled between the quartz particles so that the micro-cracks could not penetrate the entire observation area. The original "crack" is split into two pores with a diameter of about 3 μm. As for ITZ of sandstone and concrete, the interface area can be clearly observed due to the massive difference in pore size and pore structure between sandstone and concrete. There are no apparent pores on the surface of the concrete, and the shape is very dense, while the sandstone side is composed of pores and sandstone particles. The pores between the sandstone particles were covered with a cement hydration product. This result is consistent with typical laws of the aggregate influence on the micromorphology of the ITZ reported in the literature [7].

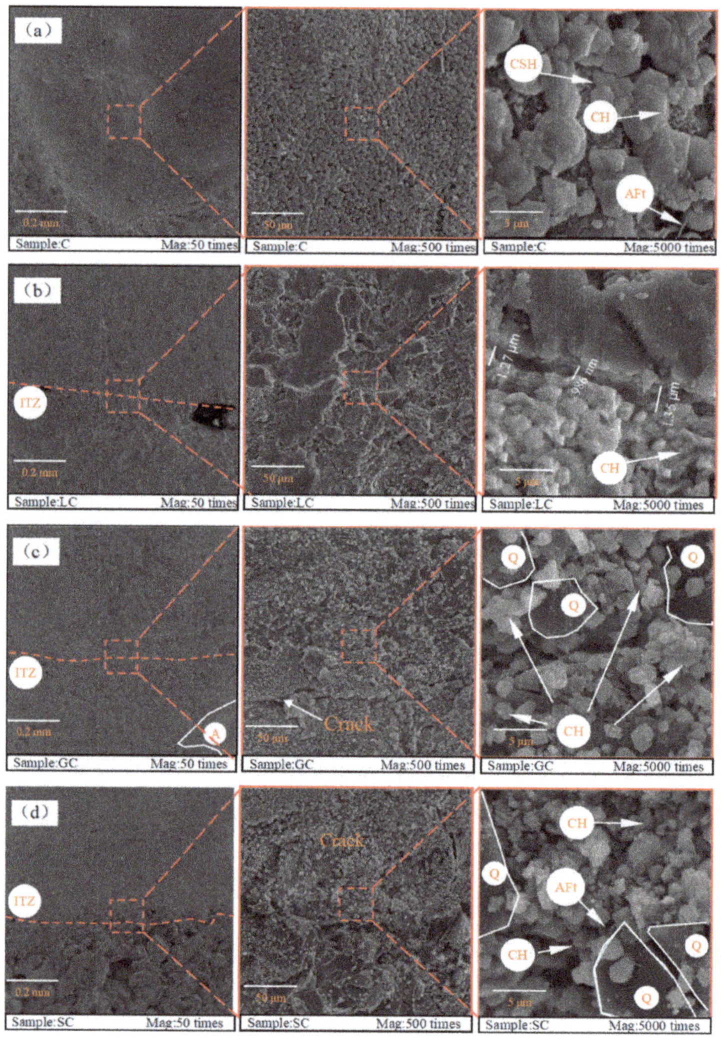

Figure 7. SEM images of ITZ microstructure at different magnifications: (**a**) C; (**b**) LC; (**c**) GC; (**d**) SC; "A"—aggregate, "Q"—quart particles.

The hydration products in granite–concrete and sandstone–concrete specimens are uniformly filled with residual pores. They form a dense microstructure, especially sandstone–concrete, which can improve the permeability of the ITZ. On the contrary, the remaining pores in limestone are more significant than other pores. The hydration products such as CSH gel and $Ca(OH)_2$ cannot fill up the remaining pore, which leads to a relatively loose microstructure. This is because the macroscopic pores inside the granite and sandstone increase the interface roughness. The rough interface restricts the shrinkage and deformation of the cement slurry during the hardening process, reducing the number of cracks in the ITZ region, and forming a better structural interface [47]. Therefore, under the three bedrocks, the sandstone–concrete interface has the tiniest cracks, followed by granite and limestone cracks developing the most.

Furthermore, the specimens with three different types of rock were measured using SEM-EDX, and the test results of Ca and Si are shown in Figure 8. The horizontal axis represents the distance from the rock to bulk cement paste in concrete, and the vertical axis is the mass element ratio. According to the linear distribution of Ca and Si elements in the ITZ region, the ITZ thicknesses of LC, GC, and SC are 155 μm, 105 μm, and 95 μm, respectively. By using granite, Xie et al. [48] observed similar results that the width of the ITZ is extended from around 60~150 μm. Moreover, the LC sample has the widest thickness. The mineral composition and surface structure of rock will affect the generation and growth of hydration products, especially for CH and Aft [45]. Combined with the ITZ XRD and SEM results, the difference in the micromorphology and phase content can lead to the difference in the thickness in the ITZ.

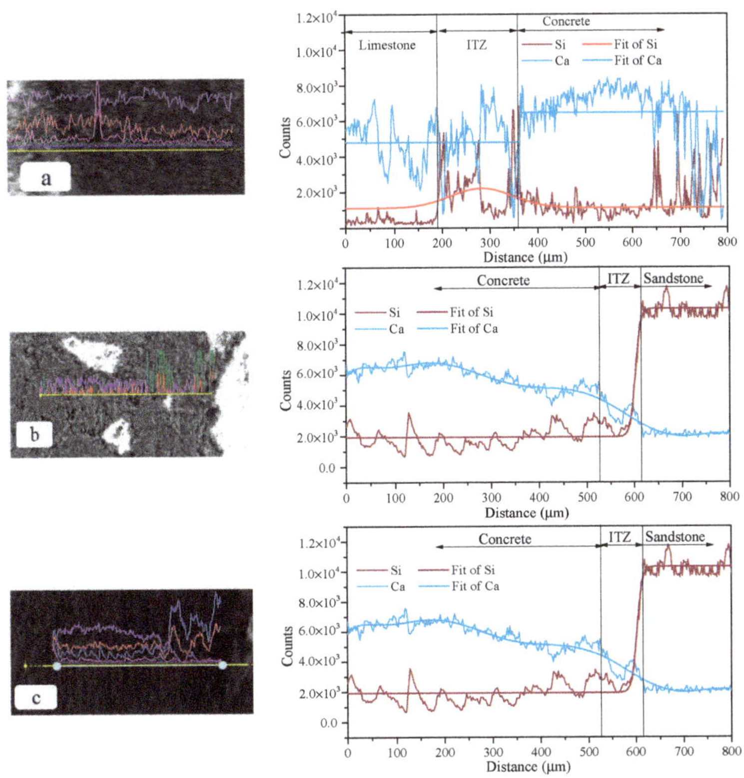

Figure 8. Comparison of SEM-EDS test results of concrete–rock interface; (**a**) limestone–concrete; (**b**) granite–concrete; (**c**) sandstone–concrete.

3.4. Pore Structure of ITZ

Referring to the research method of the ITZ pore structure in the literature [7], the pore characteristics of the mortar between 2 to 3 mm around the rock were measured by NMR imaging system to characterize the influence of different types of rock on the pore structure of the ITZ. The results are shown in Figure 9.

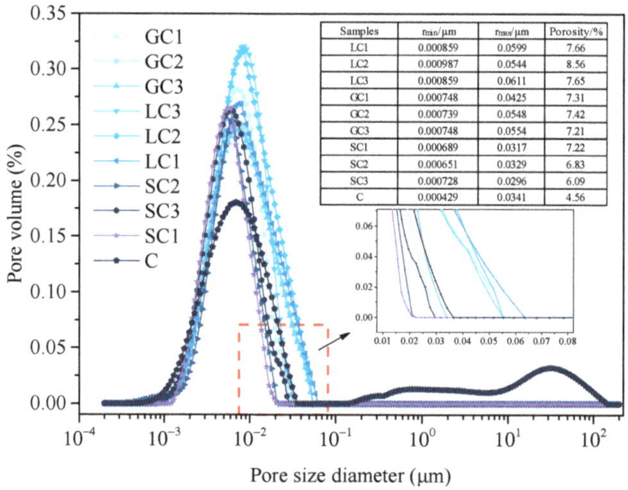

Samples	r_{min}/μm	r_{max}/μm	Porosity/%
LC1	0.000859	0.0599	7.66
LC2	0.000987	0.0544	8.56
LC3	0.000859	0.0611	7.65
GC1	0.000748	0.0425	7.31
GC2	0.000739	0.0548	7.42
GC3	0.000748	0.0554	7.21
SC1	0.000689	0.0317	7.22
SC2	0.000651	0.0329	6.83
SC3	0.000728	0.0296	6.09
C	0.000429	0.0341	4.56

Figure 9. Pore size distribution of ITZ.

The total porosity measured by NMR is summarized in the table in Figure 9. The cement slurry is composed of less porosity than the ITZ. Of all the samples examined, the ITZ between limestone and concrete had the largest porosity at 7.56~8.56%, followed by granite–concrete samples, which were between 7.21 and 7.43%, and the porosity of the sandstone–concrete samples was the smallest at 6.09~7.54%. This may be caused by the different micro-roughness of the rock foundation. Existing research shows that the micro-roughness also dramatically influences the formation of the ITZ. It is believed that the delicate pores on the surface contribute to the production of hydration product crystals, forming a state of interlaced occlusion [49]. There are many macro and micro pores between sandstone particles, which provide much space for the growth and infiltration of cement slurry hydration products, improving the physical bonding performance between rock and cement slurry, obtaining the optimal interface transition zone. As for limestone and granite, the ITZ porosity of their bedrocks is similar, and the former is slightly larger than the latter. This is because the porosity of the two bedrocks is identical, while the pore size distribution is somewhat different, and granite contains more medium and large capillary pores, which increases the micro-roughness of the bonding surface and forms a better ITZ.

Figure 9 illustrates the pore size distribution curves of the different rock ITZs. The pores of the mortar around the limestone, granite, and sandstone are distributed in the range of 0.859~61.1 nm, 0.739~55.4 nm, and 0.698~29.64 nm, respectively. These test results are similar to the pore size distribution curve measured by Wu et al. [50] for the ITZ of the aggregate–cement slurry interface, and the pore size was concentrated between 1 nm and 100 nm. Compared with the specimen of concrete, the pore size distribution curves of the ITZ around the limestone and granite shift to the right (to a larger size), the proportion of harmless pore around the rock decreases, and the pore size of the mortar around the limestone and granite are similar. However, for the cement slurry on the limestone side, with more pore content and large critical pore size, this may be attributed to the pores of mortar around the limestone and are not refined. For the cement slurry on the sandstone side, the pore structure of the transition zone was significantly improved compared to the matrix and the other two types of bedrock. The result is similar to the observation result of the light aggregate–cement slurry interface [51].

3.5. Permeability of ITZ

The permeability of the rock, concrete and concrete–rock mixed samples can be obtained through the permeability test, but the permeability of the ITZ needs further calculation. Rock and concrete are two materials with different properties, similar to the stratum composed of different thin layers. The concrete–rock samples can be regarded as a layered porous medium for the one-dimensional seepage in the seepage test. Figure 10 shows the layered porous medium. Assume that K_1, K_2, and K_3 are the permeability coefficients of the concrete layer, the rock base layer, and the bonding surface layer, respectively, and b_1, b_2, and b_3 are the corresponding layer thicknesses. Using Darcy's law to calculate the traffic Q of each layer, the total traffic should be equal to the sum of the traffic of each layer:

$$Q = \sum_{i=1}^{3} Q_i = \sum_{i=1}^{3} K_i b_i L \frac{\Delta P}{L} \tag{1}$$

where L is the seepage length, $\frac{\Delta P}{L}$ is the hydraulic gradient, constant during calculation.

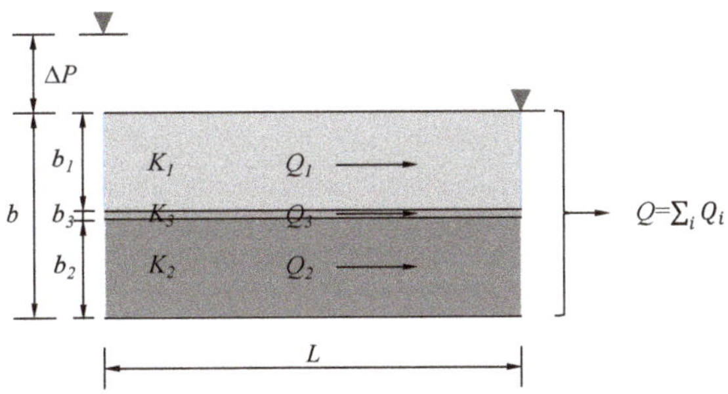

Figure 10. Schematic diagram of the flow of layered porous media along with the layer.

It is assumed that the sample has an equivalent permeability coefficient K_P, under the same hydraulic gradient, there should be the same flow through the sample:

$$Q = K_P J A = K_P b L \frac{\Delta P}{L} \tag{2}$$

Combining Equations (1) and (2), we obtained:

$$K_P = \frac{\sum_{i=1}^{3} K_i b_i}{b} \tag{3}$$

Since the permeability of sandstone is much higher than that of concrete and the ITZ, the permeable porous medium preferentially penetrates from the sandstone side during the test. The permeability of the ITZ cannot be obtained through sandstone, concrete, and sandstone–concrete binary samples. Therefore, the permeability of the ITZ in sandstone–concrete is not analyzed here. Up to now, many scholars have carried out experimental research on the permeability of granite and limestone and given the permeability test data of rock samples with different porosity, as shown in Figure 11. In the following studies, water was used as the permeable fluid. Compared with the results of this experiment, the limestone porosity range of Li et al. [52] was more extensive (1~9%), resulting in higher permeability. Selvadurai et al. [53,54] selected the limestone porosity in the range of 0.33~0.46%, and the permeability obtained in the test ranged from 1.17×10^{-20} m^2 to 6.5×10^{-20} m^2. The porosity of the limestone samples selected in this paper was 0.52~0.64%, which is similar to the limestone selected by Selvadurai, so the obtained results are also similar (3.34×10^{-20}~4.12×10^{-20} m^2). Figure 11 also plots the permeability test results of granite with different porosity (0.3~1%) by Chen et al. [55], Feng et al. [56], He [57], Jiang et al. [17], and Tian et al. [58], which were consistent with the permeability of granite in this experiment.

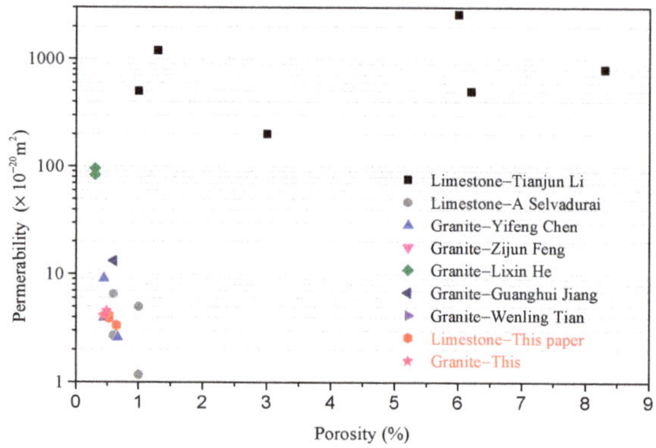

Figure 11. Comparison of the experimental data with the results of other scholars on limestone and granite.

Figure 12a presents the permeability test results for limestone, concrete, and limestone–concrete binary specimens. Under the same seepage pressure, when the permeability coefficient tends to be stable, the permeability of the limestone–concrete binary specimen is significantly larger than that of the limestone or concrete specimen alone. Considering the existence of the transition zone at the limestone–concrete interface, the overall permeability of the binary sample increases by 30.1%. Based on the above results, the permeability coefficient of the ITZ was further calculated according to Formula (3). As shown in Figure 12, the ITZ permeability was 10^{-18} m^2, two orders of magnitude larger than rock and concrete. Therefore, the limestone–concrete ITZ contributed most of the seepage channel of the sample.

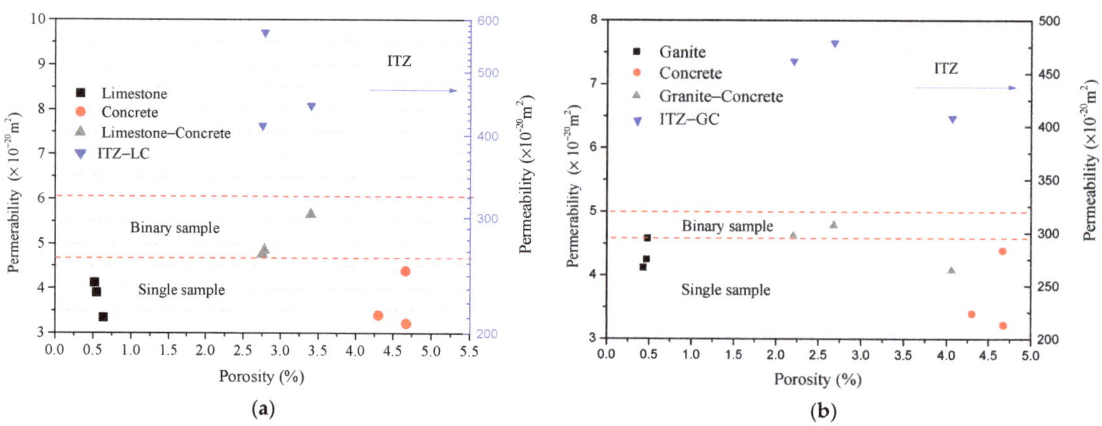

Figure 12. Comparison of permeability test results of different types of rock, concrete and binary specimens, (**a**) limestone, (**b**) granite.

Figure 12b presents the permeability test results for granite, concrete, and granite–concrete binary specimens. From the figure, it can be seen that under the same test conditions, the penetration test results of the granite–concrete binary specimen and the limestone–concrete binary specimen were similar. The permeability of the binary samples was greater than that of the granite or concrete samples alone. The ITZ penetration is much greater than the permeability of the matrix on both sides of the interface. However, compared with the limestone–concrete binary test, granite with minor porosity bonded with concrete to form samples with lower permeability. Combined with the SEM observation results, it is believed that the bonding of granite and concrete creates a better ITZ, which optimizes the impermeability of the binary body.

4. ITZ Permeability Model Based on the Fractional Theory

Most of the empirical formulas used to verify the permeability of the ITZ are established based on the fluid flow in porous media. First, they ignore that there are many micro-cracks at the concrete–rock interface, especially for the interface with poor bonding. Secondly, the median pore roar content and its connectivity are vital parameters determining permeability. The pore content in the porous media permeability model needs to be corrected. Based on the pore structure characteristics of the ITZ, we propose a dual-porosity medium model including fractures and pores. Figure 13 shows the schematic diagram of water transport in ITZ. Considering that the existing permeability models contain parameters with no apparent physical meaning, a permeability model of dual-porosity media based on the fractal theory was proposed.

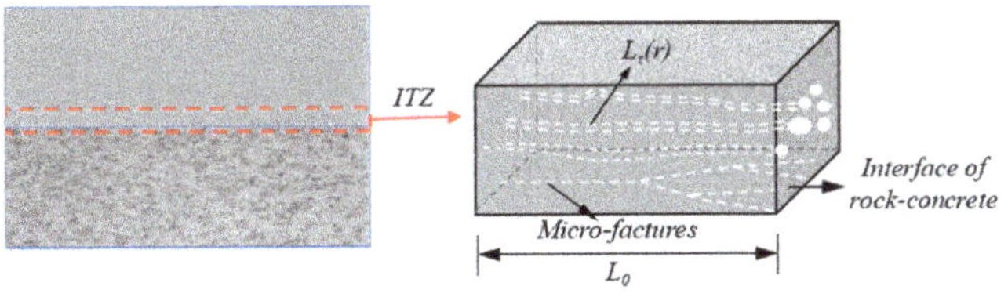

Figure 13. Schematic diagram of water transport in ITZ.

The pore size distribution was assumed to obey the following fractal power law:

$$N(\lambda \geq r) = \int_r^{r_{\max}} \mathrm{P}(\mathrm{r})dr = (r_{\max}/r)^{D_f} \tag{4}$$

where N is the number of holes, λ and r is the pore radius, D_f is the fractal dimension of pore distribution, $\mathrm{P}(\mathrm{r})$ is the pore distribution probability function:

$$\mathrm{P}(\mathrm{r}) = \left[(r_{\max}/r)^{D_f}\right]' = D_f r_{\max}^{D_f} r^{-(D_f+1)} \tag{5}$$

According to the fractal distribution of pores on the interface, the total pore area of the cross-section can be calculated by the following formula:

$$\begin{cases} A_p = \int_{r_{\min}}^{r_{\max}} \frac{\pi}{4} r^2 dN = \dfrac{\pi D_f r_{\max}^2}{4\left(2 - D_f\right)}(1-\phi) & \text{capillary} \\ A_p = \int_{r_{\min}}^{r_{\max}} \mathrm{re} dN = \dfrac{D_f e r_{\max}}{1 - D_f}\left(1 - \left(\dfrac{r_{\min}}{r_{\max}}\right)^{1 - D_f}\right) & \text{crack} \end{cases} \tag{6}$$

where ϕ is the porosity, e is the aperture of the crack

Relationship between porosity and fractal dimension of porous media:

$$D_f = d_E - \ln\phi / \ln(r_{\min}/r_{\max}) \tag{7}$$

Porous media of rock and concrete are usually defined as a structure composed of pores of different sizes, randomly distributed in space and maybe randomly connected to form curved capillary channels. The tortuosity of the capillary pores is defined as the ratio of capillary real length $L_\tau(r)$ to actual length L_0. According to Yu et al. [59]:

$$L_\tau(\mathrm{r}) = \frac{L_0^{D_T}}{r^{D_T-1}} \tag{8}$$

$$\tau(\mathrm{r}) = \frac{L_\tau(\mathrm{r})}{L_0} = \left(\frac{L_0}{R}\right)^{D_\tau - 1} \tag{9}$$

The bending fractal dimension D_T is used to describe this bending degree in the fractal model:

$$\begin{aligned} \tau_{av} &= \int_{r_{\min}}^{r_{\max}} \tau(r)P(r)dr = \frac{D_f}{(D_f + D_T - 1)}\left(\frac{N}{r_{\min}}\right)^{D_T - 1} \\ D_T &= 1 + \frac{\ln\{[\tau_{av}(D_f + D_T - 1)]/D_f\}}{\ln(N/r_{\min})} \\ \frac{N}{r_{\min}} &= \frac{r_{\max}}{r_{\min}}\sqrt{\frac{\pi}{4}\frac{D_f}{2 - D_f}\frac{(1-\phi)}{\phi}} \end{aligned} \tag{10}$$

where τ_{av} is average curvature, $\tau(\mathrm{r})$ is the curvature of a pore with a diameter of r, other parameters are consistent with the above annotations.

According to the existing tortuosity calculation model, the average tortuosity of the sample is obtained, and the tortuosity fractal parameter D_T of the porous medium is obtained by simple iterative calculation combined with Formula (10).

For two different materials, rock and concrete, due to the difference in the microstructure and phase composition of the two, a weak area will be formed near the bonding surface. Assuming that the rock base surface is composed of a series of tree-like fracture meshes, there is a fractal proportional relationship between the number of fractures and the fracture gap. According to Xu [60], the flow in the fractal tree-like crack network can be expressed as:

$$q_1 = \frac{er^{2+D_T}}{(2+D_T)2^{1+D_T}L_0^{D_T-1}}\frac{\rho g}{\mu}\frac{\Delta p}{L_0} \tag{11}$$

where ρ is the density of water, g is the gravitational acceleration, μ—the viscosity coefficient of water, Δp is the difference value of pressure.

According to the above fractal theory:

$$\begin{aligned} Q_1 &= \int_{r_w}^{r_{\max}} q(r)P(r)dr = \frac{eD_f {r_{\max}}^{D_f}}{{l_0}^{D_T-1}(2+D_T)2^{1+D_T}}\frac{\rho g}{\mu}\frac{\Delta p}{l_0}\int_{r_w}^{r_{\max}} r^{1+D_T-D_f}dr \\ &= \frac{eD_f r_{\max}^{2+D_T}}{(2+D_T)2^{1+D_T}l_0^{D_T-1}}\frac{\rho g}{\mu}\frac{\Delta p}{l_0}\frac{1}{2+D_T-D_f}\left(1-\left(\frac{r_{\min}}{r_{\max}}\right)^{2+D_T-D_f}\right) \end{aligned} \tag{12}$$

According to Darcy's Law:

$$Q_1 = k\frac{\rho g}{\mu}\frac{\Delta p}{L_0}A \tag{13}$$

Combine the above formulas:

$$k_f = \frac{\phi_1(1-D_f)r_{\max}^{1+D_T}}{(2+D_T)(2+D_T-D_f)2^{1+D_T}l_0^{D_T-1}}\frac{\left[1-\left(\frac{r_{\min}}{r_{\max}}\right)^{2+D_T-D_f}\right]}{\left[1-\left(\frac{r_{\min}}{r_{\max}}\right)^{1-D_f}\right]} \tag{14}$$

In the ideal hydrodynamic model, based on the Hagen–Poisseuille law governing steady flow, the single-width flow rate of liquid flow in a capillary is:

$$q_2 = \frac{\pi r^4}{8}\frac{\rho g}{\mu}\frac{\Delta p}{L} \tag{15}$$

For the wide range of pore size distribution in the concrete material, pores in some pore size ranges contribute in a minor or even negligible way to the material's permeability. Adopting Jullien's [41] pore size classification method, pores between 0.01 and 10 μm are capillary pores that affect material permeability. Therefore, when the permeability classification model calculates the material's permeability, the pore structure parameters (effective porosity, irreducible water ratio) need to be screened and corrected.

Calculation of effective porosity:

$$\phi' = \frac{V_{pore}}{V} = \frac{\int_{r_w}^{r_{\max}} \pi r^2 L_0 P(r) dr}{L_0 A} = \phi \frac{1 - \left(\frac{r_w}{r_{\max}}\right)^{2-D_f}}{1 - \left(\frac{r_{\min}}{r_{\max}}\right)^{2-D_f}} \tag{16}$$

where ϕ' is the effective porosity of the material, V_{pore} is the effective pore volume, V is the overall volume of the material, r_w is the minimum pore size that affects the permeability of the material.

The proportion of pores with a pore size smaller than r_w:

$$s_w = \frac{V_w}{V_{pore}} = \frac{r_w^{3-D_T-D_f} - r_{\min}^{3-D_T-D_f}}{r_{\max}^{3-D_T-D_f} - r_{\min}^{3-D_T-D_f}} \tag{17}$$

where V_w is the volume occupied by pores with a pore size smaller than r_w.

According to the pore size distribution characteristics of the concrete, the flow in a single capillary is integrated to obtain the total flow of material:

$$\begin{aligned} Q_2 &= \int_{r_w}^{r_{\max}} q(r)P(r)dr = \frac{\Delta p \rho g \pi D_f r_{\max}^{D_f}}{8\mu L_0^{D_T}} \int_{r_w}^{r_{\max}} r^{2+D_T-D_f} dr \\ &= \frac{\rho g \Delta p}{8\mu L_0^{D_T}} \frac{\pi D_f r_{\max}^{D_f}}{3+D_T-D_f} \left(r_{\max}^{3+D_T-D_f} - r_w^{3+D_T-D_f} \right) \end{aligned} \tag{18}$$

Combining Equation (17) and Equation (18), we obtained:

$$\begin{aligned} k_p &= \frac{(2-D_f) r_{\max}^{D_f-2}}{2(3+D_T-D_f) L_0^{D_T-1}} \frac{\phi'}{1-\phi'} \left(r_{\max}^{3+D_T-D_f} - r_w^{3+D_T-D_f} \right) \\ &= \frac{(2-D_f) r_{\max}^{1+D_T}}{2(3+D_T-D_f) L_0^{D_T-1}} \frac{\phi'}{1-\phi'} \left\{ 1 - \left[\left(\frac{r_{\min}}{r_{\max}}\right)^{3-D_T-D_f} + s_w \left(1 - \left(\frac{r_{\min}}{r_{\max}}\right)^{3-D_T-D_f} \right) \right]^{\frac{3+D_T-D_f}{3-D_T-D_f}} \right\} \end{aligned} \tag{19}$$

Finally:

$$\begin{aligned} k = k_f + k_p &= \frac{\phi (1-D_f) r_{\max}^{1+D_T}}{(2+D_T)(2+D_T-D_f) 2^{1+D_T} l_0^{D_T-1}} \frac{\left[1 - \left(\frac{r_{\min}}{r_{\max}}\right)^{2+D_T-D_f}\right]}{\left[1 - \left(\frac{r_{\min}}{r_{\max}}\right)^{1-D_f}\right]} + \\ & \frac{(2-D_f) r_{\max}^{1+D_T}}{2(3+D_T-D_f) L_0^{D_T-1}} \frac{\phi'}{1-\phi'} \left\{ 1 - \left[\left(\frac{r_{\min}}{r_{\max}}\right)^{3-D_T-D_f} + s_w \left(1 - \left(\frac{r_{\min}}{r_{\max}}\right)^{3-D_T-D_f} \right) \right]^{\frac{3+D_T-D_f}{3-D_T-D_f}} \right\} \end{aligned} \tag{20}$$

Table 2 lists some of the models available for the porous media permeability estimation: the fracture model proposed by Monachesi and Guarracino [61], the pore model proposed by YU B [62] and the XY model proposed by Xu and Yu [63] based on the KC model. The comparison between the calculated results of those models and the experimental data reported in Section 3.4 is shown in Figure 14. The results show that the permeability values calculated from the proposed model, pore model, KC model, and XY model were close to the experimental data, with the order 10^{-18} m^2. They were noted that the fracture model gave a worse result than others. It may be that the fracture model ignores the tortuosity of the permeable channel, assuming that the permeable channel is a simple cross model. However, the ITZ texture consists of mineral grains of various shapes and sizes, and its pore structure is highly complex.

Table 2. Some of the models for porous media permeability estimation.

Model	Equation	Reference
Fracture model	$k = \frac{\phi\left(2 - D_f\right) r_{\max}^2}{12\left(3 - D_f\right)} \frac{\left[1 - \left(\frac{r_{\min}}{r_{\max}}\right)^{3-D_f}\right]}{\left[1 - \left(\frac{r_{\min}}{r_{\max}}\right)^{2-D_f}\right]}$	Monachesi and Guarracino (2011) [61]
Pore model	$k = \frac{2 - D_f}{8\tau\left(4 - D_f\right)} \frac{1 - \phi}{\phi} r_{\max}^2$	Yu (2008) [62]
XY model	$k = \frac{\left(\pi D_f\right)^{(1-D_T)/2}\left[4\left(2 - D_f\right)\right]^{(1+D_T)/2}}{32(3 + D_T - D_f)} \left(\frac{\phi}{1 - \phi}\right)^{(1+D_T)/2} r_{\max}^2$	Kozeny (1927) [62]; Xu and Yu (2008) [63]

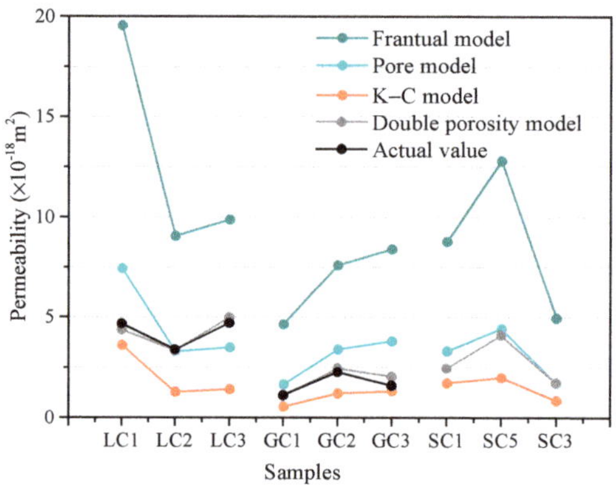

Figure 14. Comparison between measured permeability, the proposed model and other models available in the literature.

Figure 15 shows the calculated deviation results. The deviation for the proposed model, pore model, and XY model was calculated to be 0.21×10^{-18} m^2, 1.325×10^{-18} m^2, and 1.369×10^{-18} m^2, respectively. The representative comparison shows that the proposed model provides a perfect prediction with experimental data and those predicted from the other models.

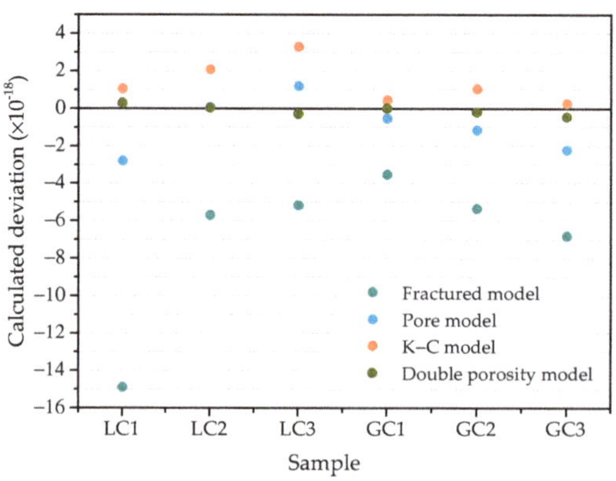

Figure 15. Comparison of calculated deviation results.

Figure 16 shows the percentage of permeability of micro-fractures and pores for samples. For all samples, the penetration caused by micro-fractures accounts for more than 50% of the total penetration, which indicates that micro-fractures have a more significant effect on fluid flow than a pore in the transition zone of the concrete–rock. For the lim-stone-concrete samples, the contribution of micro-fractures to the penetration reaches up to 82%. The calculation result supports that the bedrock with higher macropore content provides more space for cement slurry infiltration and reduces the number of microcracks. Little research compares the effects of microcracks and pores on permeability, but the above observation results are consistent with the current experimental and theoretical results [64–66].

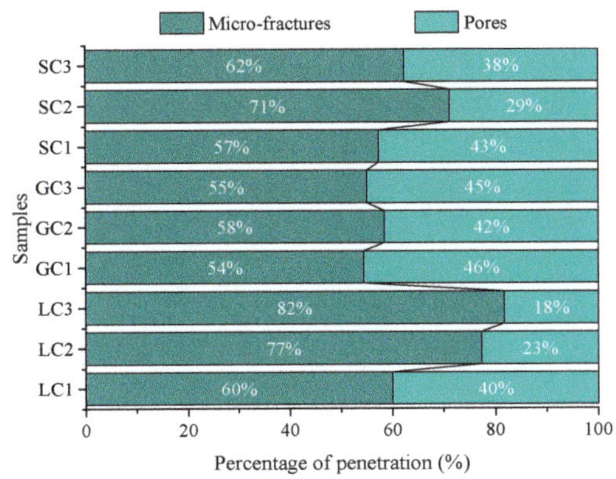

Figure 16. Percentage of permeability of different pore media.

5. Conclusions

This study investigated the effect of rock types on the calcium compounds, thickness, pore structures, and permeability of the ITZ. The main intention was to quantitatively describe the microstructure characteristics of the ITZ to understand the link between microstructure and macro seepage properties. The main conclusions are as follows:

(1) The main hydration products in the concrete–rock ITZ are CSH gel, CH, and Aft. The CSH gel content in different ITZs is between 72.5 and 75.1%, less than the cement slurry in the concrete, which is 88%. The sandstone–concrete ITZ has the highest CSH content and lowest CH content. In contrast, the limestone–concrete ITZ has the highest CH content and lowest CSH content. The CSH content is closely related to the rock particle size. The ITZ formed by coarse-grained rocks has high CSH content.

(2) The thickness of concrete–rock ITZ is between 95 and 155 μm, the porosity is between 6.09 and 8.59%, and the pore size distribution is between 0.698 and 61.6 nm. Compared with the cement slurry–aggregate ITZ, the concrete–rock ITZ has more micro-cracks and larger thickness, while its range of pore size decreases. The difference in the micromorphology and phase content leads to the difference in the thickness in the ITZ.

(3) The porosity and pore size distribution properties of bedrock have significant effects on the microstructure characteristics of the ITZ. The microstructure of the sandstone–concrete ITZ is the densest, followed by granite–concrete ITZ, and the limestone–concrete ITZ is the loosest. Macro-pores increase the roughness of the interface, which will limit the shrinkage and deformation of cement slurry in the hardening process, reduce the number of cracks in the ITZ area, and form a better structural interface.

(4) The impermeability of sandstone and concrete binary structure is not affected by the existence of a bonding interface. However, when limestone and granite (low porosity) are used as bedrocks, the ITZ has a significantly higher permeability. The

ITZ permeability between 4.08 $\times 10^{-18}$ m^2 and 5.74 $\times 10^{-18}$ m^2 was two orders of magnitude larger than the permeability coefficient of rock and concrete.

(5) The fractal permeability model in this study relates to the micropore structure. The proposed model provides a perfect prediction with experimental data and those predicted from the other models. It is pointed out that the contribution of microcracks to the permeability of the ITZ cannot be ignored.

In hydraulic engineering, the concrete–rock ITZ is often subjected to erosive ions and calcium leaching. During long-term service, the permeability of the ITZ is not constant. The evolution of the permeability of the ITZ is critical to project safety and benefits. The evolution of permeability under complex long-term conditions needs to be further studied.

Author Contributions: J.Y., J.S. and H.W. mostly contributed to the design of the manuscript. Y.H. and K.Z. carried out data collection and processing. Y.L., Q.Z. and M.Z. were involved in immersing the specimens. J.Y. revised the paper. All authors have read and agreed to the published version of the manuscript.

Funding: This research was funded by National Natural Science Foundation of China under Grant No. 51579078, 51679070, Collaborative Innovation Center for Prevention and Control of Mountain Geological Hazards of Zhejiang Province under Grant No. PCMGH-2017-Z-02, Basic Research Youth Innovation Intersection Team of Qinghai Institute of Salt Lakes, Chinese Academy of Sciences under Grant No. ISLJCTD-2022-2, and Hunan Water Conservancy Science and Technology Project under Grant No. XSKJ2021000-35, XSKJ2019081-04.

Institutional Review Board Statement: Not applicable.

Informed Consent Statement: Not applicable.

Data Availability Statement: The data that support the findings of this study are available from the corresponding author upon reasonable request.

Conflicts of Interest: The authors declare no conflict of interest.

References

1. Lizarazo, M.; Higuera, C.; Claisse, P. Measuring the effect of the ITZ on the transport related properties of mortar using electrochemical impedance. *Constr. Build. Mater.* **2014**, *52*, 9–16. [CrossRef]
2. Zhang, S.; Zhang, C.; Liao, L.; Wang, C. Numerical study of the effect of ITZ on the failure behaviour of concrete by using particle element modelling. *Constr. Build. Mater.* **2018**, *170*, 776–789. [CrossRef]
3. Zhu, Z.; Provis, J.; Chen, H. Quantification of the influences of aggregate shape and sampling method on the overestimation of ITZ thickness in cementitious materials. *Powder Technol.* **2018**, *326*, 168–180. [CrossRef]
4. Patel, R.A.; Quoc Tri, P.; Seetharam, S.C.; Perko, J.; Jacques, D.; Maes, N.; De Schutter, G.; Ye, G.; Van Breugel, K. Diffusivity of saturated ordinary Portland cement-based materials: A critical review of experimental and analytical modelling approaches. *Cem. Concr. Res.* **2016**, *90*, 52–72. [CrossRef]
5. Nemati, K.M.; Gardoni, P. Microstructural and statistical evaluation of interfacial zone percolation in concrete. *Strength Fract. Complex.* **2005**, *3*, 191–197.
6. Wang, J.; Guo, Z.; Yuan, Q.; Zhang, P.; Fang, H. Effects of ages on the ITZ microstructure of crumb rubber concrete. *Constr. Build. Mater.* **2020**, *254*, 111329. [CrossRef]
7. Qiu, J.; Zhou, Y.; Guan, X.; Zhu, M. The influence of fly ash content on ITZ microstructure of coal gangue concrete. *Constr. Build. Mater.* **2021**, *298*, 1–11. [CrossRef]
8. Nežerka, V.; Bílý, P.; Hrbek, V.; Fládr, J. Impact of silica fume, fly ash, and metakaolin on the thickness and strength of the ITZ in concrete. *Cem. Concr. Compos.* **2019**, *103*, 252–262. [CrossRef]
9. Bílý, P.; Nežerka, V.; Hrbek, V.; Fládr, J. Validation of micromechanical model for prediction of itz thickness of high-strength concrete containing secondary cementitious materials. *Mater. Sci. Forum* **2020**, *995*, 143–148. [CrossRef]
10. He, S.; Yang, E. Strategic strengthening of the interfacial transition zone (ITZ) between microfiber and cement paste matrix with carbon nanofibers (CNFs). *Cem. Concr. Compos.* **2021**, *119*, 104019. [CrossRef]
11. Hosan, A.; Shaikh, F.U.A.; Sarker, P.; Aslani, F. Nano- and micro-scale characterisation of interfacial transition zone (ITZ) of high volume slag and slag-fly ash blended concretes containing nano SiO_2 and nano $CaCO_3$. *Constr. Build. Mater.* **2021**, *269*, 121311. [CrossRef]
12. Lin, J.; Chen, H.; Zhao, Q.; Li, M. Statistical analysis of the critical percolation of ITZ around polygonal aggregates in three-phase concrete materials. *Physica A* **2021**, *572*, 125878. [CrossRef]

13. Bentz, D.P. Influence of internal curing using lightweight aggregates on interfacial transition zone percolation and chloride ingress in mortars. *Cem. Concr. Compos.* **2009**, *31*, 285–289. [CrossRef]
14. Lyu, K.; She, W. Determination of aggregate surface morphology at the interfacial transition zone (ITZ). *J. Vis. Exp.* **2019**. [CrossRef] [PubMed]
15. Skalny, J. Challenges for cement research—Are we ready. *Am. Ceram. Soc. Bull.* **1984**, *63*, 1138.
16. Wakeley, L.D.; Roy, D.M. A method for testing the permeability between grout and rock. *Cem. Concr. Res.* **1982**, *12*, 533–534. [CrossRef]
17. Jiang, G.; Zuo, J.; Ma, T.; Wei, X. Experimental investigation of wave velocity-permeability model for granite subjected to different temperature processing. *Geofluids* **2017**, *2017*, 6586438. [CrossRef]
18. Yang, C. Influence of aggregate content on the percolated interfacial transition zone of cement-based materials by using accelerated chloride migration test. *J. Chin. Inst. Eng.* **2006**, *29*, 159–164. [CrossRef]
19. Snyder, K.A.; Winslow, D.N.; Bentz, D.P.; Garboczi, E.J. Effects of interfacial zone percolation on cement-based composite transport properties. *MRS Online Proc. Libr.* **1991**, *245*, 265–270. [CrossRef]
20. Bentz, D.P.; Hwang, J.T.G.; Hagwood, C.; Garboczi, E.J.; Snyder, K.A.; Buenfeld, N.; Scrivener, K.L. Interfacial zone percolation in concrete: Effects of interfacial zone thickness and aggregate shape. *Mater. Res. Soc. Symp. Proc.* **1994**, *370*, 437. [CrossRef]
21. Zheng, J.; Zhou, X. Percolation of ITZs in concrete and effects of attributing factors. *J. Mater. Civ. Eng.* **2007**, *19*, 784–790. [CrossRef]
22. Zheng, J.; Zhou, X. Effective medium method for predicting the chloride diffusivity in concrete with ITZ percolation effect. *Constr. Build. Mater.* **2013**, *47*, 1093–1098. [CrossRef]
23. Kim, S.U.; Kim, J.S.; Kang, J.W.; Han, T.S. Construction of virtual interfacial transition zone (ITZ) samples of hydrated cement paste using extended stochastic optimization. *Cem. Concr. Compos.* **2019**, *102*, 84–93. [CrossRef]
24. Li, K.; Stroeven, M.; Stroeven, P.; Sluys, L.J. Effects of technological parameters on permeability estimation of partially saturated cement paste by a DEM approach. *Cem. Concr. Compos.* **2017**, *84*, 222–231. [CrossRef]
25. Li, K.; Stroeven, P.; Stroeven, M.; Sluys, L.J. A numerical investigation into the influence of the interfacial transition zone on the permeability of partially saturated cement paste between aggregate surfaces. *Cem. Concr. Res.* **2017**, *102*, 99–108. [CrossRef]
26. Jia, B.; Cui, X. Pore pressure dependent gas flow in tight porous media. *J. Pet. Sci. Eng.* **2021**, *205*, 108835. [CrossRef]
27. Jia, B.; Xian, C.-G. Permeability measurement of the fracture-matrix system with 3D embedded discrete fracture model. *Pet. Sci.* **2022**. [CrossRef]
28. Sun, X.; Zhang, B.; Dai, Q.; Yu, X. Investigation of internal curing effects on microstructure and permeability of interface transition zones in cement mortar with SEM imaging, transport simulation and hydration modeling techniques. *Constr. Build. Mater.* **2015**, *76*, 366–379. [CrossRef]
29. Wang, Q.; Zhang, G.; Tong, Y.; Gu, C. Prediction on permeability of engineered cementitious composites. *Crystals* **2021**, *11*, 526. [CrossRef]
30. Tran, T.T.; Pham, D.T.; Vu, M.N.; Truong, V.Q.; Ho, X.B.; Tran, N.L.; Nguyen-Sy, T.; To, Q.D. Relation between water permeability and chloride diffusivity of concrete under compressive stress: Experimental investigation and mesoscale lattice modelling. *Constr. Build. Mater.* **2021**, *267*, 121164. [CrossRef]
31. Yu, P.; Duan, Y.H.; Chen, E.; Tang, S.W.; Wang, X.R. Microstructure-based fractal models for heat and mass transport properties of cement paste. *Int. J. Heat Mass Transf.* **2018**, *126*, 432–447. [CrossRef]
32. Jin, S.; Zhang, J.; Han, S. Fractal analysis of relation between strength and pore structure of hardened mortar. *Constr. Build. Mater.* **2017**, *135*, 1–7. [CrossRef]
33. Tang, S.W.; He, Z.; Cai, X.H.; Cai, R.J.; Zhou, W.; Li, Z.J.; Shao, H.Y.; Wu, T.; Chen, E. Volume and surface fractal dimensions of pore structure by NAD and LT-DSC in calcium sulfoaluminate cement pastes. *Constr. Build. Mater.* **2017**, *143*, 395–418. [CrossRef]
34. Wang, L.; Zeng, X.; Yang, H.; Lv, X.; Guo, F.; Shi, Y.; Hanif, A. Investigation and application of fractal theory in cement-based materials: A review. *Fractal Fract.* **2021**, *5*, 247. [CrossRef]
35. Meng, Q.; Qin, Q.; Yang, H.; Zhou, H.; Wu, K.; Wang, L. Fractal characteristics of the pore structure of coral powder-cement slurry under different fractal models. *Fractal Fract.* **2022**, *6*, 145. [CrossRef]
36. Shan, J.; Zhang, Y.; Wu, S.; Lin, Z.; Li, L.; Wu, Q. Pore characteristics of pervious concrete and their influence on permeability attributes. *Constr. Build. Mater.* **2022**, *327*, 126874. [CrossRef]
37. Wang, B.; Guo, J.; Liu, J.; Wang, Q. Prediction method of permeability of textile-reinforced concrete based on fractal dimension theory. *Constr. Build. Mater.* **2022**, *327*, 126868. [CrossRef]
38. Dai, Q.; Wang, G.; Zhao, X.; Han, Z.; Lu, K.; Lai, J.; Wang, S.; Li, D.; Li, Y.; Wu, K. Fractal model for permeability estimation in low-permeable porous media with variable pore sizes and unevenly adsorbed water lay. *Mar. Pet. Geol.* **2021**, *130*, 105135. [CrossRef]
39. Li, R.; Chen, Z.; Wu, K.; Liu, X.; Dou, L.; Yang, S.; Xu, J. A fractal model for gas-water relative permeability curve in shale rocks(Article). *J. Nat. Gas Sci. Eng.* **2020**, *81*, 103417. [CrossRef]
40. Wen, F.; Fan, H.; Zhai, S.; Zhang, K.; Liu, F. Pore characteristics analysis and numerical seepage simulation of antifreeze permeable concrete. *Constr. Build. Mater.* **2020**, *255*, 119310. [CrossRef]
41. Jullien, M.; Raynal, J.; Kohler, É.; Bildstein, O. Physicochemical reactivity in clay-rich materials: Tools for safety assessment (Article). *Oil Gas Sci. Technol.* **2005**, *60*, 107–120. [CrossRef]

42. Zhu, J.; Bao, W.; Peng, Q.; Deng, X. Influence of substrate properties and interfacial roughness on static and dynamic tensile behaviour of rock-shotcrete interface from macro and micro views. *Int. J. Rock Mech. Min. Sci.* **2020**, *132*, 104350. [CrossRef]
43. Wang, Y.; Lu, B.; Hu, X.; Liu, J.; Zhang, Z.; Pan, X.; Xie, Z.; Chang, J.; Zhang, T.; Nehdi, M.L.; et al. Effect of CO2 surface treatment on penetrability and microstructure of cement-fly ash–slag ternary concrete. *Cem. Concr. Compos.* **2021**, *123*, 1–12. [CrossRef]
44. Rao, G.A.; Prasad, B.R. Influence of the roughness of aggregate surface on the interface bond strength. *Cem. Concr. Res.* **2002**, *32*, 253–257. [CrossRef]
45. Shi, Y.; Lv, X.; Zhou, S.; Liu, Z.A.; Yang, M.; Liu, C.; Lu, C. Mechanical properties, durability, and ITZ characteristics of full-grade dam concrete prepared by aggregates with surface rust stains. *Constr. Build. Mater.* **2021**, *305*, 1–14. [CrossRef]
46. Hussin, A.; Poole, C. Petrography evidence of the interfacial transition zone (ITZ) in the normal strength concrete containing granitic and limestone aggregates. *Constr. Build. Mater.* **2011**, *25*, 2298–2303. [CrossRef]
47. He, J.; Lei, D.; Xu, W. In-situ measurement of nominal compressive elastic modulus of interfacial transition zone in concrete by SEM-DIC coupled method. *Cem. Concr. Compos.* **2020**, *114*, 103779. [CrossRef]
48. Xie, Y.; Corr, D.; Jin, F.; Zhou, H.; Shah, S. Experimental study of the interfacial transition zone (ITZ) of model rock-filled concrete (RFC). *Cem. Concr. Compos.* **2015**, *55*, 223–231. [CrossRef]
49. Petrounias, P.; Giannakopoulou, P.P.; Rogkala, A.; Stamatis, P.M.; Tsikouras, B.; Papoulis, D.; Lampropoulou, P.; Hatzipanagiotou, K. The influence of alteration of aggregates on the quality of the concrete: A case study from serpentinites and andesites from central macedonia. *Geosciences* **2018**, *8*, 115. [CrossRef]
50. Wu, K.; Shi, H.; Xu, L.; Ye, G.; De Schutter, G. Microstructural characterization of ITZ in blended cement concretes and its relation to transport properties. *Cem. Concr. Res.* **2016**, *79*, 243–256. [CrossRef]
51. Liu, Y.; Shen, P.; He, Y.; Wang, F. Research progress of special aggregate concrete. *Bull. Chin. Ceram. Soc.* **2021**, *40*, 2831–2855. [CrossRef]
52. Li, T.; Huang, Z.; Feng, Y.; Chen, X.; Ma, Q.; Liu, B.; Guo, X. Reservoir characteristics and evaluation of fluid mobility in organic-rich mixed siliciclastic-carbonate sediments: A case study of the lacustrine Qiketai Formation in Shengbei Sag, Turpan-Hami Basin, Northwest China. *J. Pet. Sci. Eng.* **2020**, *185*, 106667. [CrossRef]
53. Selvadurai, A.; Głowacki, A. Estimates for the local permeability of the cobourg limestone. *J. Rock Mech. Geotech. Eng.* **2018**, *10*, 1009–1019. [CrossRef]
54. Selvadurai, A.; Głowacki, A. Stress-induced permeability alterations in an argillaceous limestone. *Rock Mech. Rock Eng.* **2017**, *50*, 1079–1096. [CrossRef]
55. Chen, Y.; Hu, S.; Wei, K.; Hu, R.; Zhou, C.; Jing, L. Experimental characterization and micromechanical modeling of damage-induced permeability variation in Beishan granite. *Int. J. Rock Mech. Min. Sci.* **2014**, *71*, 64–76. [CrossRef]
56. Feng, Z.; Zhao, Y.; Liu, D. Permeability evolution of thermally cracked granite with different grain sizes. *Rock Mech. Rock Eng.* **2021**, *54*, 1953–1967. [CrossRef]
57. He, L.; Yin, Q.; Jing, H. Laboratory investigation of granite permeability after high-temperature exposure. *Processes* **2018**, *6*, 36. [CrossRef]
58. Tian, W.; Yang, S.; Elsworth, D.; Wang, J.; Li, X.-Z. Permeability evolution and crack characteristics in granite under treatment at high temperature. *Int. J. Rock Mech. Min. Sci.* **2020**, *134*, 104461. [CrossRef]
59. Yu, B.; Cheng, P. A fractal permeability model for bi-dispersed porous media. *Int. J. Heat Mass Transf.* **2002**, *45*, 2983–2993. [CrossRef]
60. Xu, P.; Yu, B. The scaling laws of transport properties for fractal-like tree networks. *J. Appl. Phys.* **2006**, *100*, 104906. [CrossRef]
61. Monachesi, L.B.; Guarracino, L. A fractal model for predicting water and air permeabilities of unsaturated fractured rocks. *Transp. Porous Media* **2011**, *90*, 779–789. [CrossRef]
62. Yu, B. Analysis of flow in fractal porous media. *Appl. Mech. Rev.* **2008**, *61*, 0508011–05080119. [CrossRef]
63. Xu, P.; Yu, B. Developing a new form of permeability and Kozeny–Carman constant for homogeneous porous media by means of fractal geometry. *Adv. Water Resour.* **2008**, *31*, 74–81. [CrossRef]
64. Gérard, B.; Marchand, J. Influence of cracking on the diffusion properties of cement-based materials. *Cem. Concr. Res.* **2000**, *30*, 37–43. [CrossRef]
65. Wong, H.S.; Buenfeld, N.R.; Hill, J.; Harris, A.W. Mass transport properties of mature wasteform grouts. *Adv. Cem. Res.* **2007**, *19*, 35–46. [CrossRef]
66. Wong, H.S.; Zobel, M.; Buenfeld, N.R.; Zimmerman, R.W. Influence of the interfacial transition zone and microcracking on the diffusivity, permeability and sorptivity of cement-based materials after drying. *Mag. Concr. Res.* **2009**, *61*, 571–589. [CrossRef]

fractal and fractional

Article

Seepage–Fractal Model of Embankment Soil and Its Application

Xiaoming Zhao [1,2], Binbin Yang [1,3], Shichong Yuan [3], Zhenzhou Shen [4] and Di Feng [2,*]

1 School of Civil Engineering, Xuchang University, Xuchang 461000, China; 1120402012@hhu.edu.cn (X.Z.); yangbinbin@cumt.edu.cn (B.Y.)
2 College of Civil and Transportation Engineering, Hohai University, Nanjing 210098, China
3 School of Resources and Geosciences, China University of Mining and Technology, Xuzhou 221116, China; yuanshichong@cumt.edu.cn
4 Yellow River Institute of Hydraulic Research, No.45 Shunhe Road, Zhengzhou 450003, China; shenzhenzhou@hky.yrcc.gov.cn
* Correspondence: fengdi@hhu.edu.cn

Abstract: Over time and across space, the hydraulic conductivity, fractal dimension, and porosity of embankment soil have strong randomness, which makes analyzing seepage fields difficult, affecting embankment risk analysis and early disaster warning. This strong randomness limits the application of fractal theory in embankment engineering and sometimes keeps it in the laboratory stage. Based on the capillary model of porous soil, an analytical formula of the fractal relationship between hydraulic conductivity and fractal dimension is derived herein. It is proposed that the influencing factors of hydraulic conductivity of embankment soil mainly include the capillary aperture, fractal dimension, and fluid viscosity coefficient. Based on random field theory and combined with the embankment parameters of Shijiu Lake, hydraulic conductivity is discretized, and then the soil fractal dimension is approximately solved to reveal the internal relationship between hydraulic gradient, fractal dimension, and hydraulic conductivity. The results show that an increased fractal dimension will reduce the connectivity of soil pores in a single direction, increase the hydraulic gradient, and reduce the hydraulic conductivity. A decreased fractal dimension will lead to consistency of seepage channels in the soil, increased hydraulic conductivity, and decreased hydraulic gradient.

Keywords: fractal dimension; hydraulic gradient; random field; hydraulic conductivity; porosity

Citation: Zhao, X.; Yang, B.; Yuan, S.; Shen, Z.; Feng, D. Seepage–Fractal Model of Embankment Soil and Its Application. *Fractal Fract.* **2022**, *6*, 277. https://doi.org/10.3390/fractalfract6050277

Academic Editor: Wojciech Sumelka

Received: 30 March 2022
Accepted: 20 May 2022
Published: 22 May 2022

1. Introduction

The development of fractal theory provides a new theoretical basis for the study of soil with a complex internal structure [1,2]. Fractal theory is widely used in research on geotechnical materials as a theoretical method to describe the geometric characteristics of porous media [3,4]. There are extensive experimental data and theoretical research in this field showing that the pore distribution in soil has obvious fractal characteristics [5,6]. At present, the research mainly focuses on the pore–solid model, while there is less research on the seepage–fractal model.

Soil is a loose, porous three-phase medium composed of soil particles, air, and water. Soil particles form the skeleton of soil, while air and water exist in the interconnected pores between soil particles in the soil. In nature, the hydraulic parameters of soil have strong spatial variability at different scales. The pore structure and its connectivity are very complex, and pore scale ranges from millimeters to microns. The process of fluid migration depends on the multiscale pore structure of soil [7,8]. Wang [9] measured the pore size of soil with high-resolution computed tomography (CT) imaging technology, calculated the soil porosity, reconstructed the soil pores in three-dimensional space, and studied the spatial distribution law of soil pores based on their fractal characteristics. Rainfall will introduce changes to the pore structure in the soil surface, and soil particles' transportation caused by infiltration will clog soil pores and reduce the porosity of surface soil, causing a continuous change in soil pore micro characteristics consequently affecting the mechanical properties of soil [10]. The pore structure of soil is affected by many factors, and soil conditioner can significantly increase the number of macro and small pores to improve the soil structure and porosity [11].

Seepage refers to the phenomenon of pore water in the soil flowing through the interconnected pores in the soil particle skeleton under the action of water [12]. Soil permeability represents the ability to allow water or other fluids to flow through its interior. The study of soil permeability is very important in many engineering fields, including studies on the seepage instability of embankments, waterproof foundation pits, etc. [13–15]. There are interconnected pores in the soil particle skeleton, which explains the soil permeability. Therefore, the shape, size, and connection mode of pores have an important impact on the permeability of the soil. Many scholars around the world have done meaningful research on the characteristics of soil permeability; they have explored the internal connection between hydraulic conductivity and other physical parameters and tried to establish a model to determine the hydraulic conductivity [16,17].

Soil particle size distribution refers to the percentage of particles of different sizes in the total particles. It is one of the basic parameters of soil and has a strong impact on the hydraulic properties. Graded entropy can be used to represent the particle size distribution of soil. Through soil particle size, the relationship between hydraulic conductivity and graded entropy can be established. This approach can be applied to predict the hydraulic conductivity of various soils [18].

Darcy's law describes the linear function between the seepage velocity of water in saturated soil and the hydraulic gradient, also known as the linear seepage law. However, according to test results, this linear function is only applicable under certain conditions. With increased soil viscosity, the linear relationship no longer exists. Therefore, a new method is needed to calculate the hydraulic conductivity of cohesive soil. The plasticity index, average pore diameter, and particle size distribution of cohesive soil affect its permeability, resulting in changes to the hydraulic gradient and seepage velocity. The function describing the relationship between these influencing factors and hydraulic conductivity can be obtained with the single factor test. Scholars have studied the internal relationship between the physical indices and hydraulic conductivity of cohesive soil and have tried to find the optimal method to calculate the hydraulic conductivity [19].

Stress in the soil causes consolidation and deformation, which leads to changes in the pore structure and affects the distribution of hydraulic conductivity. This phenomenon is more obvious in the construction of pile foundations [20]. The pore fractal dimension refers to the size and development degree of pore channels, indicating the distribution characteristics of soil pores [21]. In the process of seepage failure in porous media, movable fine particles are continuously lost from the pore channels, and physical and mechanical parameters such as hydraulic conductivity, porosity, and the nonuniformity coefficient change, resulting in changes to the fractal dimension [22–25]. Studying the fractal characteristics of porous soil seepage and predicting and controlling the formation and development of seepage failure are significance for the long-term and safe operation of dam engineering. Capillary curvature, pressure, and wall roughness are closely related to soil hydraulic conductivity. When the capillary structure changes, the fractal dimension changes. Therefore, hydraulic conductivity can be expressed as an equation of capillary state and the fractal dimension [26]. Computer tomography and three-dimensional modeling technology can eliminate the influence of geometric size, establish relationships between soil parameters and the fractal dimension, reflect the fractal characteristics of the model in real time and reduce the error [27]. Meanwhile, the shrinkage crack caused by water loss changes the fractal characteristics, and permeability is obviously affected [28]. The expansive agent reduces the cracks of porous media, and leads to the decrease in permeability [29,30]. Soil properties can be explained by fractal theory; however, under the action of many factors, soil physical indices are not ideally fractal [31]. For complex models, the fractional Adams–

Bashforth method can quickly obtain the approximate solution, which plays a positive role in the application of fractal theory [32].

At present, the study of seepage by fractal theory is less related to the permeability of embankment soil. Research on hydraulic conductivity, hydraulic gradient, and its influencing factors is still insufficient, and the functional relationship between hydraulic conductivity, hydraulic gradient, and fractal dimension needs to be studied further. Based on the fractal characteristics of embankment soil, an analytical formula of the relationship between hydraulic conductivity, porosity, and fractal dimension is derived. Considering the influence of randomness, the seepage fractal model of embankment soil was established. Combined with engineering practice, the relationship between hydraulic conductivity, hydraulic gradient, and fractal dimension was further studied, and the internal relationship between seepage failure and fractal dimension of porous media soil is revealed.

2. Methodology

2.1. Hydraulic Conductivity Based on Pore Radius and Fractal Dimension

Soil is a typical porous medium. Spaces without a solid skeleton are called pores, which are occupied by liquid or gas, and are connected to each other to form channels. Water can flow through these pores in the soil. Therefore, the structure and scale of pores affect the hydraulic characteristics of the soil.

When there is seepage failure of soil, such as with piping, movable fine particles are continuously lost from the pore channels, resulting in changes in pore size and internal structure, and then in the fractal dimension of the soil. Under the action of seepage, some soil particles are brought to other positions by the seepage water, and the pore radius and permeability of soil change. It can be seen that pore size has an effect on the fractal dimension and hydraulic conductivity, resulting in a function that describes the relationship between them.

Hydraulic conductivity reflects the permeability of porous medium. Due to the irregularity and complexity of soil structure, a unified analytical mathematical formula cannot be used to express hydraulic conductivity at present. In engineering, some empirical values or formulas are usually used, or hydraulic conductivity is determined according to indoor geotechnical tests, in situ tests, and numerical test inversion. Considering that soil particles and pores show fractal characteristics from the atomic scale to grain size, the micro pore structure of soil is analyzed, and the analytical relationship between fractal dimension and hydraulic conductivity and porosity is deduced to further explore seepage and fractal characteristics of porous soil media.

In order to describe the characteristic of soil, a cluster of capillaries with different radiuses is introduced to simulate the pores in soil. Take a cylinder in the soil with a radius of R and a length of L. Suppose that in the cylinder, interconnected pores can be replaced by capillaries, and the radius (r) values of capillaries differ from each other. When the capillary radius satisfies $r_{\min} < r < r_{\max}$, the total number of capillaries with radius greater than r in the unit section of soil approximately obeys Equation (1) [33]:

$$M(r_c > r) = R^D \cdot r^{-D} \tag{1}$$

where D is the fractal dimension of capillaries in porous media. Since capillaries are composed of interconnected pores, their fractal dimension is equivalent to the fractal dimension of pores in porous media. R and r are the radius of the cylinder and capillary. The fractal dimension of capillaries in soil can be obtained by Equation (2):

$$D = d - \frac{\ln n}{\ln(r_{\min}/r_{\max})} \tag{2}$$

where d is the fractal dimension of Euclidean space, and n is the porosity of porous media. Obviously, the fractal dimension of a plane in Euclidean space is 2, and the fractal dimension of capillaries in soil meets $1 < D < 2$. The fractal dimension of three-dimensional Euclidean space is 3, and the fractal dimension of capillaries in soil meets $2 < D < 3$.

By differentiating Equation (1), when the capillary radius changes from r to $r + dr$, the cumulative number of capillaries can be obtained. As shown in Equation (3), it is obvious that the cumulative number of capillaries in the interval decreases with increased pore size:

$$dM(r) = -DR^D r^{-1-D} dr \tag{3}$$

According to Poiseuille's law, when fluid moves in a laminar flow in a horizontal circular pipe, such as capillary, the difference between the volume flow and the pressure at both ends of the pipe, the radius and length of the pipe, and the viscosity coefficient of the fluid meets Equation (4):

$$q(r) = \frac{1}{8\eta L}\pi r^4 (p_1 - p_2) = \frac{1}{8\eta L}\pi r^4 \gamma \Delta h \tag{4}$$

where γ and η are the gravity and viscosity coefficient of fluid, respectively; r and L are the radius and length of the pipe, respectively, and Δh is the head difference.

Assuming that the minimum and maximum capillary apertures in a section are $r_{\min}$ and $r_{\max}$, respectively, and the aperture is continuously distributed, the total flow through the section can be obtained by integrating the flow through each capillary in the section in the interval $[r_{\min}, r_{\max}]$. The total flow is shown in Equation (5):

$$Q = -\int_{r_{\min}}^{r_{\max}} q(r) dM(r) \tag{5}$$

Bringing Equations (3) and (4) into Equation (5), we obtain Equation (6):

$$Q = \frac{\pi R^D D}{8\eta} \cdot \frac{\gamma}{4-D} \cdot AJ(r_{\max}^{4-D} - r_{\min}^{4-D}) \tag{6}$$

Based on Darcy's law, the hydraulic conductivity (k) of soil is shown in Equation (7):

$$k = \frac{v}{J} = \frac{Q}{AJ} \tag{7}$$

where v is the seepage velocity, and J is the hydraulic gradient. Bringing Equation (6) into Equation (7), after sorting, hydraulic conductivity meets Equation (8):

$$k = \frac{\pi R^D D}{8\eta} \cdot \frac{\gamma}{4-D} \cdot r_{\max}^{4-D}\left[1 - \left(\frac{r_{\min}}{r_{\max}}\right)^{4-D}\right] \tag{8}$$

Soil porosity is the ratio of the pores volume to the total volume. In this study, the pores volume in soil can be expressed by the volume of capillary clusters in the interval $[r_{\min}, r_{\max}]$, and the total soil volume is equal to the volume of a hypothetical cylinder. Therefore, the soil porosity can be expressed by Equation (9):

$$n = \frac{V_p}{V} = \frac{\int_{r_{\min}}^{r_{\max}} \pi r^2 L dM(r)}{AL} \tag{9}$$

where n is the porosity, V is the total soil volume, and V_p is the pore volume in the capillary cluster. By introducing Equation (3) into Equation (9), we obtain Equation (10):

$$n = \frac{\pi R^D D}{2-D} r_{\max}^{2-D}\left[1 - \left(\frac{r_{\min}}{r_{\max}}\right)^{2-D}\right] \tag{10}$$

In nature, soil is a typical porous medium. In Equations (8) and (10), the hydraulic conductivity and porosity of soil are functions of the capillary cluster radius and fractal dimension. If the fractal dimension and pore size are different, the hydraulic conductivity and porosity are also different. Strictly speaking, due to the assumption, the relationship between hydraulic conductivity and porosity is difficult to express analytically. Considering the complexity and irregularity of soil pores, the maximum pore radius is much larger than the minimum pore radius. At this time, the relationship between maximum and minimum pore radius can be expressed by Equation (11):

$$\frac{r_{\min}}{r_{\max}} \to 0 \tag{11}$$

By introducing Equation (11) into Equations (8) and (10), the hydraulic conductivity and porosity of soil can be expressed by Equations (12) and (13):

$$k = \frac{\pi R^D D}{8\eta} \cdot \frac{\gamma}{4-D} \cdot r_{\max}^{4-D} \tag{12}$$

$$n = \frac{\pi R^D D}{2-D} \cdot r_{\max}^{2-D} \tag{13}$$

Obviously, the hydraulic conductivity of soil is affected by the fractal dimension, pore size, water viscosity coefficient, and other factors.

2.2. Randomness of Hydraulic Conductivity of Embankment Soil

Due to the differences in internal structure and pore distribution, soil hydraulic conductivity has a certain spatial variability, resulting in different values throughout the soil. In embankment engineering, due to the influence of time and space, the hydraulic conductivity of soil shows strong randomness, which cannot be ignored. It has a significant impact on the safe operation and early risk warning of embankments. Deterministic research cannot meet the requirements of the development of modern science and technology. More engineering applications need to consider the randomness of hydraulic conductivity in order to meet the needs of design and risk protection.

The randomness of hydraulic conductivity is affected by many factors, such as pore structure, plasticity index, and compactness. The relationship between these factors is complex and difficult to study at the same time. Some scholars try to use statistical methods to study the random characteristics of hydraulic conductivity, testing multiple groups of soil samples, counting the mean and variance, and obtaining an approximate expression of the probability density function through regression analysis. The concept of randomness is introduced in these studies. Soil hydraulic conductivity is regarded as a random variable that follows a certain distribution law. It is generally believed that the log normal distribution has good adaptability and can reflect the spatial structure of soil hydraulic conductivity better, and it has been applied in some studies [34–37].

In embankment engineering, when measuring the hydraulic conductivity of multiple groups of soil samples and counting the mean and variance, the mean and variance of logarithmic hydraulic conductivity can be expressed by Equation (14):

$$\begin{aligned} \sigma_{\ln k}^2 &= \ln\left(1 + \tfrac{\sigma_k^2}{\overline{\mu}_k^2}\right) \\ \overline{\mu}_{\ln k} &= -\tfrac{1}{2}\sigma_{\ln k}^2 + \ln(\overline{\mu}_k) \end{aligned} \tag{14}$$

In the discrete random field, the hydraulic conductivity of the jth soil element can be expressed by Equation (15):

$$\ln k_j = \overline{\mu}_{\ln k} + \sigma_{\ln k} G_j \tag{15}$$

where G_j is the local average of the standard normal random field in the jth soil element. There is a certain correlation between the points in the random field, which can be described by the correlation function. The exponential correlation function can represent the random characteristics of embankment soil better as shown in Equation (16):

$$\rho(\tau) = \exp\left(-\frac{2}{\theta}|\tau|\right) \tag{16}$$

The fluctuation scale (θ) can be regarded as the maximum distance beyond which points in the random field are almost irrelevant. Linear distance (τ) between any points in the random field is closely related to correlation. The randomness of hydraulic conductivity is affected by the coefficient of variation (COV) and θ. With increased COV, randomness increases. With increased θ, the correlation between points is enhanced, randomness is reduced, and the soil tends to be uniform.

As shown in Figure 1, a hydraulic conductivity random field is generated based on the random field model in this paper. The shape of the random field is a 40 m square. There are two soil layers in the random field with the same thickness. Assume that the mean of soil hydraulic conductivity is 1×10^{-5} cm/s and the fluctuation scale of the random field in any direction is fixed at 3 m. Different COV values result in changes in the dispersion degree and spatial structure of hydraulic conductivity, while the mean of hydraulic conductivity of soil is the same. In order to show this spatial distribution characteristic, different COV values were selected to generate the hydraulic conductivity random field: COV of 0.1 and 0.3 (Figure 1a) and 0.2 and 0.3 (Figure 1b). There are 1600 elements in the random field, and the element size is a 1 m square. Hydraulic conductivity is expressed in gray scale; from pure black to pure white, the scale represents hydraulic conductivity from minimum to maximum. According to the principle of minimum potential energy, groundwater mainly flows through light colored units.

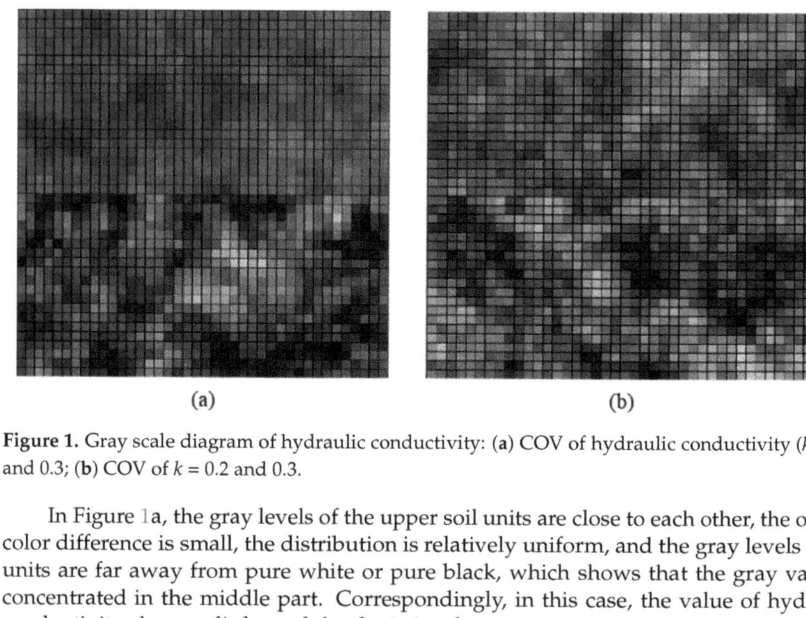

(a) (b)

Figure 1. Gray scale diagram of hydraulic conductivity: (**a**) COV of hydraulic conductivity (k) = 0.1 and 0.3; (**b**) COV of k = 0.2 and 0.3.

In Figure 1a, the gray levels of the upper soil units are close to each other, the overall color difference is small, the distribution is relatively uniform, and the gray levels of the units are far away from pure white or pure black, which shows that the gray value is concentrated in the middle part. Correspondingly, in this case, the value of hydraulic conductivity changes little, and the deviation from its mean value is small. The COV of upper soil is 0.1, which means that the deviation of unit hydraulic conductivity from its mean value is small. There are obvious differences in the gray level of subsoil units; it is close to pure black or pure white, and the gray difference increases. Based on the definition of COV, the greater the value, the higher the deviation of hydraulic conductivity from its mean value. Relatively, the unit can be assigned a larger or smaller value of hydraulic conductivity. In Figure 1a, the random field reflects the spatial distribution of hydraulic conductivity, and similar results can be seen in Figure 1b. In this model, the dispersion degree and spatial structure of hydraulic conductivity of embankment soil will change, although the mean values are equal.

Obviously, it is difficult to characterize the spatial variation of hydraulic conductivity in the fractal dimension, which limits the application of fractal theory in engineering, particularly embankment engineering. Therefore, embankment soil is divided into several basic units, the soil hydraulic conductivity random field is generated, the fractal dimension of each soil unit is calculated, and then the influence of fractal dimension on hydraulic gradient is studied, which has positive significance for engineering practice.

3. Case Study

In recent years, within the general trend of global warming, extreme weather has occurred frequently. In order to ensure the stability of lake embankments, the embankments of Shijiu and Gucheng Lakes in Nanjing City, Jiangsu Province, are reinforced. During the reinforcement process, the hydraulic conductivity of many embankment soil samples was measured. Based on this, the embankment section of Shijiu Lake was selected for analysis. It is located at K19 + 400 stake in Lishui District, near Lianhewei drainage station (shown in Figure 2). In Lishui District, the width of the embankment top is generally 5–7 m, and some embankment tops are earth roads. The lower part of the upstream slope has dry masonry and riprap toe protection, and the upper part has mortar masonry or concrete protection. The downstream slope is basically natural grassland or shrub, and there are berms on the downstream slope of some embankment sections; there is a mortar masonry wave wall approximately 0.3–1.5 m in size near the water.

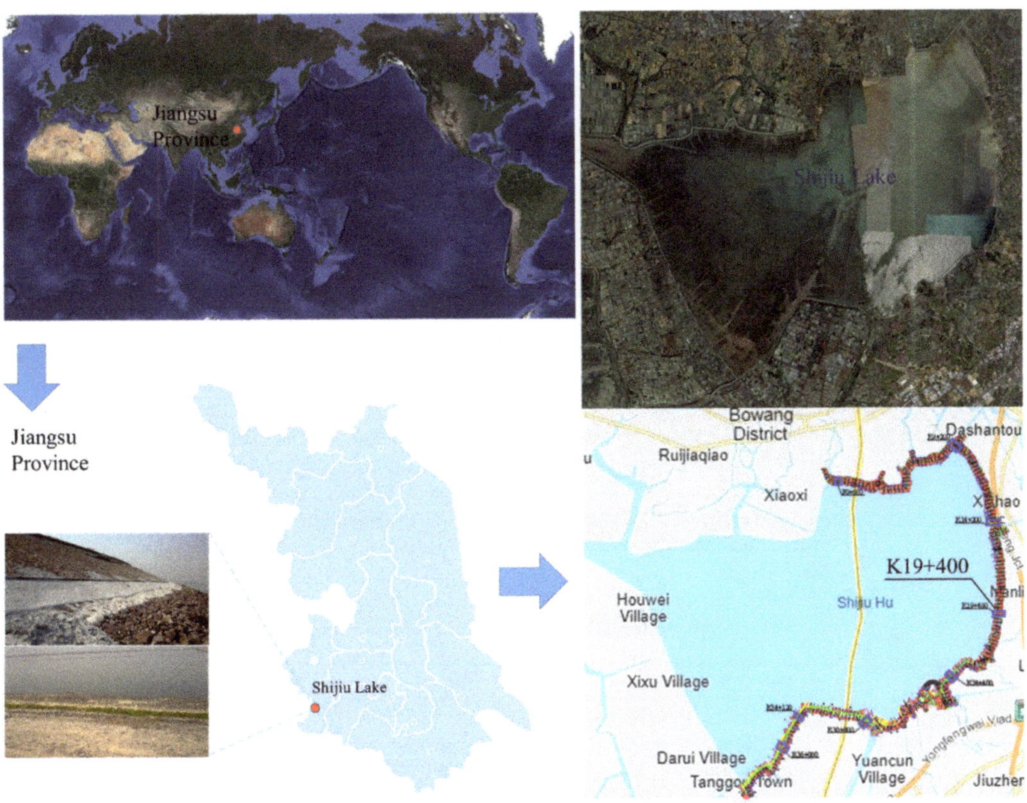

Figure 2. Location of Shijiu Lake.

Due to the length of time the embankment has been there, the upstream–downstream slopes have been eroded to varying degrees. In studying the randomness of hydraulic conductivity, the analysis model was established according to the engineering practice without considering the erosion effect on both sides. In this section, the embankment soil is not layered and is composed of heavy silty loam.

Figure 3 shows the embankment model of Shijiu Lake. According to the engineering practice, the generalized model of the lake embankment was established. The width of the embankment top and bottom was 6 and 48 m, the vertical distance from the top to the foundation was 8 m, and the model was 35 m long in the z-direction. The model was divided into 26,880 elements and 59,976 nodes. The dimensions of the elements in the y-and z-directions were 0.5 and 1 m, and the cell size was not fixed in the x-direction. Six sections were chosen to study the variation of soil hydraulic conductivity and fractal dimension: z = 5, 10, 15, 20, 25, and 30.

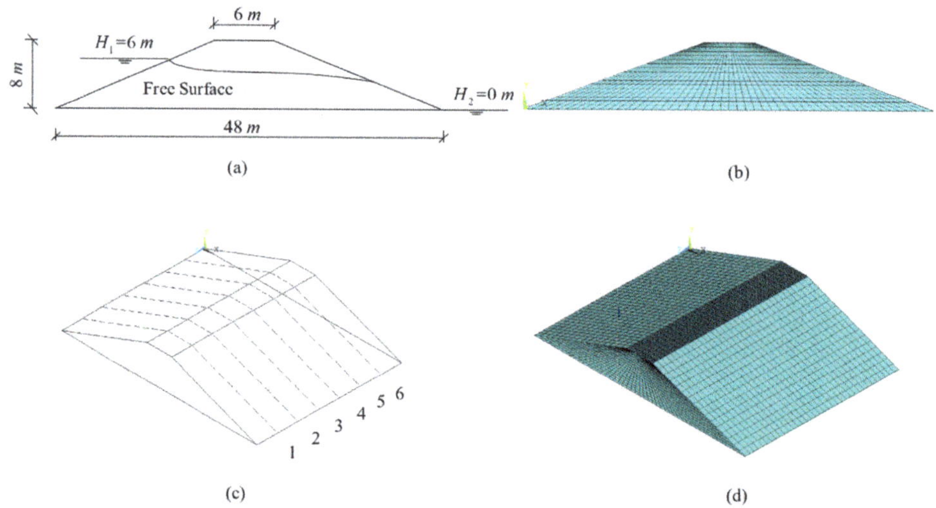

Figure 3. Embankment model of Shijiu Lake: (**a**) dimensions of section; (**b**) mesh generation of section; (**c**) location of section in embankment model; (**d**) mesh generation of embankment.

In this study, through on-site borehole sampling, indoor permeability tests were carried out on several samples of embankments in Lishui District. A total of 54 undisturbed soil samples were obtained in the soil layers inside and outside the impervious core of the embankment section. The undisturbed soil sample was 20 cm long. Each sample was measured three times, and then the hydraulic conductivity of embankment soil samples in Lishui District was obtained. The mean value measured in the test was compared with the hydraulic conductivity provided in the survey report and analyzed to determine the hydraulic conductivity of each soil layer as shown in Table 1.

Table 1. Parameters of embankment model.

Soil	k	COV	θ_x	θ_y
Heavy silty loam	3.32×10^{-6} cm/s	0.3	6 m	3 m

Based on the data in Table 1, a random field of hydraulic conductivity was generated, with a mean of 3.32×10^{-6} cm/s and COV of 0.3, and then the fractal dimension of soil element was calculated according to the relationship between soil fractal dimension and hydraulic conductivity. There were three important boundary conditions: (1) the upstream slope was permeable, and the total head was 6 m; (2) the downstream slope was permeable, and the total head was 0 m; (3) without considering the influence of unsaturated zone, the bottom of the embankment was impervious.

4. Results and Discussion

A gray scale diagram of hydraulic conductivity and fractal dimension of six sections in the embankment are shown in Figure 4. Figure 4a shows the distribution law of the hydraulic conductivity of each unit in Section 1. In the section, the gray values of the units in the middle and upper left areas are similar, showing darker colors, and the hydraulic conductivity in this area was low. On both sides of the section, the gray values of the units were close to each other and the color was light, which means the hydraulic conductivity of the soil was high. It can be seen that dark and light soil units appeared as irregular blocks. In the junction area of blocks, there was no sudden change in gray level, but an obvious transition state, gradually changing from pure white to pure black.

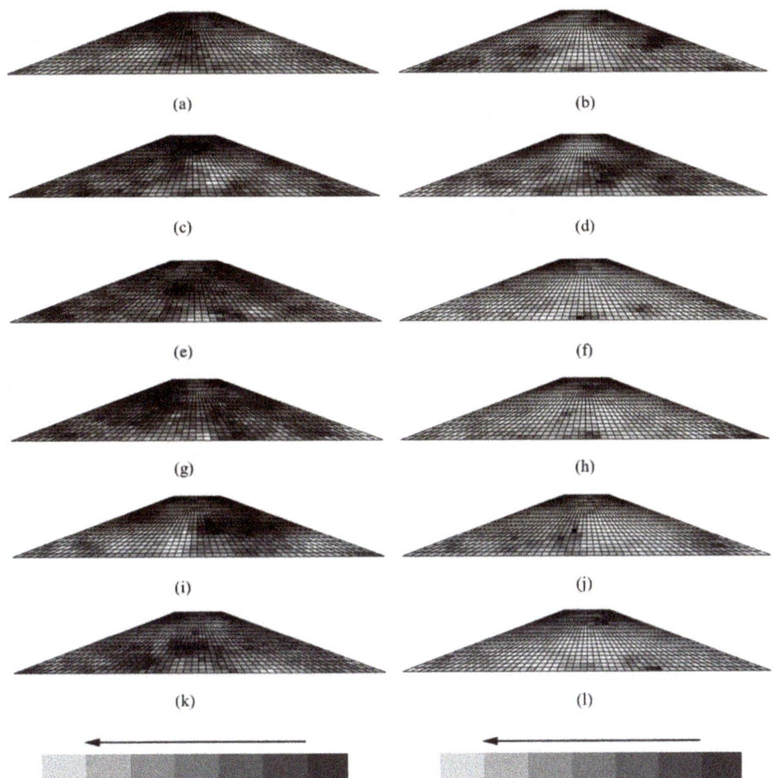

Figure 4. Distribution of hydraulic conductivity (*k*) and fractal dimension (*D*) of sections 1–6 in the embankment: (**a**) *k* in section 1; (**b**) *D* in section 1; (**c**) *k* in section 2; (**d**) *D* in section 2; (**e**) *k* in section 3; (**f**) *D* in section 3; (**g**) *k* in section 4; (**h**) *D* in section 4; (**i**) *k* in section 5; (**j**) *D* in section 5; (**k**) *k* in section 6; (**l**) *D* in section 6.

Figure 4b shows the distribution law of the fractal dimension of each unit in Section 1. In this section, the gray values of the units in the middle and upper left regions were similar, showing lighter colors. The unit fractal dimension in this region was large; according to the statistics, the soil fractal dimension was mainly distributed in the interval (1.7, 1.8), which was close to two. In a plane, the soil fractal dimension was distributed in the interval (1, 2);

therefore, in this part of the region, it was large. On both sides of the section, the gray values of the units were also close to each other, but the color in this part was dark. According to the statistics, in this part of the area, the soil fractal dimension was mainly distributed in the interval (1.55, 1.6), which means that it was small. At the same time, it can be seen that the dark and light units appeared as irregular blocks. In the junction area of the blocks, there was no sudden change in the gray, but an obvious shift from pure white to pure black.

Comparing Figure 4a,b, the unit distribution area with dark color in Figure 4a was close to the unit distribution area with light color in Figure 4b, and the unit distribution area with light color in Figure 4a was close to the unit distribution area with medium dark color in Figure 4b, indicating that the unit gray levels of hydraulic conductivity and fractal dimension present opposite characteristics. In Figure 4a, the soil units with low hydraulic conductivity were represented by dark color, and the units with high hydraulic conductivity were represented by light color. In Figure 4b, the soil units with small fractal dimension were represented by dark color, and the units with large fractal dimension were represented by light color. In a unit at the same point, when the hydraulic conductivity was low, the corresponding fractal dimension was large, and when the hydraulic conductivity was high, the corresponding fractal dimension was small.

The reasons for this phenomenon are as follows: The fractal dimension of two-dimensional Euclidean space is 2. For the soil distributed in this space, considering the influence of pore structure, the fractal dimension is $1 < D < 2$. Assuming that the structure of pores in the soil tends to be consistent, a single-direction seepage channel is formed, and groundwater can only flow through this channel. Based on the meaning of fractal, the fractal dimension of soil is infinitely close to 1. In this case, the seepage path of water is a straight line. In the flow process, it is rarely hindered by soil particles or soil particle agglomerates, the flow rate is high, and the hydraulic conductivity of soil can reach the maximum value.

Correspondingly, it is assumed that there is no rule governing the structure of pores. At any point, the seepage channel can point in any direction and water can flow in any direction. In this case, the fractal dimension of the soil is infinitely close to 2, and the seepage path of water is no longer a straight line but a curve related to the head difference. In the process of flow, soil particles or soil particle agglomerates form the greatest obstruction to seepage, the hydraulic conductivity and flow rate are reduced and can reach the minimum value. In nature, the two extreme working conditions are almost impossible. In the section, when the fractal dimension of soil increases, the hydraulic conductivity decreases, and when the fractal dimension decreases, the hydraulic conductivity increases.

Figure 4c,d show similar laws. Comparing Figure 4a,c, due to the randomness of hydraulic conductivity, although the laws are consistent, its distribution in the two sections differs. In Figure 4e,g, the dark areas are widely distributed, which indicates that most of the unit hydraulic conductivity is low in the two sections. In Figure 4f,h, there are many units with light color, which means that the fractal dimension of most units is large. Consistent results can also be seen in Figure 4i–l. Obviously, hydraulic conductivity can be explained by fractal theory; however, it is not an ideal fractal, which leads to the fractal dimension of soil properties always being greater than 1.5 [31,38]. Similar explanations can be found in this study.

Figure 5 shows the variations in hydraulic conductivity with fractal dimension in different sections. In the sections, hydraulic conductivity decreases with increased fractal dimension. For the same fractal dimension, some of the unit hydraulic conductivity is not unique, and its value changes within a range. The reason is that the unit permeability is affected by the maximum pore diameter and porosity, and these are random variables, which will affect the fractal dimension and hydraulic conductivity values.

The fractal dimension values in each section are in the interval [1.5, 1.8], which is far from 1.0. In this condition, it is difficult to form a single-direction seepage channel in the soil, and water can flow in multiple directions. The fractal dimension is close to 2.0, which means that the seepage channel can point in almost any direction in the plane, and the water can flow in any direction in the soil. The seepage path of water is not a straight line but is in a complex and changeable state.

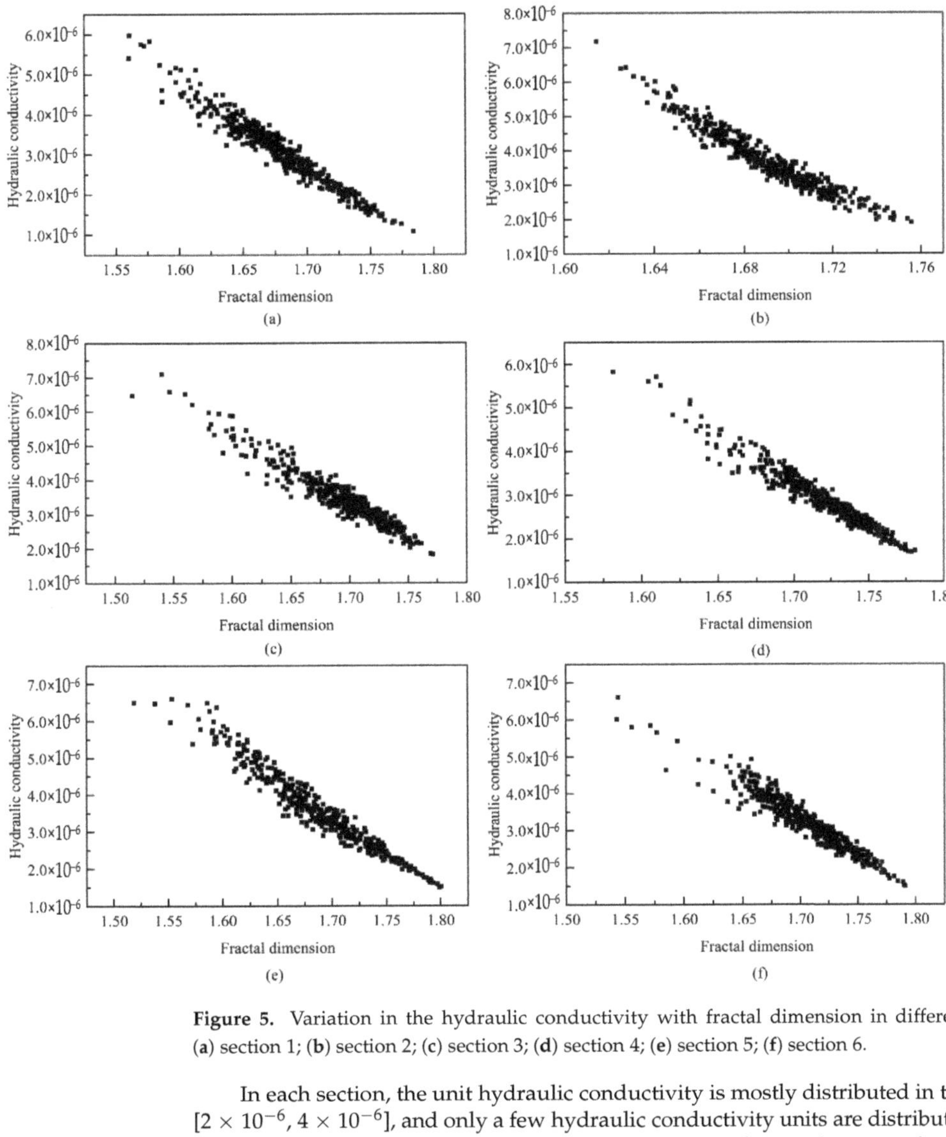

Figure 5. Variation in the hydraulic conductivity with fractal dimension in different sections: (**a**) section 1; (**b**) section 2; (**c**) section 3; (**d**) section 4; (**e**) section 5; (**f**) section 6.

In each section, the unit hydraulic conductivity is mostly distributed in the interval $[2 \times 10^{-6}, 4 \times 10^{-6}]$, and only a few hydraulic conductivity units are distributed outside this interval. This phenomenon conforms to the objective law. In this research, the mean of hydraulic conductivity is 3.32×10^{-6} cm/s, COV is 0.3, and the scale of fluctuations is 6 and 3 m, respectively. When the hydraulic conductivity random field is discretized, COV is within a certain range, and the value after discretization deviates little from the mean value. With increased COV, the value after discretization will be distributed in a larger range. In Figure 5b,c,e, the discrete points show a slight downward bending phenomenon, which means that the unit permeability may decrease slowly with increased fractal dimension.

In order to study the relationship between the fractal dimension and hydraulic gradient of soil, four soil units were selected near the upstream slope angle of each section, which were in the same position. The random seepage field of the embankment was calculated by

ABAQUS software, and the hydraulic gradient at each point was obtained. The variation in hydraulic gradient with fractal dimension in different sections is shown in Figure 6. With increased fractal dimension, the hydraulic gradient increased accordingly, because the soil fractal dimension was inversely proportional to the hydraulic conductivity; when the fractal dimension increased the unit permeability decreased, which means that water flows the same distance in the soil, encounters more resistance, and needs more potential energy, resulting in increased hydraulic gradient.

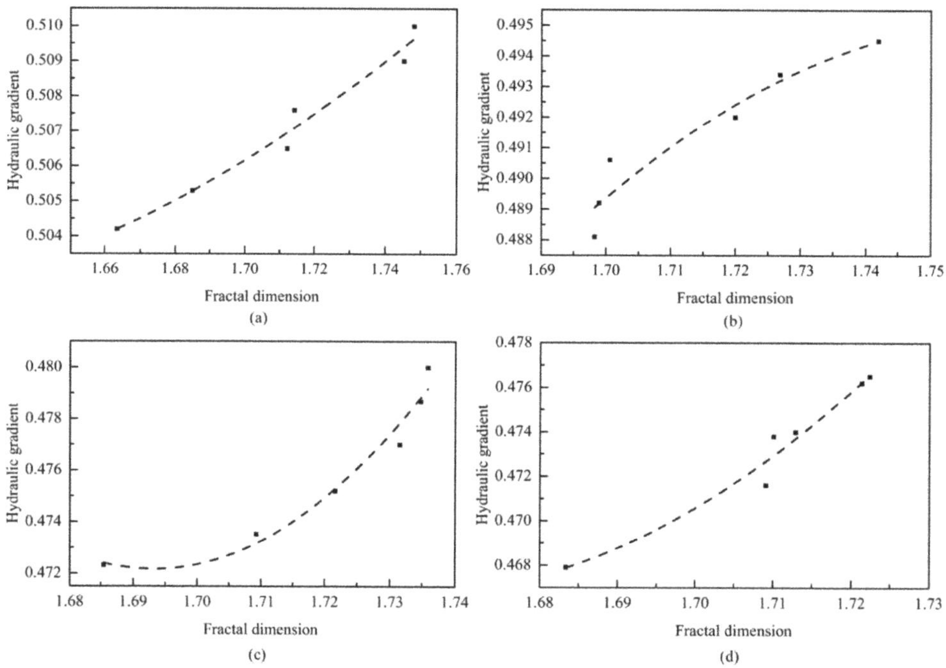

Figure 6. Variation in hydraulic gradient with fractal dimension in different sections: (**a**) point 1; (**b**) point 2; (**c**) point 3; (**d**) point 4.

5. Conclusions

This study focused on constructing and applying a seepage–fractal model of embankment soil. The analytical formula of fractal dimension was deduced, and a random field of hydraulic conductivity was generated to calculate the fractal dimension of soil units. With the seepage–fractal model, the internal relationship between embankment hydraulic conductivity, hydraulic gradient, and fractal dimension was revealed. There are four main conclusions:

(1) The proposed seepage–fractal model of embankment soil is suitable for porous soil media under laminar flow. The influencing factors of hydraulic conductivity mainly include pore size, fractal dimension, and fluid viscosity coefficient, and fractal dimension is the main factor;

(2) Hydraulic conductivity is inversely proportional to fractal dimension. Increased fractal dimension will reduce the connectivity of soil pores in a single direction, increase the seepage resistance of water, and reduce the hydraulic conductivity. Decreased fractal dimension will lead to consistency of seepage channels in the soil, limited seepage direction of water, decreased resistance in the seepage direction, and increased hydraulic conductivity;

(3) Increased fractal dimension leads to decreased hydraulic conductivity, increased potential energy consumption through the same seepage path, and increased hydraulic

gradient. When the seepage resistance increases further, the seepage path changes, and the water will bypass units with high fractal dimension and flow through units with low fractal dimension;

(4) In the seepage–fractal model of embankment soil, the fractal dimension and hydraulic conductivity had significant field characteristics. Units with similar attributes formed agglomerates within which the soil interacts, and the attributes tended to be consistent. On the outside, the interaction between clusters shows obvious transition.

Author Contributions: Investigation, methodology, writing—original draft preparation, X.Z.; data curation, B.Y.; conceptualization, S.Y.; funding acquisition, Z.S.; writing—review and editing, project administration, D.F. All authors have read and agreed to the published version of the manuscript.

Funding: This research was funded by [Key Scientific Research Project of Colleges and Universities in Henan Province] grant number [22B570002 and 21A410003] And [National Natural Science Foundation of China] grant number [U2243210] And [Natural Science Foundation of Henan] grant number [222300420281].

Institutional Review Board Statement: Not applicable.

Informed Consent Statement: Not applicable.

Data Availability Statement: The datasets generated during the current study are available from the corresponding author on reasonable request.

Acknowledgments: The authors would like to thank all the anonymous referees for their constructive comments and suggestions.

Conflicts of Interest: No conflict of interest exits in the submission of this manuscript, and the manuscript was approved by all authors for publication.

References

1. Peli, T. Multiscale fractal theory and object characterization. *JOSA A* **1990**, *7*, 1101–1112. [CrossRef]
2. Perfect, E.; Kay, B.D. Fractal theory applied to soil aggregation. *Soil Sci. Soc. Am. J.* **1991**, *55*, 1552–1558. [CrossRef]
3. Wang, L.; Zeng, X.; Yang, H.; Lv, X.; Guo, F.; Shi, Y.; Hanif, A. Investigation and application of fractal theory in cement-based materials: A review. *Fractal Fract.* **2021**, *5*, 247. [CrossRef]
4. Dıaz-Zorita, M.; Perfect, E.; Grove, J.H. Disruptive methods for assessing soil structure. *Soil Tillage Res.* **2002**, *64*, 3–22. [CrossRef]
5. Tyler, S.W.; Wheatcraft, S.W. Fractal scaling of soil particle-size distributions: Analysis and limitations. *Soil Sci. Soc. Am. J.* **1992**, *56*, 362–369. [CrossRef]
6. Romero, E.; Simms, P.H. Microstructure investigation in unsaturated soils: A review with special attention to contribution of mercury intrusion porosimetry and environmental scanning electron microscopy. *Geotech. Geol. Eng.* **2008**, *26*, 705–727. [CrossRef]
7. Gimenez, D.; Perfect, E.; Rawls, W.J.; Pachepsky, Y. Fractal models for predicting soil hydraulic properties: A review. *Eng. Geol.* **1997**, *48*, 161–183. [CrossRef]
8. Ju, X.; Jia, Y.; Li, T.; Gao, L.; Gan, M. Morphology and multifractal characteristics of soil pores and their functional implication. *Catena* **2021**, *196*, 104822. [CrossRef]
9. Wang, J.; Qin, Q.; Guo, L.; Feng, Y. Multi-fractal characteristics of three-dimensional distribution of reconstructed soil pores at opencast coal-mine dump based on high-precision CT scanning. *Soil Tillage Res.* **2018**, *182*, 144–152. [CrossRef]
10. Yang, M.; Fu, Y.; Li, G.; Ren, Y.; Li, Z.; Ma, G. Microcharacteristics of soil pores after raindrop action. *Soil Sci. Soc. Am. J.* **2020**, *84*, 1693–1704. [CrossRef]
11. Yang, Y.; Wu, J.; Zhao, S.; Han, Q.; Pan, X.; He, F.; Chen, C. Assessment of the responses of soil pore properties to combined soil structure amendments using X-ray computed tomography. *Sci. Rep.* **2018**, *8*, 1–10. [CrossRef]
12. Chang, D.S.; Zhang, L.M. Extended internal stability criteria for soils under seepage. *Soils Found.* **2013**, *53*, 569–583. [CrossRef]
13. Liu, Q.Q.; Li, J.C. Effects of water seepage on the stability of soil-slopes. *Procedia IUTAM* **2015**, *17*, 29–39. [CrossRef]
14. Yang, Y.; Sun, G.; Zheng, H. Modeling unconfined seepage flow in soil-rock mixtures using the numerical manifold method. *Eng. Anal. Bound. Elem.* **2019**, *108*, 60–70. [CrossRef]
15. Wilson, G.V.; Periketi, R.K.; Fox, G.A.; Dabney, S.M.; Shields, F.D.; Cullum, R.F. Soil properties controlling seepage erosion contributions to streambank failure. *Earth Surf. Processes Landf. J. Br. Geomorphol. Res. Group* **2007**, *32*, 447–459. [CrossRef]
16. Hu, S.; Tian, C.; Gan, Y. Determination and calculation of soil permeability coefficient. *Trans. Chin. Soc. Agric. Eng.* **2011**, *27*, 68–72.
17. Elhakim, A.F. Estimation of soil permeability. *Alex. Eng. J.* **2016**, *55*, 2631–2638. [CrossRef]
18. O'Kelly, B.C.; Nogal, M. Determination of soil permeability coefficient following an updated grading entropy method. *Geotech. Res.* **2020**, *7*, 58–70. [CrossRef]

19. Kozlowski, T.; Ludynia, A. Permeability coefficient of low permeable soils as a single-variable function of soil parameter. *Water* **2019**, *11*, 2500. [CrossRef]
20. Lu, M.M.; Xie, K.H.; Guo, B. Consolidation theory for a composite foundation considering radial and vertical flows within the column and the variation of soil permeability within the disturbed soil zone. *Can. Geotech. J.* **2010**, *47*, 207–217. [CrossRef]
21. Dathe, A.; Eins, S.; Niemeyer, J.; Gerold, G. The surface fractal dimension of the soil–pore interface as measured by image analysis. *Geoderma* **2001**, *103*, 203–229. [CrossRef]
22. Ghanbarian-Alavijeh, B.; Millán, H.; Huang, G. A review of fractal, prefractal and pore-solid-fractal models for parameterizing the soil water retention curve. *Can. J. Soil Sci.* **2011**, *91*, 1–14. [CrossRef]
23. Sun, X.; She, D.; Wang, H.; Fei, Y.; Gao, L. Modelling soil hydraulic properties with an improved pore-solid fractal (PSF) model through image analysis. *Eur. J. Soil Sci.* **2022**, *73*, e13156. [CrossRef]
24. Wise, W.R. A new insight on pore structure and permeability. *Water Resour. Res.* **1992**, *28*, 189–198. [CrossRef]
25. Wang, L.; Song, X.; Yang, H.; Wu, B.; Mao, W. Pore structural and fractal analysis of the effects of MgO reactivity and dosage on permeability and F–T resistance of concrete. *Fractal Fract.* **2022**, *6*, 113. [CrossRef]
26. Su, H.; Zhang, Y.; Xiao, B.; Huang, X.; Yu, B. A fractal-monte carlo approach to model oil and water two-phase seepage in low-permeability reservoirs with rough surfaces. *Fractals* **2021**, *29*, 2150003. [CrossRef]
27. Chang, B.; Du, C.; Sun, M.; Lin, Y.; Wang, Y.; Chu, X.; Zhang, L.; He, J. Mesoscopic Seepage Simulation and Analysis of Unclassified Tailings Pores Based on 3D Reconstruction Technology. *ACS Omega* **2021**, *6*, 14309–14316. [CrossRef]
28. Ye, W.; Hu, J.; Ma, F. Centrifuge model study on the influence of desiccation cracks on the seepage behavior of upstream clay anti-seepage system subjected to abrupt flood. *Bull. Eng. Geol. Environ.* **2021**, *80*, 5075–5090. [CrossRef]
29. Wang, L.; Li, G.; Li, X.; Guo, F.; Tang, S.; Xiao, L.; Hanif, A. Influence of reactivity and dosage of MgO expansive agent on shrinkage and crack resistance of face slab concrete. *Cem. Concr. Compos.* **2022**, *126*, 104333. [CrossRef]
30. Yang, H.M.; Zhang, S.M.; Lei, W.; Chen, P.; Shao, D.K.; Tang, S.W.; Li, J.Z. High-ferrite Portland cement with slag: Hydration, microstructure, and resistance to sulfate attack at elevated temperature. *Cem. Concr. Compos.* **2022**, *130*, 104560. [CrossRef]
31. Burrough, P.A. Multiscale sources of spatial variation in soil. I. The application of fractal concepts to nested levels of soil variation. *J. Soil Sci.* **1983**, *34*, 577–597. [CrossRef]
32. Shah, K.; Arfan, M.; Mahariq, I.; Ahmadian, A.; Salahshour, S.; Ferrara, M. Fractal-Fractional Mathematical Model Addressing the Situation of Corona Virus in Pakistan. *Results Phys.* **2020**, *19*, 103560. [CrossRef] [PubMed]
33. Yu, B. Analysis of Flow in Fractal Porous Media. *Appl. Mech. Rev.* **2008**, *61*, 050801. [CrossRef]
34. Freeze, R.A. A stochastic-conceptual analysis of one-dimensional groundwater flow in nonuniform homogeneous media. *Water Resour. Res.* **1975**, *11*, 725–741. [CrossRef]
35. Hoeksema, R.J.; Kitanidis, P.K. Analysis of the spatial structure of properties of selected aquifers. *Water Resour. Res.* **1985**, *21*, 563–572. [CrossRef]
36. Sudicky, E.A. A natural gradient experiment on solute transport in a sand aquifer: Spatial variability of hydraulic conductivity and its role in the dispersion process. *Water Resour. Res.* **1986**, *22*, 2069–2082. [CrossRef]
37. Huang, H.; Hu, B.X.; Wen, X.H.; Shirley, C. Stochastic inverse mapping of hydraulic conductivity and sorption partitioning coefficient fields conditioning on nonreactive and reactive tracer test data. *Water Resour. Res.* **2004**, *40*, 1–16. [CrossRef]
38. Shou, D.; Fan, J.; Ding, F. A difference-fractal model for the permeability of fibrous porous media. *Phys. Lett. A* **2010**, *374*, 1201–1204. [CrossRef]

fractal and fractional

MDPI

Article

Simple Graphical Prediction of Relative Permeability of Unsaturated Soils under Deformations

Gaoliang Tao [1,2], Qing Wang [1], Qingsheng Chen [1,*], Sanjay Nimbalkar [3], Yinjie Peng [1] and Heming Dong [4]

1 School of Civil, Architectural and Environmental Engineering, Hubei University of Technology, Wuhan 430068, China; beining@hbut.edu.cn (G.T.); 101910581@hbut.edu.cn (Q.W.); pyjamazing123@163.com (Y.P.)
2 School of Intelligent Construction, Wuchang University of Technology, Wuhan 430223, China
3 School of Civil and Environmental Engineering, University of Technology Sydney (UTS), 15 Broadway, Ultimo, NSW 2007, Australia; sanjay.nimbalkar@uts.edu.au
4 Shiyan City Transportation Investment Co., Ltd., Shiyan 442000, China; dhm.wayne@hotmail.com
* Correspondence: 20181085@hbut.edu.cn

Abstract: At present, there are only a few existing models that can be used to predict the relative permeability of unsaturated soil under deformations, and the calculation process is relatively complex. In order to fit the measured value of the relative permeability coefficient of unsaturated soil before deformation, this work employs the simplified unified model of the relative permeability coefficient of unsaturated soil, and it obtains the index λ before deformation of the soil. In addition, the value of index λ remains unchanged before and after deformation. Based on the actual measured value of the soil–water characteristic curve before deformation, the air-entry value prediction model is used to predict the air-entry value of soil with different initial void ratios. The relative permeability coefficient of unsaturated soil is then conveniently predicted using the graphical method in combination with the simplified unified model. The method is validated by using the test data of silt loam, sandy loam, and unconsoildated sand. The results show that the predicted results are consistent with the measured values. The prediction method in this paper is simple and overcomes the limitations associated with the determination of the index λ. It expands the application range of the unsaturated relative permeability coefficient model while improving the accuracy of predictions.

Keywords: deformation conditions; relative permeability coefficient; air-entry value; void ratio

Citation: Tao, G.; Wang, Q.; Chen, Q.; Nimbalkar, S.; Peng, Y.; Dong, H. Simple Graphical Prediction of Relative Permeability of Unsaturated Soils under Deformations. *Fractal Fract.* **2021**, *5*, 153. https://doi.org/10.3390/fractalfract5040153

Academic Editor: Norbert Herencsar

Received: 25 August 2021
Accepted: 29 September 2021
Published: 5 October 2021

1. Introduction

At present, there are many studies on the mechanical properties and hydraulic properties of soil [1–3]. Accurate prediction of permeability coefficient is an important prerequisite for studying soil water conservancy characteristics, and it plays an important role in the impact of groundwater on buildings, slope instability caused by rainfall, seepage and diffusion of pollutants in landfills, and seepage prevention of soil dams. Considering the high prevalence with which unsaturated soils are encountered in these projects, the permeability coefficient is a critical parameter for the research of unsaturated soils and has garnered widespread attention from scholars both at home and abroad [4–6]. Unsaturated soil is composed of soil particles, water, and air. Water and air occupy the pore volume in the soil, so its permeability coefficient is easily affected by saturation, even in the near-saturated state (suction range is 0~1 kPa). A small change in suction can cause a great influence in the permeability coefficient [7]. This renders the task of measuring the permeability of unsaturated soil difficult to achieve. In particular, the measurement of the permeability coefficient at low saturation is time demanding. Therefore, it is crucial to propose a method to predict the unsaturated permeability coefficient. The current classic prediction models mainly include the CCG model [8], the Burdine model [9], and the Mualem model [10], among which the Mualem model is widely used in practice. Additionally, Tao and Kong developed a novel methodology for forecasting the permeability coefficient [11]. Tao et al. [12] derived the above four models using fractal theory and analyzed the applicable scope of each model. They then proposed a simplified unified model for predicting the unsaturated permeability coefficient and verified its effectiveness using a large number of measured values. However, these models do not consider the influence of deformation on the variation in permeability coefficient of unsaturated soil.

Mitchell et al. [13] applied the deformable saturated soil permeability theory proposed by Taylor to unsaturated soil and conducted corresponding experiments with compacted clay to evaluate the permeability coefficient of deformable soils. Chang et al. [14] considered the influence of saturation and void ratio on the permeability coefficient and proposed a more reasonable empirical equation. Zhang et al. [15] applied the theories of fluid mechanics and probability to study the change of pore structure parameters under deformation conditions and put forward an expression of the permeability coefficient of unsaturated soil considering pore changes. Hu et al. [16] introduced new parameters to reflect the influence of the change of pore structure under effective saturation and deformation conditions and proposed an unsaturated soil permeability coefficient prediction model considering deformation conditions. Gao et al. [17] used the normalized soil water retention curves (SWRCs) method to investigate the permeability and hydraulic hysteresis behavior of unsaturated soils with varying bulk densities or void ratios and proposed a simple and effective method for simulating the effect of bulk density or void ratio on the relative permeability coefficient of unsaturated soils.

Based on the simplified unified model [12] of relative permeability coefficient of unsaturated soil and air-entry value prediction model [18], a new approach is proposed to readily obtain the relative permeability coefficient of unsaturated soil under deformation conditions by using the graphical technique method. The proposed approach is simple to implement and widely applicable.

2. Prediction Method

2.1. Unified Model

Tao et al. [18] proposed a fractal form of the soil-water characteristic curve (SWCC) expressed in terms of gravimetric water content on the basis of fractal theory. Based on the relationship between volumetric and gravimetric water content and saturation, this is translated into the fractal form of the soil-water characteristic curve expressed in terms of volumetric water content and saturation. Then, these fractal forms are combined with the known CCG model, Burdine model, Mualem model, and Tao–Kong model and their unsaturated permeability coefficient fractal form is deduced. The results found that they have similarities, and then a simplified unified model was proposed [12]. Finally, the effectiveness and applicable scope of the unified model was verified.

This is expressed by the following equation:

$$k_r(\psi) = (a + bS_r) \cdot (\frac{\psi_a}{\psi})^{\lambda} \tag{1}$$

where k_r is the relative permeability coefficient, ψ_a is the air-entry value, ψ is matric suction, S_r is the degree of saturation, a and b are coefficients related to the fractal dimension D, and λ is model parameters. The relationship between a, b, λ, and D is shown in Table 1.

Table 1. The relationship between model coefficients, λ and D.

Model	a Coefficient	b Coefficient	Relationship between λ and D
CCG	$a = (8 - 2D)/(3 - D)$	$b = (D - 5)/(3 - D)$	$\lambda = 5 - D$
Mualemm	$a = 1$	$b = 0$	$\lambda = 9.5 - 2.5D$
Burdine	$a = 1$	$b = 0$	$\lambda = 11 - 3D$
Tao–Kong	$a = 1$	$b = 0$	$\lambda = 5 - D$

For different types of soil, the model parameters are different. The reference for the specific parameter selection is related to the fractal dimension D. The CCG model and the Tao–Kong model predict better when the range of D is $2.8 \leq D < 3$. Within the range of $2.6 \leq D < 2.8$, the Tao–Kong model results into good predictions as well. The Burdine model and the Mualem model predict better when $D < 2.6$. The coefficients $a = (8 - 2D)/(3 - D)$ and $b = (D - 5)/(3 - D)$ in the CCG model are more difficult to evaluate, while the Tao–Kong model also yields good prediction for $2.8 \leq D < 3$. The coefficients $a = 1$ and $b = 0$, as applicable to three models (i.e., Mualemm, Burdine, Tao–Kong), are easier to calculate. Therefore, for the sake of computational simplicity, the CCG model is not considered in this paper. Equation (1) can be simplified as follows:

$$k_r = \left(\frac{\psi_a}{\psi}\right)^{\lambda} \tag{2}$$

2.2. Relationship between Parameter λ and Fractal Dimension D

Since the 1980s, many scholars have discovered that the distribution of internal pores and particles for many types of rock and soil have similar characteristics in their natural state. Therefore, the fractal theory has been applied to a variety of soils and achieved promising results [19–22]. Fractal dimension is a physical parameter used to describe the irregularity of fractal geometry, which is closely related to the pore structure of the soil. With different fractal dimensions, the soil shows great difference in mechanical properties and hydraulic properties. Therefore, the fractal dimension is a very important parameter in the study of soil permeability coefficient. Tao et al. [18] analyzed SWCC variation under different starting void ratio conditions and discovered that when the initial void ratio is changed, the air-entry value is the primary component affecting the change in SWCC. In a study [18], the data after the air-entry value were chosen to calculate the fractional dimension, so it could be assumed that the soil fractal dimension does not change significantly at different void ratios. The relationship between the index λ and the fractal dimension D in the simplified unified model of the unsaturated soil permeability coefficient are described in the previous section, shown in Table 1. It can be obtained that the index λ of the unified model also does not change significantly under different initial pore ratios.

However, the relationship between index λ and fractal dimension D in the above four models cannot cover all the cases relevant to practice. In some instances, the relationship between λ and D is beyond the scope of the above expressions. This paper suggested that, in order to predict the unsaturated permeability coefficient under deformation conditions, the unsaturated permeability coefficient of the soil should be measured first, and the λ value obtained by fitting the experimental data of the unsaturated relative coefficient and air-entry value prior to deformation. From this, the λ value after deformation can be obtained.

2.3. Law of Relative Permeability Coefficient of Unsaturated Soil under Deformation Conditions

Tao and Kong [11] established the relationship model between the SWCC and saturated/unsaturated permeability coefficient based on the SWCC, capillary, and fluid mechanics theories from the microscopic pore channel, which has ideal effects. The total pore channels are assumed to have n grades, and now only the channels of grade 1 m are filled with water (m < n). Combined with the soil–water characteristic curve prediction method of deformed soil proposed by Tao et al. [18], a novel expression of unsaturated relative permeability coefficient is proposed as follows:

$$k_r(\theta_{i=\mathrm{m}}) = \sum_{i=1}^{i=\mathrm{m}} \frac{\Delta\theta_i}{{\psi_i}^2} \Big/ \sum_{i=1}^{i=\mathrm{n}} \frac{\Delta\theta_i}{{\psi_i}^2} \tag{3}$$

where the water content in the above formula requires the form of volumetric water content; ψ_i is the matrix suction of the corresponding pore channel; θ_i is the volumetric water content corresponding to the matrix suction ψ_i; n is the total number of pore channels; and m is the number of pore channels filled with water (m < n).

The model is used to predict the unsaturated relative permeability coefficient of Wuhan clay soil under different deformations. It is found that the variation law of the unsaturated relative permeability coefficient with the matrix suction under different deformation conditions can be described by the "brush type distribution" under double logarithmic coordinates. That is: before air-entry value ψ_a, unsaturated relative permeability coefficient is 1; after air-entry value, ψ_a, the unsaturated relative permeability coefficient decreases with the increase in matrix suction, and the corresponding slopes are approximately equal under different initial void ratios, as shown in Figure 1.

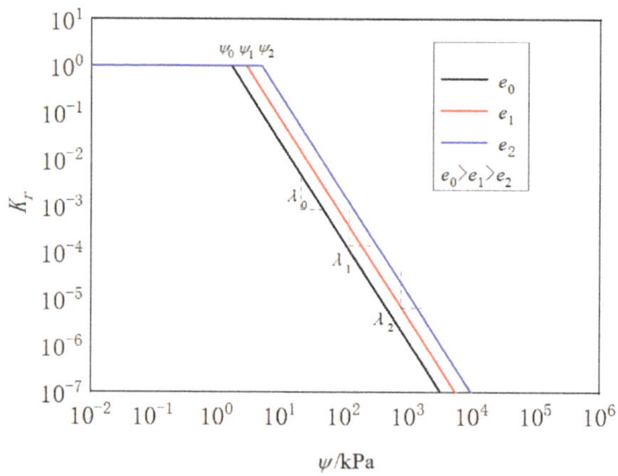

Figure 1. Law of permeability coefficient of unsaturated soil under deformation conditions.

2.4. Simple Graphing Method

According to the variation law for the relative permeability coefficient of unsaturated soil under deformation conditions, this article presents a simple graphing method for predicting the relative permeability coefficient of unsaturated soil with varying initial void ratios. This method is based on the measured value of the unsaturated permeability coefficient and the air-entry value prior to deformation; the relevant model is not confined, unlike the Tao–Kong, Burdine, or Mualem models, and is applicable to a broader range of situations.

2.4.1. Measure the SWCC and Unsaturated Relative Permeability Coefficient before Deformation

The unsaturated permeability coefficient is calculated by multiplying the saturated permeability coefficient by the unsaturated relative permeability coefficient. After determining the soil's unsaturated and saturated permeability coefficients, the measured value of the soil's unsaturated relative permeability coefficient can be computed by dividing the unsaturated permeability coefficient by the saturated permeability coefficient. Currently, the primary methods for determining the SWCC are the salt solution method, the filter paper method, and the tensiometer approach.

2.4.2. Solve the Fractal Dimension

The literature [18] uses the fractal theory to express the fractal form of the SWCC by gravimetric water content:

$$\begin{cases} w = \frac{e}{G_s}\left(\frac{\psi_a}{\psi}\right)3-D & \psi \geq \psi_a \\ w = \frac{e}{G_s} & \psi < \psi_a \end{cases} \tag{4}$$

where w is the gravimetric water content, e is the void ratio, ψ is matric suction, Gs is the relative density, ψ_a is air-entry value, and D is the fractal dimension.

Equation (4) can be transformed into the fractal form of the SWCC concerning saturation, as follows:

$$\begin{cases} S_r = \left(\frac{\psi_a}{\psi}\right)3 - D & \psi \geq \psi_a \\ S_r = 1 & \psi < \psi_a \end{cases} \tag{5}$$

where S_r is the degree of saturation. The method of solving the fractal dimension D is as follows: Using—$ln\psi$ as the horizontal coordinate and $ln\ w$, $ln\ \theta$, or $ln\ S_r$ as the vertical coordinate, a scatter plot is drawn and a straight line fit is then made; assuming a slope of k, the number of fractional dimensions $D = 3 - k$.

2.4.3. Predict the Air-Entry Value after Deformation

A previous study [18] analyzed the fractal dimension and the change law of air-entry value under different initial void ratio conditions, and then gave the following formula for predicting air-entry value under deformation conditions:

$$\psi = \frac{\psi_{a0}}{\left(\frac{e_1}{e_0}\right)^{1/(3-D_0)}}, \tag{6}$$

where ψ_{a0} is the air-entry value corresponding to the maximum initial void ratio e_0, which can be obtained through the fitting of Equation (4), and D_0 is the fractal dimension, which is almost unchanged under deformation conditions. The fractal dimension D_0 can be calculated from the test results of the SWCC at e_0. According to Equation (6), the air-entry value under any initial void ratio can be predicted.

2.4.4. Determine λ According to the Measured Value of the Unsaturated Relative Permeability Coefficient before Deformation

Based on the relative permeability coefficient of the unsaturated soil before deformation, the air-entry value ψ_{a0} of the soil before deformation is obtained by fitting Equation (4) against experimental data, and the two are substituted into the simplified unified model Equation (2) to obtain the index λ before deformation.

2.4.5. Determine the Deformed λ

As shown in Figure 1, if the index is λ_0 at the maximum initial porosity ratio e_0, and after obtaining the index λ_0 at e_0 before deformation, it is known that λ will not change significantly under deformation conditions, then $\lambda_0 = \lambda_1 = \lambda_2$ for e_1, e_2 ($e_0 > e_1 > e_2$) for any deformation condition can be taken.

2.4.6. Simple Drawing

As shown in Figure 1, based on the relative permeability coefficient and air-entry value of unsaturated soil at e_0 before deformation mentioned above, the parameter λ can be obtained by fitting with Equation (2) and then drawing the relationship between the measured relative permeability coefficient and matrix suction at e_0 before deformation. The air-entry value under void ratio e_1 and e_2 after deformation can be predicted using Equation (6). Then, starting with the air-entry value, a straight line can be drawn to the right. The slope of the straight line and the initial void ratio e_0 after the air-entry value are the same. The slope of the straight line is the same, so the relationship between the relative coefficient of the unsaturated soil and the matrix suction after deformation can be quickly obtained, and the purpose of predicting the relative coefficient of the unsaturated soil under arbitrary deformation conditions can be achieved.

3. Method Verification

To verify the rationality of the above method, this paper uses the soil–water characteristic curve test data and the measured values of the unsaturated relative permeability coefficients of silt loam, sandy loam, and unconsoildated sand in the literature [23] to verify the rationality of the proposed method.

3.1. Calculation of Fractal Dimension

To predict the air-entry value of the soil after deformation by Equation (6), firstly, it is required to calculate the fractal dimension D_0 of the soil before deformation. The solution method is as mentioned above. Figure 2 shows the fractal dimensions of three types of soil.

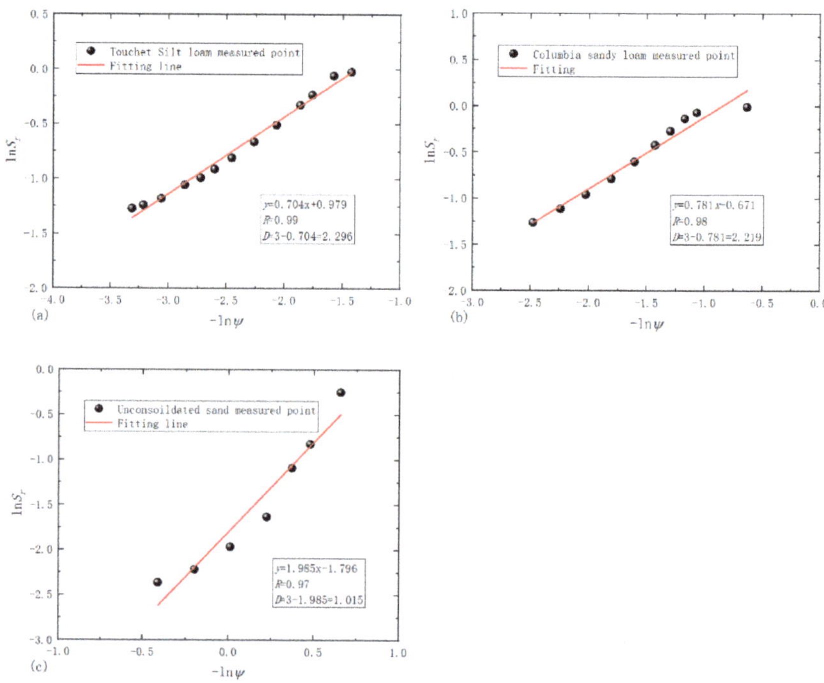

Figure 2. Fitting to obtain the fractal dimension of three soil types: (**a**) Touchet Silt loam; (**b**) Columbia sandy loam; (**c**) unconsoildated sand.

In Figure 2a–c, it is shown that the fitting correlation coefficient R is above 0.97, indicating that the fractal characteristic of soil is obvious. For different types of soils, this method has better applicability when solving fractal dimensions. This demonstrates the significance of the proposed approach.

3.2. Prediction of Air-Entry Value under Deformation Conditions

Based on the obtained fractal dimensions of the soil and the air-entry value ψ_{a0} of the maximum initial void ratio of the three soils obtained by Equation (4), substituting it into Equation (6), the air-entry values after deformation are predicted and listed in Table 2.

It can be seen from Table 2 that the predicted values of air-entry value under different void ratios are all very close to the measured values, which indicates that the prediction effect is good.

Table 2. Air-entry value predicted value.

Soil Type	Void Ratio	Air-Entry Value/kPa	
		Measured	Prediction
Touchet silt loam [23]	1.012	4.13	4.13
	0.916	5.07	4.76
	0.815	6.35	5.62
	0.733	7.56	6.53
	0.653	8.95	7.70
Columbia sandy loam [23]	1.268	2.65	2.65
	1.114	3.34	3.13
	0.942	4.52	3.88
	0.890	5.07	4.17
	0.815	5.87	4.67
Unconsoildated sand [23]	0.852	0.49	0.49
	0.825	0.52	0.50
	0.799	0.54	0.51
	0.767	0.56	0.52
	0.715	0.59	0.54

3.3. Determine λ According to the Measured Value of the Unsaturated Relative Permeability Coefficient before Deformation

Based on the measured values of the relative permeability coefficients of the three kinds of soils before deformation, the air-entry value ψ_{a0} of the soil before deformation is obtained by Equation (4). Substituting the two parameters into the simplified unified model Equation (2), the model indices λ_0 before deformation are obtained, which are 4.96, 4.83, and 12.87 (the data before the air-entry value is discarded during fitting).

In Figure 3a–c, the correlation coefficient R^2 is higher than 0.96, which indicates that the λ index obtained by the fitting process has high accuracy. As illustrated in the figure, with the increase in matrix suction, its fitting correlation increases, which indicates the high accuracy of the model. It shows that, for different types of soil, the method of obtaining the parameter λ by using the unified model to fit the measured values and air-entry values of the unsaturated relative permeability coefficients of the soils before deformation has good applicability.

3.4. Determine the Deformed λ

It is apparent from the relationship between parameter λ and fractal dimension D that the index λ does not change significantly under deformation conditions. The λ of Touchet silt loam, Columbia sandy loam, and unconsoildated sand are the same irrespective of deformation, and are 4.96, 4.83, and 12.87, respectively.

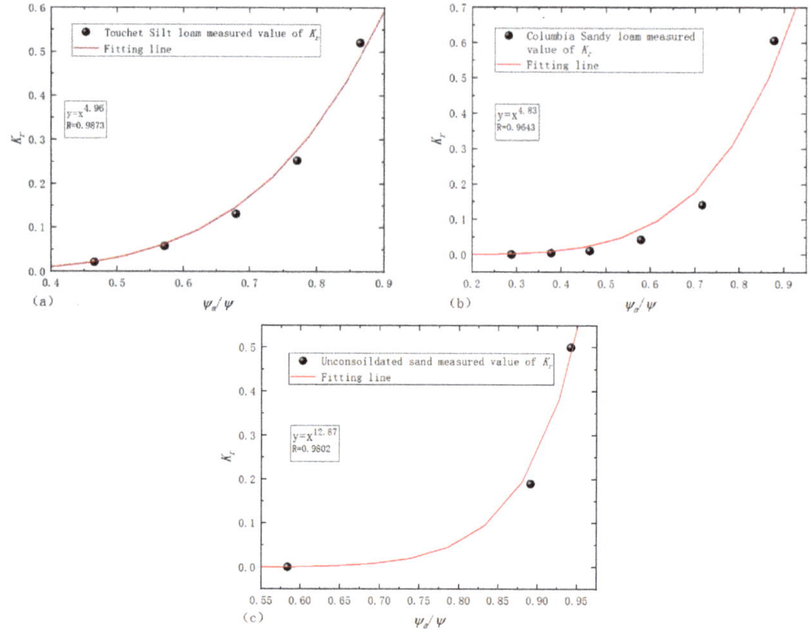

Figure 3. Fitting values of λ for three types of soil: (**a**) Touchet silt loam; (**b**) Columbia sandy loam; (**c**) unconsoildated sand.

3.5. Simple Drawing

The relationship between the measured value of the relative permeability coefficient of the unsaturated soil at the maximum initial void ratio and the matrix suction data is obtained. Then, the air-entry value of each void ratio after deformation is predicted using Equation (6). Then, starting with the air-entry value, a straight line is drawn to the right. When the suction value is after the air-entry value, the slope of the straight line at each void ratio after deformation is the same as the slope of the straight line at the initial void ratio. Therefore, the relationship between the relative coefficient of the unsaturated soil and the matrix suction after deformation can be quickly obtained, as shown in Figures 4–6.

In Figures 4–6, the graphs marked (a) depict the relationship between the unsaturated relative permeability coefficient and the matrix suction at the maximum initial void ratio. The graphs marked (b) to (e) present the relationships between the unsaturated relative permeability coefficient and the matrix suction under different void ratios. It is apparent that the predicted value of the three types of soil unsaturated relative permeability coefficient and the measured value coincide, indicating that the prediction effect is better. It is worth noting that, using the prediction method in this paper, unsaturated relative permeability coefficient of soil under arbitrary deformation conditions can be effectively and quickly predicted.

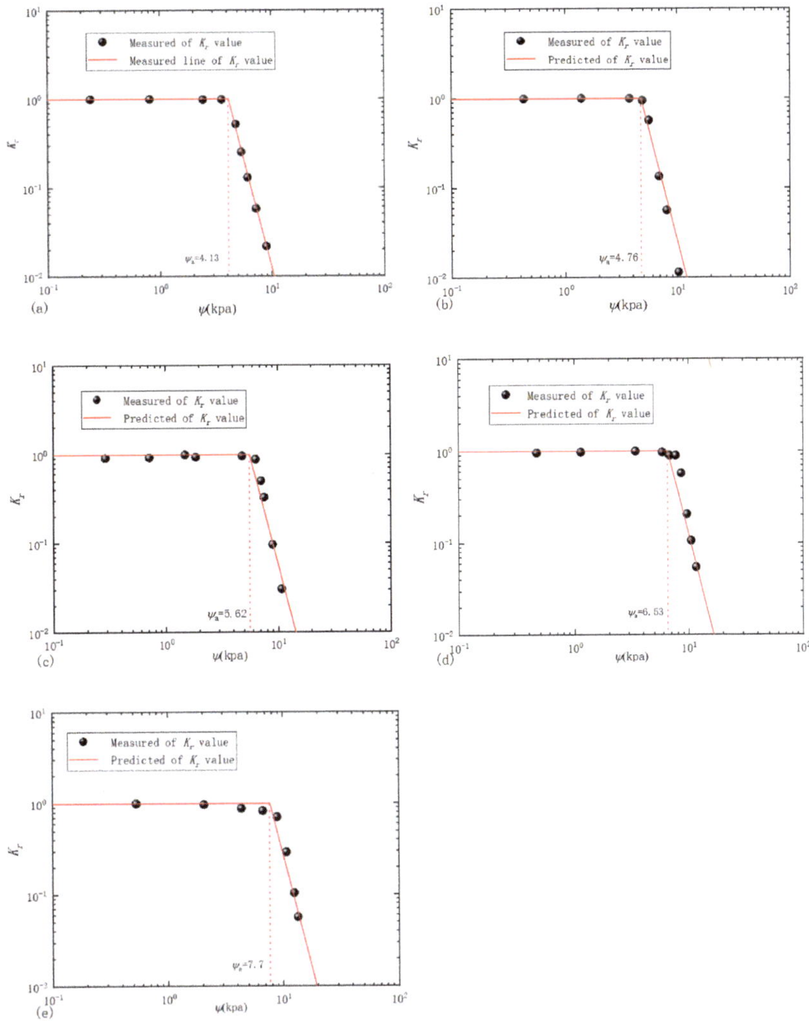

Figure 4. Touchet silt loam's different porosity ratio unsaturated relative permeability coefficient prediction results: (**a**) $e = 1.012$; (**b**) $e = 0.916$; (**c**) $e = 0.815$; (**d**) $e = 0.733$; (**e**) $e = 0.653$.

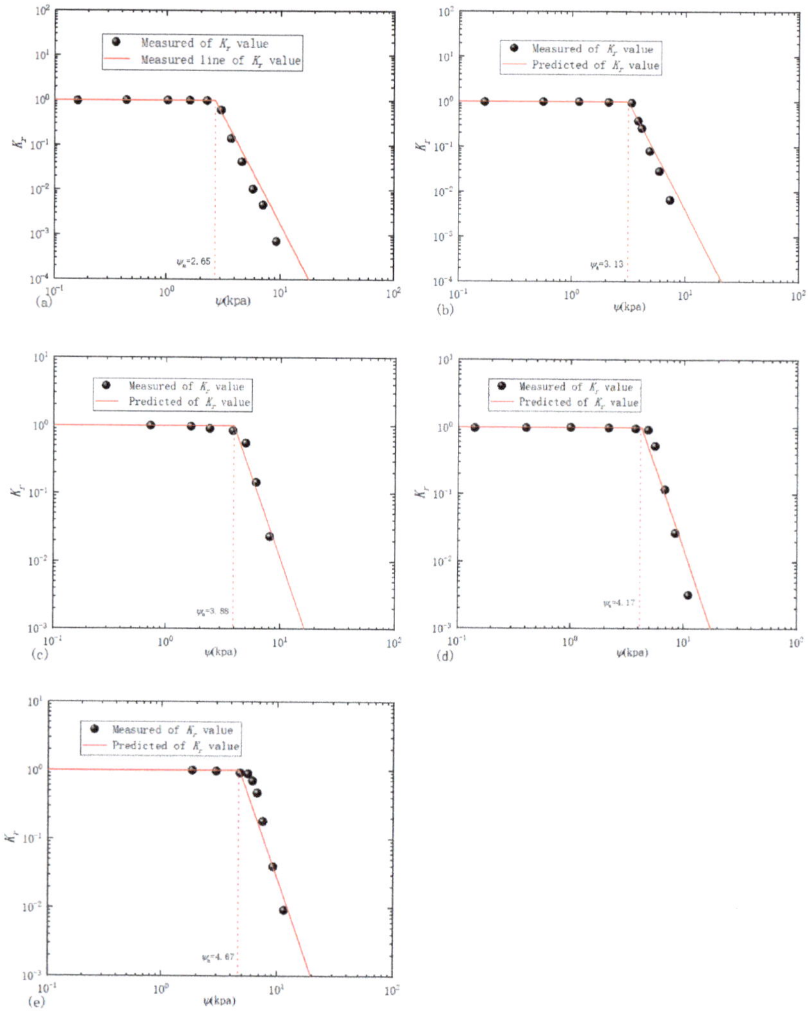

Figure 5. Prediction results of the unsaturated relative permeability coefficients of Columbia sandy loam with different void ratios: **(a)** $e = 1.268$; **(b)** $e = 1.114$; **(c)** $e = 0.942$; **(d)** $e = 0.89$; **(e)** $e = 0.815$.

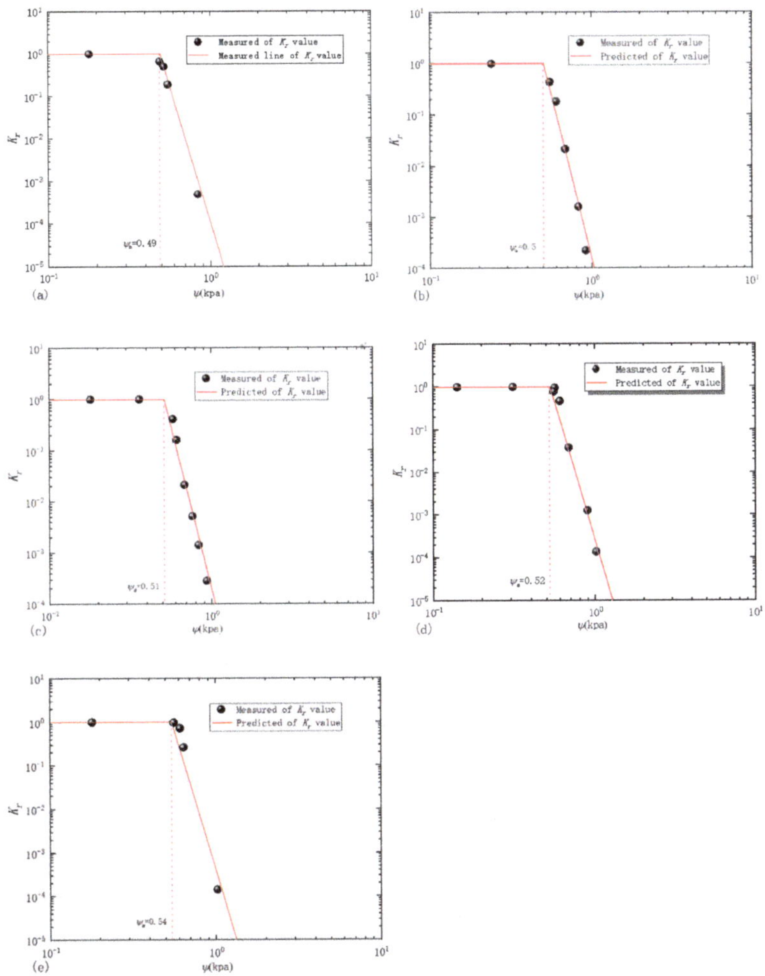

Figure 6. The prediction results of the unsaturated relative permeability coefficients of unconsoildated sand with different porosity ratios: **(a)** $e = 0.852$; **(b)** $e = 0.825$; **(c)** $e = 0.799$; **(d)** $e = 0.767$; **(e)** $e = 0.715$.

4. Discussion

The deformation of soil will cause changes in the pore ratio, pore size distribution, and pore shape of soil, for example, large pores will be compressed into small pores [24]. Tao et al. conducted experiments on Wuhan clay through mercury intrusion test, nuclear magnetic resonance test, and scanning electron microscope (SEM) test, and found that the cumulative pore volume distribution per unit particle mass showed a "broom-type" distribution under different compression deformation conditions. The soil–water characteristic curve represented by the mass moisture content also presents a similar distribution, the change law of the two is relatively consistent. Therefore, it is reasonable to use simple mapping method to predict the permeability coefficient of unsaturated soil. Tao and Zhu [25] derive a fractal model of the relative permeability coefficient of unsaturated soil based on fractal theory and seepage theory. The expression is as follows:

$$k_r(\psi) = \left(\frac{\psi_a}{\psi}\right)^{5-D} \tag{7}$$

By combining Equations (6) and (7), we can obtain the air-entry value for various void ratios and, eventually, the unsaturated relative permeability coefficients for soils under deformation conditions. The model index is $5 - D$, which may not apply to some soils. Although the fractal forms of the CCG model, Burdine model, and Mualem model derived in the simplified unified model give other possible forms of the relationship between λ and D, as illustrated in Table 1, these relationships cannot cover all cases. In some instances, λ and D cannot be related using the any of above expressions. Therefore, this paper recommends first measuring the unsaturated relative permeability coefficient and SWCC of the soil before deformation and fitting the value of λ based on the measured data before deformation, that is the λ value after deformation. This paper takes the Touchet silt loam in the literature [23] as an example to compare and analyze the difference between the prediction method proposed by Tao and Zhu [25], and the predicted results are shown in Figure 7.

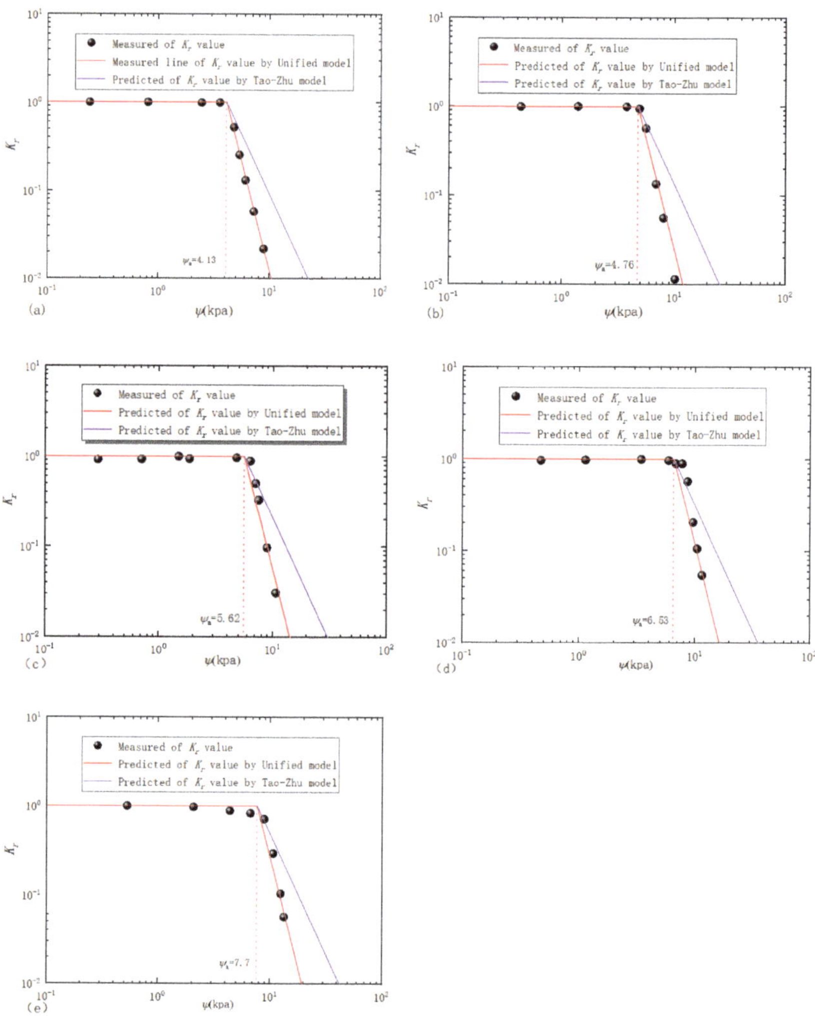

Figure 7. Comparison of Tao and Zhu models with different void ratios: **(a)** e = 1.012; **(b)** e = 0.916; **(c)** e = 0.815; **(d)** e = 0.733; **(e)** e = 0.653.

As illustrated in Figure 7, the predicted values of the unsaturated relative permeability coefficient obtained using the simplified unified model are better to those obtained using the Tao and Zhu model and are more in line with the measured value. It demonstrates that the simple graphing method used to predict the relative permeability coefficient of unsaturated soil under deformations has a broader range of applicability.

5. Conclusions

This study proposed a unified model to determine index λ based on the laboratory measured value of the unsaturated relative permeability coefficient and the air-entry value before deformation. Additionally, the value of index λ remains unchanged before and after deformation. Based on the SWCC obtained before deformation, the fractal dimension and air-entry value could be obtained, which then could be used to predict the air-entry value after deformation. Combined with the simplified unified model, the relationship between the relative permeability coefficient and matrix suction under arbitrary deformations conditions can be readily predicted using the proposed graphing method. The graphical procedure is simple and easy to implement. The method is based on the measured values of the unsaturated permeability coefficient before deformation and overcomes the defect that calculating the index λ through a specific relationship in the unsaturated relative permeability fractal model. The proposed model does not have limitations like the Tao–Kong model, the Burdine model, and the Mualem model, and it can be adopted for a wider range with improved accuracy.

Author Contributions: G.T. and Q.W. contributed to the conceptualization and methodology; Q.C. and Y.P. conducted the analyses; Q.W., S.N. and H.D. wrote and edited the manuscript. All authors discussed the results. All authors have read and agreed to the published version of the manuscript.

Funding: The work in this paper is supported by grants from Natural Science Foundation of China (No. 51978249) and the Innovation Group Project of Hubei Science and Technology Department (No. 2020CFA046), China.

Institutional Review Board Statement: Not applicable.

Informed Consent Statement: Not applicable.

Data Availability Statement: We acknowledge source data from the literature "Hydraulic Properties of Disturbed Soil Materials Affected by Porosity. *Soil. Sci. Soc. Am. J.* **1967**, *31*, 451–454" [23], published by Laliberte, G.E. and Brooks, R.H.

Conflicts of Interest: The authors declare no conflict of interest.

References

1. Sun, Y.; Sumelka, W. Multiaxial stress-fractional plasticity model for anisotropically overconsolidated clay. *Int. J. Mech. Sci.* **2021**, *205*, 106598. [CrossRef]
2. Sun, Y.; Sumelka, W.; Gao, Y.; Nimbalkar, S. Phenomenological fractional stress-dilatancy model for granular soil and soil-structure interface under monotonic and cyclic loads. *Acta. Geotech.* **2021**, *16*, 3115–3132. [CrossRef]
3. Yang, Y.; Lin, Y.; Ji, X. Hydrodynamic characteristics of flow over emergent vegetation in a strongly curved channel. *J. Hydraul. Res.* **2021**, 1–18. [CrossRef]
4. Topp, G.C.; Miller, E.E. Hysteretic Moisture Characteristics and Hydraulic Conductivities for Glass-Bead Media1. *Soil Sci. Soc. Am. J.* **1966**, *30*, 156. [CrossRef]
5. Shafiee, A. Permeability of compacted granule-clay mixtures. *Eng. Geol.* **2008**, *97*, 199–208. [CrossRef]
6. Leung, A.K.; Coo, J.L.; Ng, C.W.W.; Chen, R. New transient method for determining soil hydraulic conductivity function. *Can. Geotech. J.* **2016**, *53*, 1332–1345. [CrossRef]
7. Jarvis, N.J.; Messing, I. Near-saturated hydraulic conductivity in soils of contrasting texture as measured by tension infiltrometers. *Soil Sci. Soc. Am. J.* **1995**, *59*, 27–34. [CrossRef]
8. Childs, E.C.; Collis-George, G.N. The permeability of porous materials. *Proc. R. Soc. A-Math. Phys.* **1950**, *201*, 392–405.
9. Burdine, N.T. Relative permeability calculation from poresize distribution data. *J. Pet. Technol.* **1953**, *5*, 71–78. [CrossRef]
10. Mualem, Y. A new model for predicting the hydraulic conductivity of unsaturated porous media. *Water Resour. Res.* **1976**, *12*, 513–522. [CrossRef]

11. Tao, G.; Kong, L. Permeability model of saturated/unsaturated soil based on microscopic pore channels and its application. *J. Hydraul. Eng.* **2017**, *48*, 702–709. (In Chinese)
12. Tao, G.; Wu, X.; Xiao, H.; Chen, Q.; Cai, J. A unified fractal model for permeability coefficient of unsaturated soil. *Fractals* **2019**, *27*, 1940012. [CrossRef]
13. Mitchell, J.K.; Hooper, D.R.; Campanella, R.G. Permeability of compacted clay. *Soil Mech. Found. Div.* **1965**, *91*, 41–65. [CrossRef]
14. Chang, C.S.; Duncan, J.M. Consolidation analysis for partly saturated clay by using an elastic-plastic effective stress-strain model. *Int. J. Numer. Anal. Met.* **1983**, *7*, 39–55. [CrossRef]
15. Zhang, X.; Zhao, C.; Liu, Y. Probability-based model for influence of deformation on hydraulic conductivity function of unsaturated soils. *J. Eng. Geol.* **2010**, *18*, 132–139. (In Chinese)
16. Hu, R.; Chen, Y.; Liu, H.; Zhou, C. A water retention curve and unsaturated hydraulic conductivity model for deformable soils: Consideration of the change in pore-size distribution. *Géotechnique* **2013**, *63*, 1389–1405. [CrossRef]
17. Gao, Y.; Li, Z.; Sun, D.; Yu, H. A simple method for predicting the hydraulic properties of unsaturated soils with different void ratios. *Soil Tillage Res.* **2021**, *209*, 104913. [CrossRef]
18. Tao, G.; Chen, Y.; Kong, L.; Xiao, H.; Chen, Q.; Xia, Y. A Simple Fractal-Based Model for Soil-Water Characteristic Curves Incorporating Effects of Initial Void Ratios. *Energies* **2018**, *11*, 1419. [CrossRef]
19. Tyler, S.W.; Wheatcraft, S.W. Fractal scaling of soil particle-size distributions: Analysis and Limitations. *Soil Sci. Soc. Am. J.* **1992**, *56*, 362–369. [CrossRef]
20. Xu, Y.; Matsuoka, H.; Sun, D. Swelling characteristics of fractal-textured bentonite and its mixtures. *Appl. Clay Sci.* **2003**, *22*, 197–209. [CrossRef]
21. Brid, N.R.A.; Perrier, E.; Rieu, M. The water retention function for a model of soil structure with pore and solid fractal distributions. *Eur. J. Soil Sci.* **2000**, *51*, 55–63. [CrossRef]
22. Xu, Y.; Huang, Y. Fractal-textured soils and their unsaturated mechanical properties. *Chin. J. Geotech. Eng.* **2006**, *28*, 635–638. (In Chinese)
23. Laliberte, G.E.; Brooks, R.H. Hydraulic Properties of Disturbed Soil Materials Affected by Porosity. *Soil. Sci. Soc. Am. J.* **1967**, *31*, 451–454. [CrossRef]
24. Zhou, C.; Chen, R. Modelling the water retention behaviour of anisotropic soils. *J. Hydrol.* **2021**, *599*, 126361. [CrossRef]
25. Tao, G.; Zhu, X.; Cai, J.; Xiao, H.; Chen, Q.; Chen, Y. A Fractal Approach for Predicting Unsaturated Hydraulic Conductivity of Deformable Clay. *Geofluids* **2019**, *2019*, 8013851. [CrossRef]

fractal and fractional

MDPI

Article

The Fractal Characteristics of Soft Soil under Cyclic Loading Based on SEM

Bowen Kong [1,2,3,*], Chen-Xiang Dai [4], Haibo Hu [4], Jianzhong Xia [1] and Shao-Heng He [4,*]

1 School of Civil Engineering and Architecture, Zhejiang University of Science & Technology, Hangzhou 310023, China; 197002@zust.edu.cn
2 School of Physical and Mathematical Sciences, Nanjing Tech University, Nanjing 211816, China
3 Zhejiang Wuzhou Engineering Project Management Co., Ltd., Hangzhou 310053, China
4 Research Center of Coastal and Urban Geotechnical Engineering, Zhejiang University, Hangzhou 310058, China; 21912017@zju.edu.cn (C.-X.D.); huhaibo@zju.edu.cn (H.H.)
* Correspondence: kbw@zust.edu.cn (B.K.); heshaoheng@zju.edu.cn (S.-H.H.); Tel.: +86-137-3547-6567 (B.K.)

Abstract: Cyclic loading always results in great damage to the pore structure and fractal characteristics of soft soil. Scanning electron microscope (SEM) can help collect data to describe the microstructure of soft soil. This paper conducted a series of SEM tests to interpret the effect of consolidation confining pressure, circulating dynamic stress ratios and overconsolidation ratio on soil's micro-pore structure and fractal characteristics. The results demonstrate that fractal dimension can well represent the complex characteristics of the microstructure of the soil; the larger the consolidation confining pressure, the greater the cyclic dynamic stress ratio, and the greater the overconsolidation ratio, the smaller the fractal dimension number of soil samples. Finally, an empirical fitting formula for cumulative strain considering microstructure parameters is established through data fitting.

Keywords: soft soil; fractal characteristics; SEM; microstructure; cumulative strain

Citation: Kong, B.; Dai, C.-X.; Hu, H.; Xia, J.; He, S.-H. The Fractal Characteristics of Soft Soil under Cyclic Loading Based on SEM. *Fractal Fract.* **2022**, *6*, 423. https://doi.org/10.3390/fractalfract6080423

Academic Editor: Wojciech Sumelka

Received: 31 May 2022
Accepted: 26 July 2022
Published: 30 July 2022

1. Introduction

Marine-saturated silty soft clay is widely distributed in coastal areas, and is prone to settlement under long-term cyclic loads. Severe uneven settlement will cause damage to the superstructure and bring huge economic losses. Controlling the settlement of soft clay foundations is a key issue that needs to be solved urgently. Clarifying the deformation characteristics and microscopic mechanism of soft soil under dynamic load has great significance for the settlement control of soft soil foundation, so it is necessary to conduct systematic basic theoretical research.

Fractal theory is mainly used to study objects with self-similarity characteristics, which refer to the shape, function and feature information of the object. The plastic deformation parameters of fractional geomaterials have been studied extensively [1,2]. To explore the complexity of material composition, fractal theory was gradually widely used in the quantitative study of soil material composition [3–6]. In the past study of soil composition, fractal theory studies were mostly used to study the particle shape of sandy soil [7,8]. Delage R and Lefebvre G analyzed the distribution and arrangement of pores in the soil [9]. Peyton et al. and Zeng et al. used fractal theory to propose a microscopic difference conjecture that used fractal dimensions combined with fractal non-uniformity to describe soil structure [10,11]. He S et al. used fractal theory to study the distribution of soil pores, and the results showed that the pore structure in soil has fractal characteristics, which can be used to quantitatively describe the geometric characteristics of pore structure in the medium [12]. Wang P et al. explored the influence of electron microscopy factors on the quantitative study of microscopic information through a series of scanning electron microscopy pictures [13]. Kong B et al. carried out research on the fractal characteristics of soft soil under different temperatures and pressures [14].

However, existing research can mostly assist in explaining the reason of deformation causes and cannot quantify the relationship between macroscopic deformation and microscopic characteristic parameters. In order to find the precise quantitative relationship between micro and macro deformation, in this paper, the deformation of Hangzhou original soft clay was analyzed from a microscopic point of view under cyclic loading. With the help of SEM, the microstructure parameters of soil samples can be obtained [15]. Based on the existing empirical model of cumulative plastic strain index, the cumulative plastic strain model considering the microstructural parameters was established. Then, the correlation between microstructure parameters and cumulative plastic strain was calculated. The research results can be used as a reliable basis for predicting foundation deformation under cyclic loading by fractal dimension information.

2. Materials and Methods

2.1. Materials

The original soft soil samples were taken from the tunnel section of Zhejiang University's Zijingang Station and Sanba Station, which belongs to Hangzhou Metro Line 2. The maximum burial depth of the soil layer in this interval is 11.8 m, and the groundwater level is 1–2 m. The soil sample used in this test was about 5–5.5 m depth, which belongs to soft soil (based on USCS classification). The soil specimen was 40 mm in diameter and 40 mm in thickness (showed as Figure 1). The soil extraction processes were carried out in strict accordance with the relevant requirements, so the quality of soil samples could be effectively guaranteed.

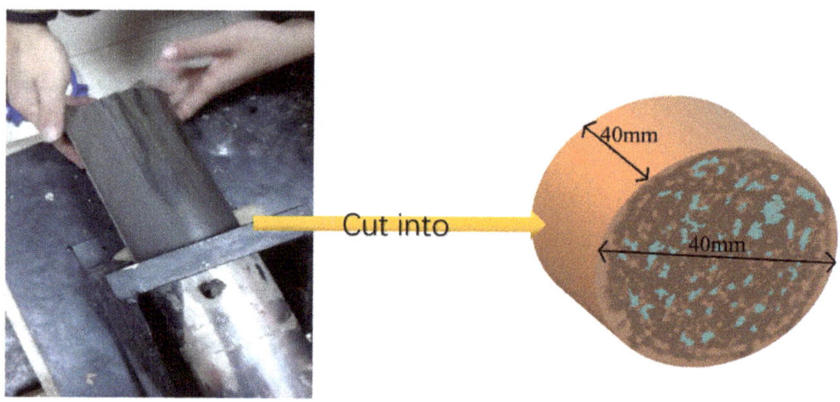

Figure 1. Photograph of undisturbed soft soil sample.

Basic physical and mechanical parameters of soil obtained through indoor geotechnical tests are shown in Table 1.

Table 1. Physical and mechanical parameters of soil.

Soil Type	Unit Weight/kN/m^3	Water Content/%	Specific Gravity	Plastic Limit/%	Liquid Limit/%	Plasticity Index	Liquid Index
Soft soil	15.7	62.47	2.74	27	44.6	17.6	2.01

2.2. Methods

The research includes two parts: firstly, the analysis of the macroscopic parameters (cumulative plastic strain) and microstructure parameters by SEM (fractal dimension and probability entropy). Secondly, based on the existing empirical model of cumulative plastic strain, the cumulative plastic strain model considering the microstructural parameters is established.

In this paper, the one-dimensional circulating load drainage consolidation test from Dai et al. [16] was adopted to provide new insight into the relationship between the microstructure and cumulative deformation. Consolidation conditions and cyclic loading conditions both have great influence on the soil fractal structure [17–19]. After determining the pre-consolidation pressure of the soil sample, the specimens were divided into three groups, namely Group A, Group B and Group C, which corresponded to different consolidation surrounding pressures P_0 (100, 200, 300, 400, 500 kPa); circulating dynamic stress ratios ζ (3, 6, 8, 10, 15); and overconsolidation ratios *OCR* (1, 3, 6, 9). Among them, specimens A_0, B_0, and C_0 represent the samples before loading (parallel specimens). In order to comprehensively explore the relationship between microporous structure and macroscopic mechanical properties, high stresses above 300 kPa were also considered. The experimental program is detailed in Table 2, where p_0 is pressure; ζ is cyclic dynamic stress ratio; N_{max} is the largest number of vibrations and *OCR* is overconsolidation ratio.

Table 2. Experimental program.

Group	Number	p_0/(kPa)	ζ	N_{max}	*OCR*
A	A0		not loaded		
	A1	100	3	20,000	1
	A2	200	3	20,000	1
	A3	300	3	20,000	1
	A4	400	3	20,000	1
	A5	500	3	20,000	1
B	B0		not loaded		
	B1	100	3	20,000	1
	B2	100	6	20,000	1
	B3	100	8	20,000	1
	B4	100	10	20,000	1
	B5	100	15	20,000	1
C	C0		not loaded		
	C1	100	3	20,000	1
	C2	300	3	20,000	3
	C3	600	3	20,000	6
	C4	900	3	20,000	9

Soil samples were saturated before the tests in vacuum equipment: air was sucked out to maintain negative pressure for 3 h, then airless water was added and samples were soaked for 12 h. The specimens were then consolidated under the consolidation surrounding pressure p_0. After the consolidation was completed, a one-dimensional cyclic load was applied to the specimen. Using a semi-sine wave as a stress waveform, the dynamic stress amplitude is p_f; the cyclic dynamic stress ratio ζ indicates the difference between the dynamic stress amplitude p_f, and the consolidation confining pressure p_0 ratio to the consolidation confining pressure p_0, and $\zeta = (p_f - p_0)/p_0$. According to Table 1, the amplitude of cyclic dynamic stress pf can be obtained by determining the consolidation confining pressure p_0 and the cyclic dynamic stress ratio ζ. N refers to the number of loaded half-sine waves experienced, and the total number of vibrations is recorded as N_{max} (set N_{max} = 20,000). Vibration frequency in f refers to the reciprocal of each half-sine wave time course (set f = 0.1 Hz). The loading and waveform diagram is shown in Figure 2.

Scanning electron microscopy (SEM) was used to observe the marine soft soil under different dynamic load conditions (based on ASTM standard), as shown in Figure 3. Under the premise of ensuring that the observation surface was not disturbed, the sample was cut and polished into a 2 mm × 2 mm × 4 mm microscopic sample, and the loose float particles on the observation section were blown away with the ear ball. Along the height direction of the sample, a horizontal and vertical section were selected every 2 cm as the observation area. Due to the poor electrical conductivity of soft soil, in order to ensure the quality of microscopic images the surface of the dried sample was sprayed with 20–50 nm gold film as a conductive material. The SEM test used 8000 times magnification. The microstructure parameters were extracted using the software Image Pro-Plus 6.0 (referred to as IPP). The microstructure parameters were analyzed under different dynamic load conditions, and the contour boundaries of the soil sample shape were extracted to output pore characteristic data, including area, diameter, angle, orientation frequency, probabilistic entropy and fractal dimension.

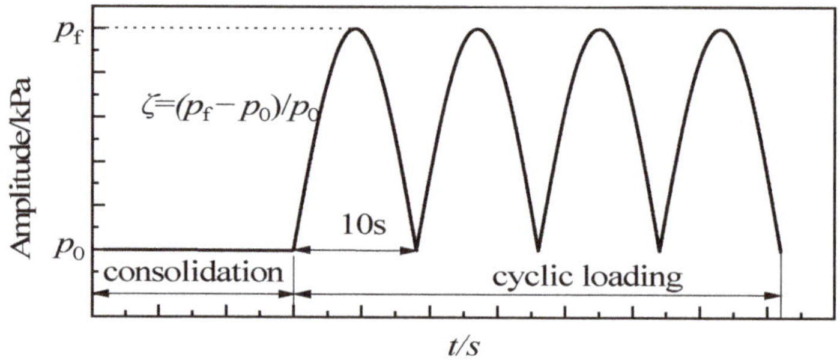

Figure 2. Loading and waveform diagram.

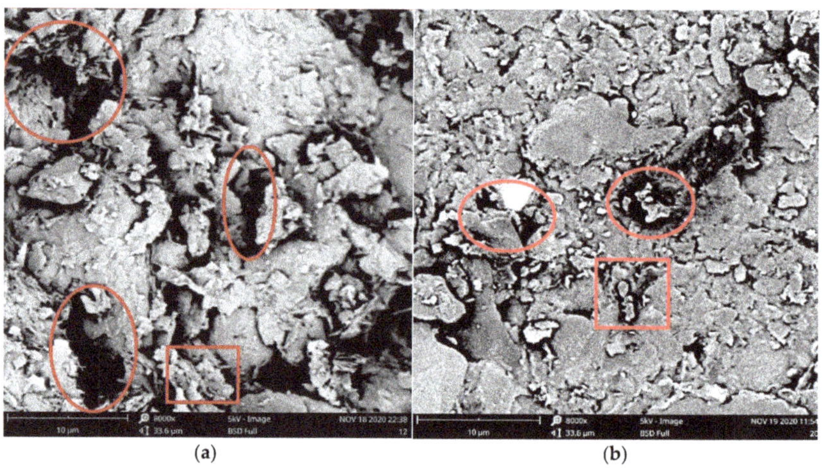

Figure 3. Microscopic scanning images before and after loading [16]: (**a**) before loading; (**b**) after loading. *OCR* = 9 identifications of porosity units. (The circle was marked as a pore, and the square was marked as a flocculent structure).

The analysis has three steps [16]: first of all, a reference plane scale was created to convert the pixel units into length units; secondly, the images were binarized into black and white parts, where the black part represents soil grains and the white part represents pores (Figure 4a); finally, after the image was binarized, the system processing function of the IPP software was used to automatically collect image data (Figure 4b).

Based on the research results of Voss et al. [20], the fractal theory was introduced [21]. It was proposed that the area of particles in sandy soil images has the following relationship with the equivalent perimeter:

$$Log(Perimeter) = \frac{D_d}{2} \times Log(Area) + C \quad (1)$$

where Perimeter represents the equivalent perimeter of any geometric polygon in the scanned image; Area represents the equivalent area of the corresponding polygon; D_d

represents the fractal dimension of the particle shape of the soil corresponding to the scanned image; and C is a constant influenced by the microstructure characteristics of the material. The fractal information can be used both in microstructure and numerical simulation experiments [22].

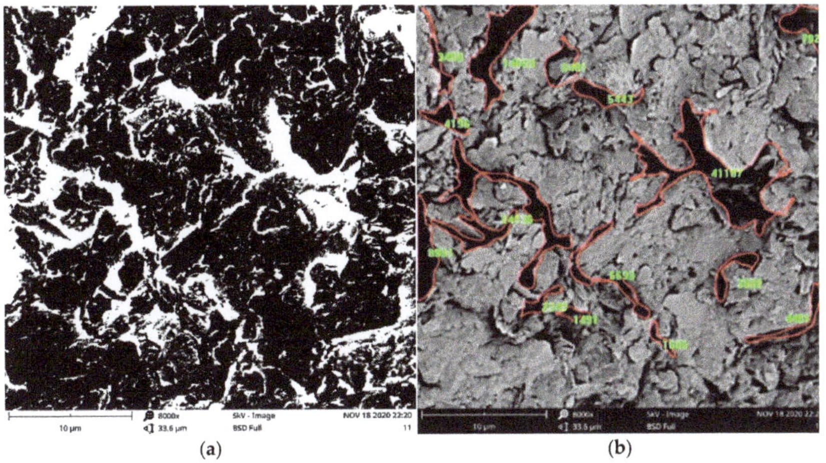

Figure 4. IPP picture processing [16]: (a) binary; (b) identification of porosity units (The pore contour and area are marked).

Shi Bin (1996) used modern systems theory to introduce the concept of probabilistic entropy into the study of soil's structure characteristics, and used H_m to describe the orderliness of the arrangement of soil microstructure units, defined as:

$$H_m = -\sum_{i=1}^{n} P_i(\alpha)\frac{\ln P_i(\alpha)}{\ln n} \tag{2}$$

where H_m represents probability entropy, which indicates the orderliness of the element body distribution, and P represents the probability of occurrence. The larger the H_m, the lower the order of the element arrangement.

3. Results

Figure 5a–c shows the cumulative strain curve of soil samples from Dai et al. [16]. It can be seen that in the early stage of cyclic load loading, an obvious squeezing effect occurs because the internal pores of the soil are subjected to load. When the vibration number increases to 10,000, the specimen is gradually compacted, and the strain accumulation rate is gradually slowed down, showing a slow growth trend. It can be inferred that in the process of vibration, the pore water is gradually discharged, resulting in the hardening of the soil structure. When the number of vibrations reaches a certain time (about 1000 times for the soil in this paper), the soil reaches a stable state and the cumulative strain tends to be stable.

With the continuous increase in the consolidation confining pressure and cyclic dynamic stress ratio, the final cumulative strain of the specimen gradually increases (Figure 5a,b). Under the same cyclic loads, high pressure or stress ratio were more conducive to the compaction of the soil, resulting in a faster strain accumulation rate, and a lager final cumulative strain. When the P_0 pressure reached 500 kPa, the cumulative strain value was twice of that at 100 kPa. The reason for this is that when the consolidation confining pressure and cyclic dynamic stress ratio are larger, the damage of the soil structure is aggravated.

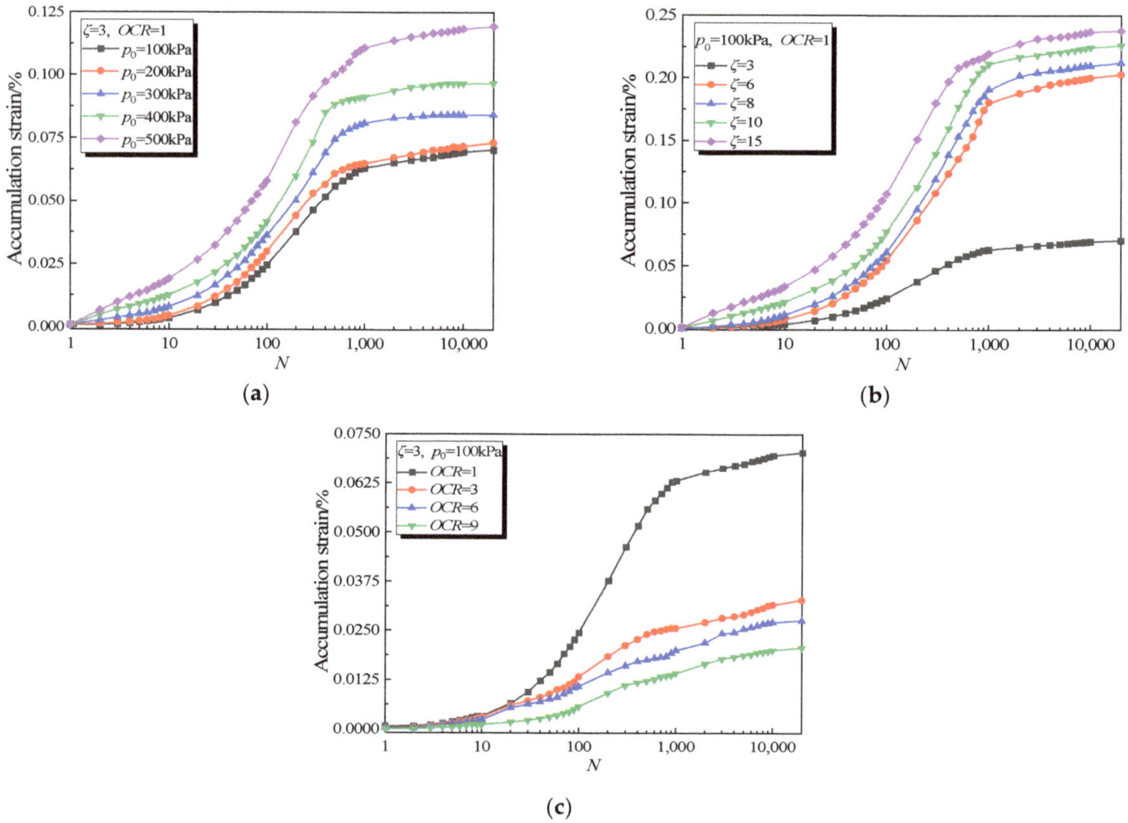

Figure 5. Cumulative strain-vibration curve in different dynamic load modes [16]: (**a**) group A; (**b**) group B; (**c**) group C.

It is worth noting that as the overconsolidation ratio continues to increase, the final cumulative strain of the specimen gradually decreases (Figure 5c). The reason is that under a high overconsolidation ratio the soil has a stronger resistance to deformation, resulting in a smaller final cumulative strain. This indicates that the increase in the overconsolidation ratio has an inhibitory effect on the cumulative strain.

Figures 6–8 show the *log* (*Perimeter*)-*log* (*Area*) curves of soft soil under different test conditions, which have obvious linear relationships. If the linear relationships are expressed as y = ax + b, then a in the linear relationships corresponds to D_d, and b corresponds to the perimeter in logarithmic coordinates. Therefore, it can be considered that the particle shape in soft clay is fractal. The number of fractal dimensions before loading is the largest (1.531 for the undisturbed soil sample), and the fractal dimension of soft soil is always between 1 and 2. After loading, the fractal dimension reaches 1. The results illustrate that the orderliness and directionality of the microstructure of soft soil are worst in the original state. After loading, with the reduction in the fractal dimension number, the pores become more orderly at any loading condition.

It can be seen that, with the increase in confine pressure, dynamic stress and ratio overconsolidation ratio, the fractal dimension value D_d decreased. The essence of the change of fractal dimension information is the change of the stress of the soil in equilibrium state under cyclic loading, which is caused by the recombination of particles in the soil and the dislocation movement of soil particles. In the early stage of loading, the soil particles moved in dislocation under cyclic loading, thus forming new pore structures. The accumulated energy of plastic strain dissipated, and the soil began to deform. With the increase in vibration N, the cumulative viscous energy dissipation rate gradually exceeds the cumulative plastic strain energy dissipation rate, the deformation tends to be stable, and the pore structure trends to be regular.

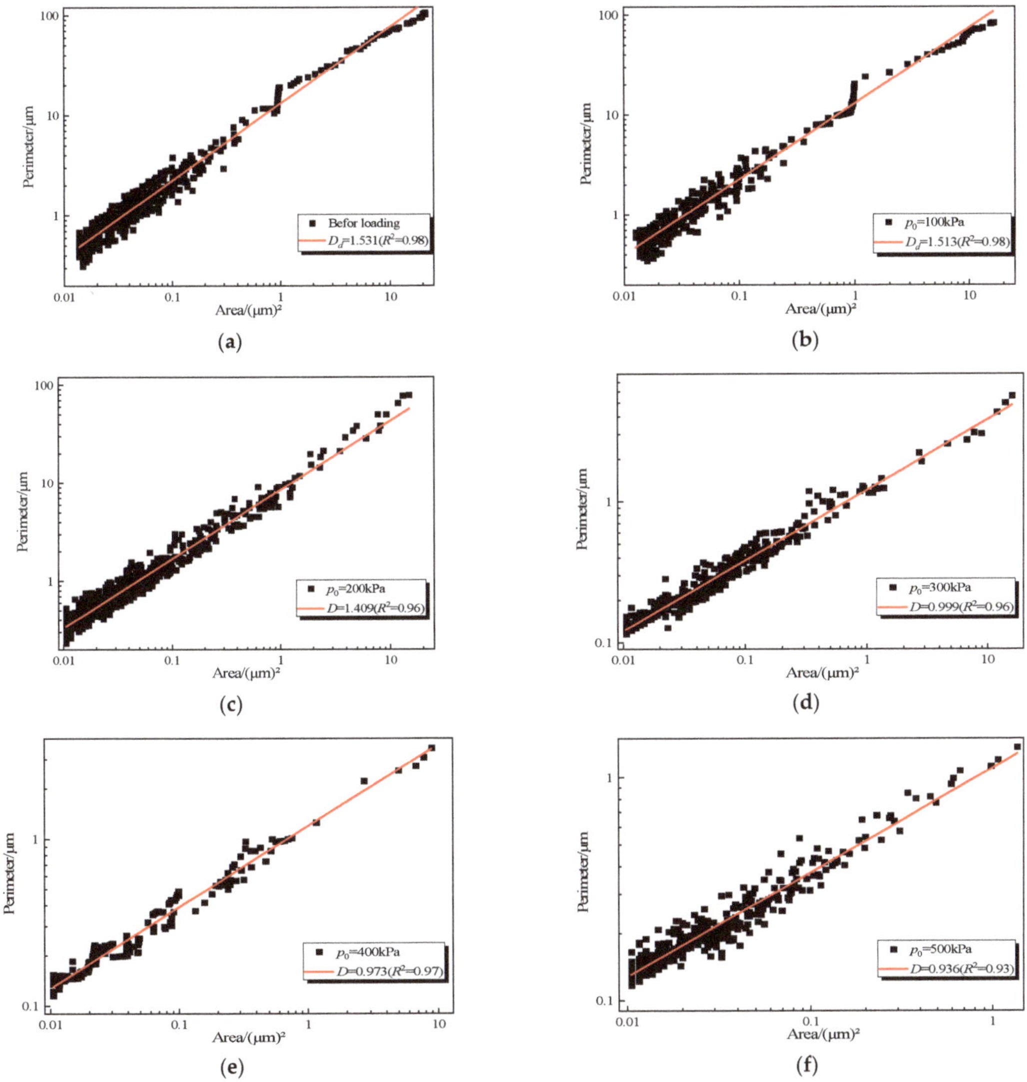

Figure 6. Fractal dimension of soil samples under different pressure: (**a**) before loading; (**b**) p_0 = 100 kPa; (**c**) p_0 = 200 kPa; (**d**) p_0 = 300 kPa; (**e**) p_0 = 400 kPa; (**f**) p_0 = 500 kPa.

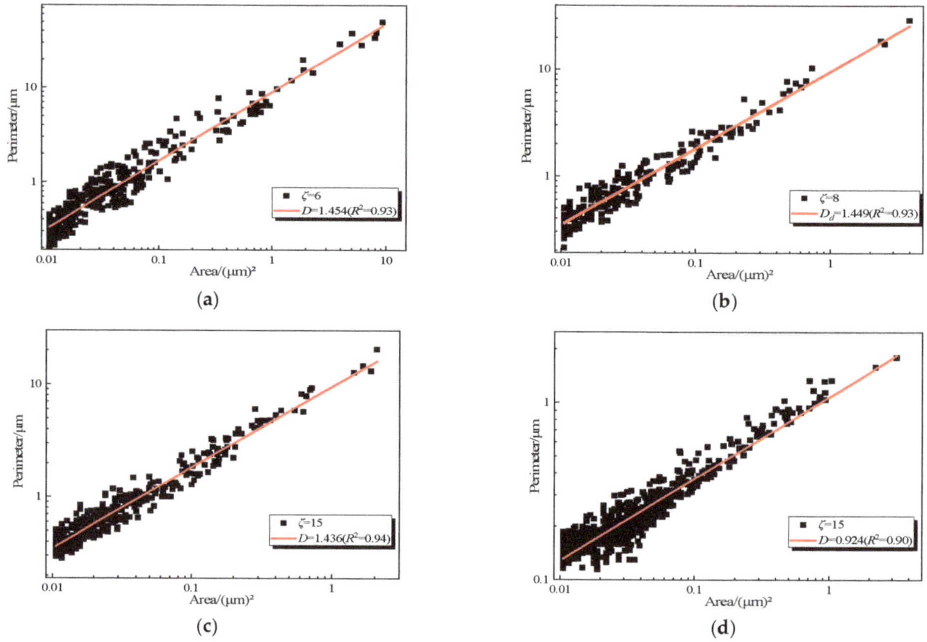

Figure 7. Fractal dimension of soil samples under different cyclic dynamic stress ratio: (**a**) $\zeta = 6$; (**b**) $\zeta = 8$; (**c**) $\zeta = 10$; (**d**) $\zeta = 15$.

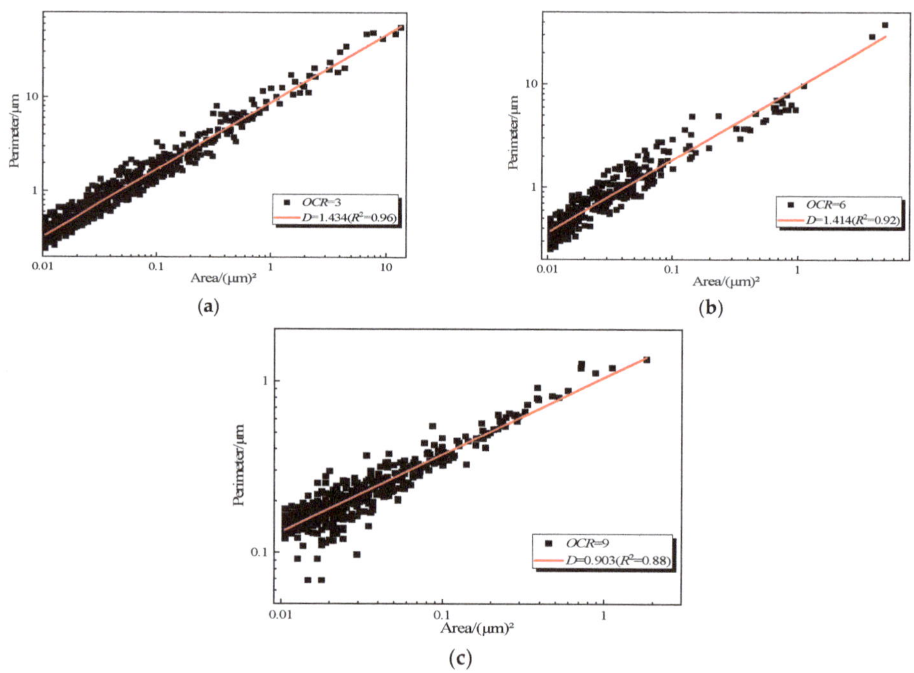

Figure 8. Fractal dimension of soil samples under different overconsolidation ratios: (**a**) $OCR = 3$; (**b**) $OCR = 6$; (**c**) $OCR = 9$.

4. Discussion

The macroscopic mechanical behavior of soil is closely related to the microinternal structure, so the microstructure changes of soft soil during cyclic loading can reveal the fundamental causes of its deformation.

At present, there are two main methods for calculating the cumulative plastic strain of soil under dynamic load: establishing a constitutive model and an empirical fitting formula. When calculating cumulative strain by constitutive model, it is difficult to determine the calculation parameters and apply in practical engineering. In contrast, the concise empirical fitting formulas are widely used in practical engineering.

Based on the Monismith exponential empirical model $\varepsilon = A \times N^b$ and the existing creep model [23], this paper introduces constant A_0 to reflect the stress σ and strain of soft soil ε:

$$\varepsilon(t) = A_0 + A \times t^b \tag{3}$$

where A_0, A, b represents the fitting constant.

Consider that A_0, A, b are parameters related to the stress state and the microstructural characteristics of the soil. Based on Formula (3), the cumulative plastic strain considering the microstructural parameters can be expressed as:

$$\varepsilon(t) = A_0(\sigma_i, C) + A(\sigma_i, C) \times t^{b(\sigma_i, C)} \tag{4}$$

where C is a series of microstructure characteristics of soft soil.

Since the shape of the cumulative strain curve at different stress levels also tends to be consistent; thus, $b(\sigma_i, C)$ can be considered as a constant β. The above formula can be simplified to:

$$\varepsilon(t) = E_0(\sigma_i, C) + E(\sigma_i, C) \times t^{\beta} \tag{5}$$

In Equation (5), $E_0(\sigma_i, C)$ is called the microstructure function, the value of which is related to the microstructure state of the soil at specific stress level. Therefore, if the probability entropy and fractal dimension are used as representatives of the microstructure parameters of soft soil, Equation (5) can be written as:

$$\varepsilon(t) = E_0(\sigma_i, H_m) + E(\sigma_i, H_m) \times t^{\beta m} \tag{6}$$

$$\varepsilon(t) = E_0(\sigma_i, D_d) + E(\sigma_i, D_d) \times t^{\beta d} \tag{7}$$

Let:

$$E(\sigma_i, H_m) = \left(\frac{\sigma_i}{A_m}\right)^{f_m(H_m)} \tag{8}$$

And

$$E(\sigma_i, D_d) = \left(\frac{\sigma_i}{A_d}\right)^{f_d(D_d)} \tag{9}$$

The exponential fitting empirical Equations (6) and (7) can be used to represent the orientation parameters of the soil structure. The calculation formula of cumulative plastic strain can be expressed as:

$$\varepsilon(t) = \left(\frac{\sigma_i}{A_{m0}}\right)^{f_m(H_m)} + \left(\frac{\sigma_i}{A_m}\right)^{f_m(H_m)} \times t^{\beta_m} \tag{10}$$

$$\varepsilon(t) = \left(\frac{\sigma_i}{A_{d0}}\right)^{f_d(D_d)} + \left(\frac{\sigma_i}{A_d}\right)^{f_d(D_d)} \times t^{\beta_d} \tag{11}$$

where A_m, A_{m0}, A_{d0} and A_d is the coefficient to be determined, $f_m(H_m)$, $f_d(D_d)$ is the probability entropy and fractal dimension function. Since the value of the function should be greater than 1 and $H_m \in [0,1]$, $D_d \in [0,1]$, the probability entropy and fractal dimension function can be expressed as:

$$f_m(H_m) = \frac{1}{H_m} \tag{12}$$

$$f_d(H_d) = \frac{1}{D_d} \tag{13}$$

where H_m, D_d is the probability entropy and fractal dimension. According to the change curve of probability entropy and fractal dimension with stress, the relationship between probability entropy and fractal dimension and stress can be expressed as follows:

$$H_m = a_m\sigma^2 + b_m\sigma + c_m \tag{14}$$

$$D_d = a_d\sigma^2 + b_d\sigma + c_d \tag{15}$$

where a_m, a_d, c_m, b_m, b_d and c_d are the fitted values of the probability entropy H_m and fractal dimension D_d and stress σ.

Then, Equations (12) and (13) can be expressed as:

$$f_m(H_m) = \frac{1}{a_m\sigma^2 + b_m\sigma + c_m} \tag{16}$$

$$f_d(D_d) = \frac{1}{a_d\sigma^2 + b_d\sigma + c_d} \tag{17}$$

Based on Equations (16) and (17), Equations (10) and (11) can be written as:

$$\varepsilon(t) = \left(\frac{\sigma_i}{A_{m0}}\right)^{\frac{1}{a_m\sigma^2+b_m\sigma+c_m}} + \left(\frac{\sigma_i}{A_m}\right)^{\frac{1}{a_m\sigma^2+b_m\sigma+c_m}} \times t^{\beta_m} \tag{18}$$

$$\varepsilon(t) = \left(\frac{\sigma_i}{A_{d0}}\right)^{\frac{1}{a_d\sigma^2+b_d\sigma+c_d}} + \left(\frac{\sigma_i}{A_d}\right)^{\frac{1}{a_d\sigma^2+b_d\sigma+c_d}} \times t^{\beta_d} \tag{19}$$

According to the experimental data of the probabilistic entropy and fractal dimension above, the value of parameters can be obtained in Table 3.

Table 3. Summary of parameters.

Parameter	a_m	b_m		c_m	a_d		b_d	c_d
numeric value	-1.279×10^{-7}	-4267×10^{-5}		0.9834	3831×10^{-7}		−0.0016	1.597
Parameter	A_{m0}	A_m	β_m	R^2	A_{d0}	A_d	β_d	R^2
numeric value	9.451×10^5	1.871×10^7	10.98	0.8396	7.106×10^4	981.5	−1.76	0.9629

As shown in Figure 9, the formula of cumulative plastic strain with fractal dimension as the microstructural parameter has a higher correlation coefficient (0.9629) than the formula with probabilistic entropy (0.8396) as the microstructure parameter. Consequently, the fractal dimension can describe the cumulative plastic strain law of Hangzhou's soft soil more accurately. In future studies, other microstructural parameters could be investigated to make the fitted formula curve closer to the experimental data.

Therefore, this paper introduces the fractal dimension as a microstructure parameter into the cumulative strain formula:

$$\varepsilon(t) = \left(\frac{\sigma_i}{A_{d0}}\right)^{\frac{1}{a_d\sigma^2+b_d\sigma+c_d}} + \left(\frac{\sigma_i}{A_d}\right)^{\frac{1}{a_d\sigma^2+b_d\sigma+c_d}} \times t^{\beta_d} \tag{20}$$

where a_d, b_d, c_d, A_{d0}, A_d and β_d are test fitting parameters.

Overall, fractal dimension information from soft soil provides a reliable method to obtain strain, which is helpful when predicting the physical and mechanical properties. The impact of various confine pressures, cyclic dynamic stress ratio and *OCR* were obtained from the analysis of the test results, which describe the fractal characteristics of soft soil. The mathematical relationship between the fractal characteristics and deformation characteristics deserves further investigation.

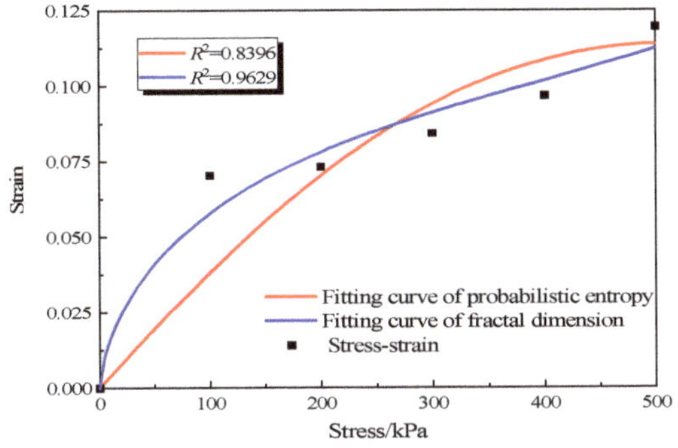

Figure 9. Fitting results by probability entropy and fractal dimension.

5. Conclusions

Using SEM technology, the microstructural change in soft soil under loading was observed, and the fractal dimension of pores was calculated by image processing techniques. Then, the relationship between microporous structure and macrodynamic characteristics were studied by using correlation and empirical model analysis methods. The relationship between the microstructure parameters obtained by SEM and the cumulative strain was discussed. Finally, the following main conclusions were drawn:

(1) The soft soil of Hangzhou marine has a flocculation structure. The distribution of soil particles in the microstructure of soft clay conforms to the fractal characteristics, and the fractal dimension number is between 1 and 2.

(2) After loading, the pores become more orderly. The larger the consolidation confining pressure, the greater the cyclic dynamic stress ratio and the greater the overconsolidation ratio, the smaller the fractal dimension number of soil samples.

(3) Fractal dimension and probabilistic entropy are closely related to cumulative strain. Based on the empirical fitting formula of strain index, an empirical fitting formula for cumulative strain considering microstructure parameters was established.

(4) The trends from the fractal dimension and probabilistic entropy are consistent with each other. The accuracy of the predicted probabilistic entropy strain is 0.83, whereas this value is 0.96 for the fractal dimension. Then, a new way to predict subsidence based on fractal dimension information was obtained for soft soil.

Author Contributions: Conceptualization, B.K. and C.-X.D.; methodology, B.K.; validation, C.-X.D.; formal analysis, B.K. and H.H.; investigation, C.-X.D.; resources, S.-H.H.; data curation, S.-H.H.; writing—original draft preparation, B.K.; writing—review and editing, B.K. and H.H.; visualization, S.-H.H.; supervision, J.X.; project administration, J.X.; funding acquisition B.K. and J.X. All authors have read and agreed to the published version of the manuscript.

Funding: National Natural Science Foundation of China (No. 52008373); Zhejiang Science and Technology Plan Project (2022C35026).

Institutional Review Board Statement: Not applicable.

Informed Consent Statement: Not applicable.

Data Availability Statement: The data are available from the corresponding author upon request.

Acknowledgments: The authors are very grateful to Yifei Sun from Ruhr University Bchum for his kind guidance in writing this article.

Conflicts of Interest: The authors declare no conflict of interest.

References

1. Sun, Y.F.; Sumelka, W. Bounding surface model refined with fractional dilatancy relation for sand. *Soils Found.* **2022**, *62*, 101149. [CrossRef]
2. Sumelka, W.; Blaszczyk, T. Fractional continua for linear elasticity. *Arch. Mech.* **2014**, *66*, 147–172.
3. Li, Y.; Zhang, H.; Huang, M.; Yin, H.; Jiang, K.; Xiao, K.; Tang, S. Influence of Different Alkali Sulfates on the Shrinkage, Hydration, Pore Structure, Fractal Dimension and Microstructure of Low-Heat Portland Cement, Medium-Heat Portland Cement and Ordinary Portland Cement. *Fractal Fract.* **2021**, *5*, 79. [CrossRef]
4. Pfeifer, P.; Avnir, D. Chemistry in noninteger dimensions between two and three. I. Fractal theory of heterogeneous surfaces. *J. Chem. Phys.* **1983**, *79*, 7. [CrossRef]
5. Zhao, X.; Yang, B.; Yuan, S.; Shen, Z.; Feng, D. Seepage–Fractal Model of Embankment Soil and Its Application. *Fractal Fract.* **2022**, *6*, 277. [CrossRef]
6. Fu, X.; Ding, H.; Sheng, Q.; Zhang, Z.; Yin, D.; Chen, F. Fractal Analysis of Particle Distribution and Scale Effect in a Soil–Rock Mixture. *Fractal Fract.* **2022**, *6*, 120. [CrossRef]
7. Zhou, S.; Liu, D.; Cai, Y.; Yao, Y. Fractal characterization of pore–fracture in low-rank coals using a low-field NMR relaxation method. *Fuel* **2016**, *181*, 218–226. [CrossRef]
8. Sun, W.; Zuo, Y.; Wu, Z.; Liu, H.; Xi, S.; Shui, Y.; Wang, J.; Liu, R.; Lin, J. Fractal analysis of pores and the pore structure of the Lower Cambrian Niutitang shale in northern Guizhou province: Investigations using NMR, SEM and image analyses. *Mar. Pet. Geol.* **2018**, *99*, 416–428. [CrossRef]
9. Delage, P.; Lefebvre, G. Study of the structure of a sensitive Champlain clay and of its evolution during consolidation. *Can. Geotech. J.* **1984**, *21*, 21–35. [CrossRef]
10. Peyton, R.L.; Gantzer, C.J.; Anderson, S.H.; Haeffner, B.A.; Pfeifer, P. Fractal dimension to describe soil macropore structure using X ray computed tomography. *Water Resour. Res.* **1994**, *30*, 691–700. [CrossRef]
11. Zeng, Y.; Payton, R.L.; Gantzer, C.J.; Anderson, S.H. Fractal Dimension and Lacunarity of Bulk Density Determined with X-ray Computed Tomography. *Soil Sci. Soc. Am. J.* **1996**, *60*, 1718–1724. [CrossRef]
12. He, S.-H.; Ding, Z.; Hu, H.-B.; Gao, M. Effect of Grain Size on Microscopic Pore Structure and Fractal Characteristics of Carbonate-Based Sand and Silicate-Based Sand. *Fractal Fract.* **2021**, *5*, 152. [CrossRef]
13. Wang, P.; Jiang, Z.; Ji, W.; Zhang, C.; Yuan, Y.; Chen, L. Heterogeneity of intergranular, intraparticle and organic pores in long-maxi shale in Sichuan basin, south China: Evidence from SEM digital images and fractal and multifractal geometries. *Mar. Pet. Geol.* **2016**, *72*, 122–138. [CrossRef]
14. Kong, B.; He, S.-H.; Tao, Y.; Xia, J. Pore Structure and Fractal Characteristics of Frozen–Thawed Soft Soil. *Fractal Fract.* **2022**, *6*, 183. [CrossRef]
15. Yang, R.; He, S.; Yi, J.; Hu, Q. Nano-scale pore structure and fractal dimension of organic-rich Wufeng-Longmaxi shale from Jiaoshiba area, Sichuan Basin: Investigations using FE-SEM, gas adsorption and helium pycnometry. *Mar. Pet. Geol.* **2015**, *7*, 27–45. [CrossRef]
16. Dai, C.-X.; Zhang, Q.-F.; He, S.-H.; Zhang, A.; Shan, H.-F.; Xia, T.-D. Variation in micro-pores during dynamic consolidation and compression of soft marine soil. *J. Mar. Sci. Eng.* **2021**, *9*, 750. [CrossRef]
17. Cai, Y.Q.; Hao, B.B.; Gu, C.; Wang, J.; Pan, L.Y. Effect of anisotropic consolidation stress paths on the undrained shear behavior of reconstituted Wenzhou clay. *Eng. Geol.* **2018**, *242*, 23–33. [CrossRef]
18. Yuan, B.; Li, Z.; Chen, W.; Zhao, J.; Lv, J.; Song, J.; Cao, X. Influence of Groundwater Depth on Pile–Soil Mechanical Properties and Fractal Characteristics under Cyclic Loading. *Fractal Fract.* **2022**, *6*, 198. [CrossRef]
19. Bohaienko, V.; Bulavatsky, V. Fractional-Fractal Modeling of Filtration-Consolidation Processes in Saline Saturated Soils. *Fractal Fract.* **2020**, *4*, 59. [CrossRef]
20. Voss, R.F.; Laibowitz, R.B.; Alessandrini, E.I. *Fractal Geometry of Percolation in Thin Gold Films*; Springer: New York, NY, USA, 1991.
21. Moore, C.A.; Donaldson, C.F. Quantifying soil microstructure using fractals. *Geotechnique* **1995**, *45*, 105–116. [CrossRef]
22. Sumelka, W.; Nowak, M. Non-normality and induced plastic anisotropy under fractional plastic flow rule: A numerical study. *Int. J. Numer. Anal. Methods Geomech.* **2016**, *40*, 651–675. [CrossRef]
23. Zhang, M.J.; Zhang, L.P. The study on the licrostructure parameters and rheological constitutive relations of structural soft soils. *J. Shenyang Jianzhu Univ. Nat. Sci. Ed.* **2006**, *22*, 177–180.

fractal and fractional

MDPI

Article

Effect of Grain Size on Microscopic Pore Structure and Fractal Characteristics of Carbonate-Based Sand and Silicate-Based Sand

Shao-Heng He [1,2], Zhi Ding [1,2,*], Hai-Bo Hu [2] and Min Gao [2]

1 Department of Civil Engineering, Zhejiang University City College, Hangzhou 310015, China; heshaoheng@zju.edu.cn
2 Research Center of Coastal and Urban Geotechnical Engineering, Zhejiang University, Hangzhou 310058, China; huhaibo@zju.edu.cn (H.-B.H.); 21812013@zju.edu.cn (M.G.)
* Correspondence: dingz@zucc.edu.cn

Abstract: In this study, a series of nuclear magnetic resonance (NMR) tests was conducted on calcareous sand, quartz sand, and glass bead with a wide range of grain sizes, to understand the effect of grain size on the micro-pore structure and fractal characteristics of the carbonate-based sand and silicate-based sand. The pore size distribution (PSD) of the tested materials were obtained from the NMR T_2 spectra, and fractal theory was introduced to describe the fractal properties of PSD. Results demonstrate that grain size has a significant effect on the PSD of carbonate-based sand and silicate-based sand. As grain size increases, the PSD of sands evolves from a binary structure with two peaks to a ternary structure with three peaks. The increase in the grain size can cause a remarkable increase in the maximum pore size. It is also found that the more irregular the particle shape, the better the continuity between the large and medium pores. In addition, grain size has a considerable effect on the fractal dimension of the micro-pore structure. The increase of grain size can lead to a significant increase in the heterogeneity and fractal dimension in PSD for calcareous sand, quartz sand and glass bead.

Keywords: nuclear magnetic resonance; pore size distribution; fractal dimension; grain size; calcareous sand; quartz sand

Citation: He, S.-H.; Ding, Z.; Hu, H.-B.; Gao, M. Effect of Grain Size on Microscopic Pore Structure and Fractal Characteristics of Carbonate-Based Sand and Silicate-Based Sand. *Fractal Fract.* **2021**, *5*, 152. https://doi.org/10.3390/fractalfract5040152

Academic Editor: Wojciech Sumelka

Received: 3 September 2021
Accepted: 1 October 2021
Published: 4 October 2021

1. Introduction

Carbonate-based sands (i.e., sands with the main component of calcium carbonate) and silicate-based sands (i.e., sands with the main component of quartz) exist extensively in nature, and they have long been used as the two most important granular geomaterials in the construction of geotechnical engineering. Specifically, siliceous-based sand is mainly distributed on the continent [1], which is applied to terrigenous geotechnical projects, such as roads, foundation pits, slopes, tunnels, etc. Calcareous-based sand, formed by the geological deposition of marine biological debris, is widely distributed onshore and offshore. It is commonly used in land reclamation projects. For example, it has been adopted as the filling material for the foundations of wharves, roads, and airport runways and in island reefs. Carbonate-based sand has more complex mechanical properties than silicate-based sand because its particle shape is more irregular [2–15]. The microstructure of soils, especially the pore-size distribution (i.e., PSD), has been extensively acknowledged to be related to mechanical properties, such as shear strength, compressibility, and water-retention ability [16–22]. Therefore, it is essential to investigate the micropore structure of carbonate-based sand and silicate sand for a better understanding of their macroscopic mechanical behaviors.

In the last decade, numerous efforts have been devoted to revealing the pore size distribution of soils, mainly based on several experimental measurement techniques, such as scanning electron microscope (SEM), mercury intrusion porosimetry (MIP), transmission electron microscopy (TEM), X-ray micro-tomography (XR-uCT), and nuclear magnetic resonance (NMR) [10,11,16,23–27]. The previous studies demonstrated that the microscopic pore structure of soils was highly complicated, highlighting the necessity of additional experimental research in this field. Moreover, previous works have provided sufficient evidence that the pore structures of porous media are born with fractal characteristics and could be studied by fractal theory [17,28–32]. Through a series of SEM and NMR tests, Shao et al. [33] observed a rather irregular pore size distribution in tight sandstone which consists of quartz and clay minerals. They found that the pore network of tight sandstones could be recognized as a dual-scale pore system based on fractal theory, and the fractal dimensions could reflect the physical properties. Sun et al. [34] carried out NMR tests on the Lower Cambrian Niutitang formation shale in northern Guizhou, China, and they investigated the relationships between fractal dimensions of micro-pore structure and mineral compositions. Li et al. [35] conducted NMR tests on the organic-rich marine shales of the Lower Cambrian Niutitang Formation in northern Guizhou, China, to explore the pore structures and fractal characteristics. It was found that shale samples with a larger surface fractal dimension have a higher methane adsorption capacity. Although some experimental investigations have been performed on sandstone and shale to investigate the micro-pore structure, few attempts have been made to quantify the microscopic pore structure and fractal characteristics of granular material, such as calcareous sand and siliceous sand. Another challenging problem is that the sands in the field are always heterogeneous with different grain sizes rather than with a single grain size. However, the effect of the grain size on the microstructure of granular material is still poorly understood. Consequently, further investigations are needed to eliminate the gap surrounding the effect of grain size on the microscopic pore structure and fractal characteristics of carbonate-based sand and silicate-based sand.

In this work, a series of nuclear magnetic resonance (NMR) tests was conducted on calcareous sand, quartz sand, and glass bead with a wide range of grain sizes. The effect of grain size on the micropore structure and fractal characteristics of the carbonate-based sand and silicate-based sand was analyzed in detail. Fractal theory was then introduced to describe the fractal properties of PSD, where the fractal dimension was obtained. The relationship between the fractal dimension and grain size was deeply discussed. The present findings might help to achieve a better interpretation and prediction of the macroscopic behavior of sands with different grain sizes from a microscopic structure view.

2. Materials and Methods

2.1. Materials

The carbonate-based sand used in this study was the natural calcareous sand from the South China Sea (as shown in Figure 1a). The South China Sea has a typical dual-structure coral reef stratum due to its complex seabed landform type. The upper layer of coral islands and reefs in the South China Sea is carbonate-based sand, and the lower layer is reef limestone. The carbonate-based sand adopted in this study was obtained from the upper layer of the coral island reef in the South China Sea, which is a good construction filler material widely used in marine reclamation projects. Through additional X-ray diffraction tests already conducted on this carbonated-based sand [2], it was found that this sand is mainly composed of aragonite and calcite, and the calcium carbonate ($CaCO_3$) content of this sand is about 92%. The silicate sands adopted in this study were quartz sand (see Figure 1b) and glass bead (see Figure 1c) used for comparison experiments. The standard quartz sand was obtained from Fujian, China. It is extensively distributed in various strata in Fujian, China, and it is the most common geotechnical material in the local area. Through further X-ray diffraction tests, this sand was found to have a high silicon dioxide (SiO_2) content of around 98%. The glass bead was a commercial glass bead with artificial round particles and the same particle size as sands. In the sampling process of this study, homogeneous calcareous sand and quartz sand were selected as much as possible in their respective identical locations. Moreover, additional sieving was performed before the NMR test to further ensure the uniformity of the experimental materials. Figure 2 shows the scanning electron microscope (SEM) images of calcareous sand and quartz sand. As can be seen, compared with quartz sand, the particles of calcareous sand have more corners, rougher surfaces and more irregular shapes.

(a) (b) (c)

Figure 1. Photographs of the tested materials: (**a**) Calcareous sand; (**b**) Quartz sand; (**c**) Glass bead.

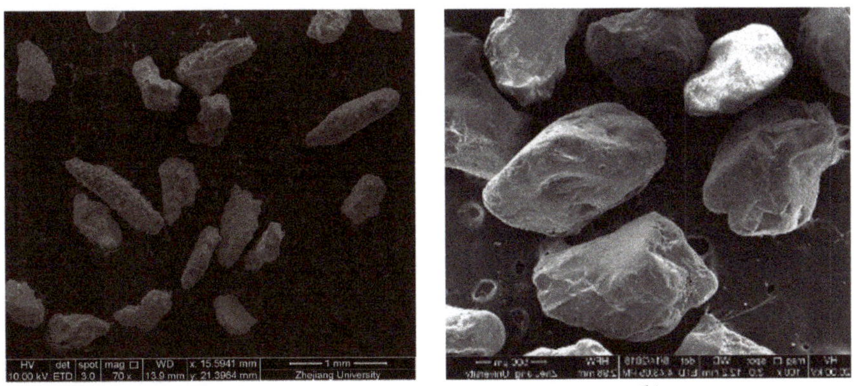

(a) (b)

Figure 2. SEM images of calcareous sand and quartz sand: (**a**) Calcareous sand; (**b**) Quartz sand.

2.2. Test Equipment and Specimen Preparation

The NMR technique has been used to capture the pore size distribution of the adopted specimens. The NMR technique is fast, comprehensive, non-destructive and accurate, which makes it different than the other existing methods like MIT, SEM. Unlike MIT and SEM scanning, NMR can accurately detect the whole pore structures of the sand assembly based on the reliable theoretical relationship between relaxation time and pore size. The device used in this study is the low-field nuclear magnetic resonance device (MesoMR23-060H-I, Shanghai Niumag Corporation, Shanghai, China), as shown in Figure 3. The nuclear magnetic resonance device is composed of five major systems, namely the temperature control system, the magnet system, the spectrometer system, the radiofrequency system and the data acquisition system. The magnet system used 0.5T magnetic field strength, and the magnet temperature was controlled at 32 °C ± 0.01 °C.

The NMR specimens were prepared on a triaxial device. Standard cylindrical specimens with diameters of 50 mm and heights of 100 mm were employed in this study [36–38], as shown in Figure 4. The approach of air pluviation was adopted to prepare the specimen to prevent sand particles from breaking during specimen preparation [39]. The prepared specimens were then saturated by passing carbon dioxide and deaired water through the specimens. Subsequently, the specimen was saturated at a back pressure of 200 kPa. After the B-value was detected to be greater than 95%, the saturation operation was stopped. The saturated sample was removed from the triaxial device and then transferred to the sample tube of the NMR device under the protection of the mold. To investigate the effect of grain size on the microstructure and its fractal dimension, NMR tests were carried out on calcareous sand with six single grain sizes (i.e., 0.1–0.25 mm, 0.5–0.71 mm, 0.71–1 mm, 1–2 mm, 2–3 mm, and >5 mm), quartz sands with three single grain sizes (0.1–0.25 mm, 0.5–0.71 mm, and 1–2 mm), and glass beads with four single grain sizes (i.e., 0.5–0.71 mm, 0.71–1.0 mm, 1–2 mm, and 2–3 mm) for comparison. The detailed test program is summarized in Table 1.

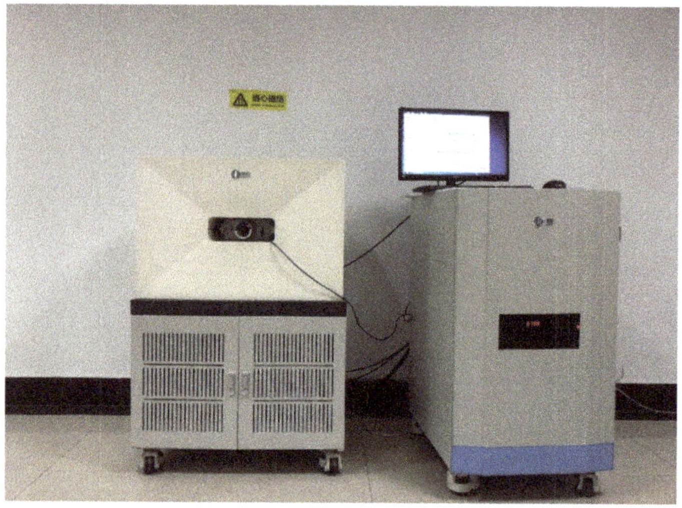

Figure 3. Photograph of the NMR device.

Figure 4. Photograph of the tested specimens.

Table 1. Test program of this study.

No.	Material	Grain Size (mm)	Void Ratio
C-D-1	Calcareous sand	0.1–0.25	1.143
C-D-2		0.5–0.71	1.142
C-D-3		0.71–1	1.146
C-D-4		1–2	1.145
C-D-5		2–3	1.144
C-D-6		>5	1.143
S-D-1	Quartz sand	0.1–0.25	0.507
S-D-2		0.5–0.71	0.508
S-D-3		1–2	0.509
G-D-1	Glass sand	0.5–0.71	0.631
G-D-2		0.71–1.0	0.630
G-D-3		1–2	0.632
G-D-4		2–3	0.633

2.3. Calculation Method for Determining Pore Size Distribution Using NMR

After the main magnetic field (B_0) is applied to the nuclei, the spin direction of some nuclei will become the same as the applied magnetic field, while most of the other spin directions are still chaotic. When the applied magnetic field is removed, the non-equilibrium protons at a high energy level will return to the equilibrium state of a low energy level. This process of returning the protons from a high energy level to the equilibrium state is called relaxation. From a macro point of view, the maximum magnetization (M_0) will tend to return to the initial state parallel to B_0 (i.e., the smallest energy and the most stable state). Therefore, attenuation occurs in the xy plane and recovery occurs in the z-direction. The Bloch equation can be used to describe the relaxation phenomenon, as follows:

$$\frac{\mathrm{d}\vec{M}}{\mathrm{d}t} = \vec{M} \times \gamma \vec{B} - \frac{(M_z - M_0)\vec{k}}{T_1} - \frac{M_x \vec{i} + M_y \vec{j}}{T_2} \tag{1}$$

which can be transformed into:

$$\frac{\mathrm{d}}{\mathrm{d}t}\begin{pmatrix} M_x \\ M_y \\ M_z \end{pmatrix} = \gamma \begin{pmatrix} \vec{i} & \vec{j} & \vec{k} \\ M_x & M_y & M_z \\ B_x & B_y & B_z \end{pmatrix} - \frac{(M_z - M_0)\vec{k}}{T_1} - \frac{M_x \vec{i} + M_y \vec{j}}{T_2} \tag{2}$$

The solution of the equation with the specific B_0 is given by:

$$\begin{aligned} \frac{\mathrm{d}}{\mathrm{d}t}\begin{pmatrix} M_x \\ M_y \\ M_z \end{pmatrix} &= \gamma \begin{pmatrix} \vec{i} & \vec{j} & \vec{k} \\ M_x & M_y & M_z \\ 0 & 0 & B_0 \end{pmatrix} - \frac{(M_z - M_0)\vec{k}}{T_1} - \frac{M_x \vec{i} + M_y \vec{j}}{T_2} \\ &= \gamma\left(M_z B_0 \vec{i} - M_x B_0 \vec{j}\right) - \frac{(M_z - M_0)\vec{k}}{T_1} - \frac{M_x \vec{i} + M_y \vec{j}}{T_2} \\ &= \begin{pmatrix} -\frac{1}{T_2} & \gamma B_0 & 0 \\ -\gamma B_0 & -\frac{1}{T_2} & 0 \\ 0 & 0 & -\frac{1}{T_1} \end{pmatrix}\begin{pmatrix} M_x \\ M_y \\ M_z \end{pmatrix} + \begin{pmatrix} 0 \\ 0 \\ \frac{M_0}{T_1} \end{pmatrix} \end{aligned} \tag{3}$$

where M is the magnetization vector, B is the magnetic field vector, t is the time, γ is the rotation ratio, T_1 is the longitudinal relaxation time, and T_2 is the transverse relaxation time.

T_1 relaxation is the process of energy exchange between hydrogen nuclei and the surrounding environment, that is, T_1 relaxation (i.e., longitudinal relaxation) is spin-lattice relaxation; T_2 relaxation (i.e., transverse relaxation) is the process of energy exchange between hydrogen nuclei and other atomic nuclei, that is, T_2 relaxation is spin-spin relaxation. Figure 5 shows the longitudinal and transverse components of the relaxation process versus time. The equation of longitudinal relaxation is given by:

$$\frac{\mathrm{d}M_z}{\mathrm{d}t} = -\frac{M_z - M_0}{T_1} \quad (4)$$

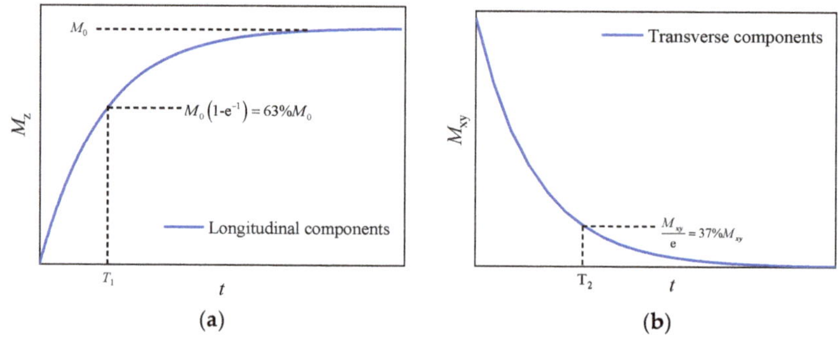

Figure 5. Schematic diagram of the changes in the longitudinal and transverse components in the relaxation process: (**a**) Longitudinal components; (**b**) Transverse components.

Then, the solution can be expressed as:

$$M_z(t) = M_0 + (M_z(0) - M_0)\mathrm{e}^{-t/T_1} \quad (5)$$

The equation of transverse relaxation is given by:

$$\frac{\mathrm{d}M_{xy}}{\mathrm{d}t} = -\frac{M_{xy}}{T_2} \quad (6)$$

Then, the solution can be given by:

$$M_{xy}(t) = M_{xy}(0)\mathrm{e}^{-t/T_2} \quad (7)$$

Because the completion time of T_2 relaxation is much faster than T_1 relaxation, T_2 relaxation is commonly used to study the pore distribution characteristics of the sample in porous fluid media. According to the relaxation mechanism of low-field NMR, there are three different relaxation mechanisms for fluid in porous media. Therefore, the relaxation time of the pore fluid can be expressed as follows [40]:

$$\frac{1}{T_2} = \frac{1}{T_{2\mathrm{B}}} + \frac{1}{T_{2\mathrm{S}}} + \frac{1}{T_{2\mathrm{D}}} \quad (8)$$

where T_2 is the transverse relaxation time of pore fluid collected by Carr-Purcell-Meiboom-Gill (CPMG) pulse sequence; $T_{2\mathrm{B}}$ is the free transverse relaxation time of the liquid, which is determined by the physical properties of the liquid (such as viscosity and chemical composition); $T_{2\mathrm{S}}$ is the transverse relaxation time caused by surface relaxation; $T_{2\mathrm{D}}$ is the transverse relaxation time of pore fluid caused by diffusion under magnetic field gradient. For water in porous media, the influence of the first and third terms on the equation can be ignored, and the surface relaxation plays the main role as follows:

$$\frac{1}{T_2} = \frac{1}{T_{2\mathrm{S}}} = \rho_2(\frac{S}{V}) \quad (9)$$

where ρ_2 is the surface relaxation coefficient, which is a constant and is not affected by temperature and pressure; the method to determine the ρ_2-value of the sand-gravel mixture is described in detail in Feng, Xu, Chai and Li [23]; S/V is the ratio between the pore surface area (S) to the pore volume (V), which is related to the pore shape. Tian et al. [40] represented that the pore shape in the soil can be approximated as a cylindrical tube shape. Therefore, Equation (9) can be written as such:

$$R = 2\rho_2 T_2 \tag{10}$$

2.4. Calculation Method of Fractal Dimension Using Fractal Theory

Fractal dimension reflects the validity of space occupied by complex shapes, and it is a measure of the irregularity of complex shapes. According to previous works, the pore structures of porous media show fractal characteristics and can be studied by fractal theory [41]. The different fractal dimensions (D) resulted in the mechanisms of pore generation. In recent years, fractal theory has been widely used in quantitatively describing and studying the geometric characteristics of the PSD. The fractal property for the pore size distribution is given by [42]:

$$N(R > \mathrm{Ra}) \propto R^{-D} \tag{11}$$

where D is the fractal dimension; N ($R > R_a$) is the number of pores whose size R is greater than the specific measurement scale R_a. For most geomaterials, counting the number of pores of a given size is not feasible. Instead, it seems more appropriate to use mass to interpret the pore size distribution, which is a more readily measurable quantity. The definition of mass M ($R > R_a$) is suggested as [43]:

$$M(R > R_a) = \rho_P C_m [1 - (R/\lambda_m)^{3-D}] \tag{12}$$

where C_m and λ_m are constants related to pore shape and size, and ρ_P is the pore density under the assumed conditions. The total mass M_T can be determined by setting $R = 0$ in Equation (12), as given by:

$$M_T = \rho_P C_m \tag{13}$$

Substitution of Equation (13) into Equation (12) yields:

$$M(R > R_a)/M_T = 1 - (R/\lambda_m)^{3-D} \tag{14}$$

The constant λ_m is equal to the maximum pore size R_{max} for the fractal behavior in the case of M ($R > R_a$)/M_T = 0 with setting $R = R_{max}$. For the fractal nature of soil pore size distribution, the fractal dimension D is strictly limited to the range of $0 < D < 3$ [44].

On account of the fact M ($R > R_a$)/M_T + M ($R < R_a$)/M_T = 1, Equation (8) can be rearranged to display the pore size distribution in a finer percentage by weight, that is, M ($R < R_a$) by noting that:

$$M(R < R_a)/M_T = (R/R_{max})^{3-D} \tag{15}$$

According to the geometric fractal theory [41], the model of fractal dimension D in PSD corresponding to the pore size R obtained by NMR is as follows:

$$\lg S_v = (3 - D)\log R + (D - 3)\log R_{max} \tag{16}$$

where S_v is the percentage of the pore volume. It is calculated by the pore size smaller than the specific pore diameter R to the total pore volume; D is the fractal dimension, and R_{max} is the maximum pore size.

3. Results and Discussion

3.1. Effect of Grain Size on the Pore Size Distribution

Figure 6a shows the pore size distribution curve of calcareous sand with different grain sizes obtained from the T_2 relaxation time. It can be observed that the grain size significantly affects the pore size distribution curve of calcareous sand. For specimens with a grain size of 0.1 mm–0.25 mm, the pore size distribution belongs to a binary type with two peaks. With the increase of the grain size, the binary pore size distribution gradually evolves into a ternary pore size distribution with three peaks. The pore diameter corresponding to the turning point where the two peaks evolve into three peaks is around 100 μm. From the pore size distribution curves of the calcareous sands with different grain sizes, it can be inferred that the pore size distribution curve of the specimen with a grain size less than 0.25 mm belongs to a binary type. However, for the specimen with grain size larger than 1 mm, the pore size distribution curve belongs to a ternary type. The pore size distribution curve of the specimen with grain size around 0.25 mm–1 mm is between the binary type and ternary type. In addition to the number of main peaks in the pore size distribution curve of calcareous sand, the grain size also significantly affects the range of the pore size distribution curve. As shown in Figure 6a, the larger the grain size, the larger the maximum pore size and the wider the pore size range.

Figure 6b shows the pore size distribution curves of quartz sands with different grain sizes. It can be seen that, similar to calcareous sand, for quartz sand specimens with a grain size of 0.1 mm–0.25 mm, the pore size distribution also belongs to a binary type with two peaks. As the grain size increases, the pore size distribution curve of quartz sand gradually evolves into a ternary type with three peaks. Figure 6c shows the pore size distribution curves of glass beads. The results show the pore size distribution curves of glass beads all have three peaks. Note that in the pore size range between 100 μm and 1000 μm, carbonate-based sand has a continuous pore size distribution, whereas the pore size distribution of silicate-based sand is discontinuous. This indicates that the more irregular the particle shape, the better the continuity between the large and medium pores. Moreover, for quartz sand and glass bead, as the particle size increases, the maximum pore size also increases significantly.

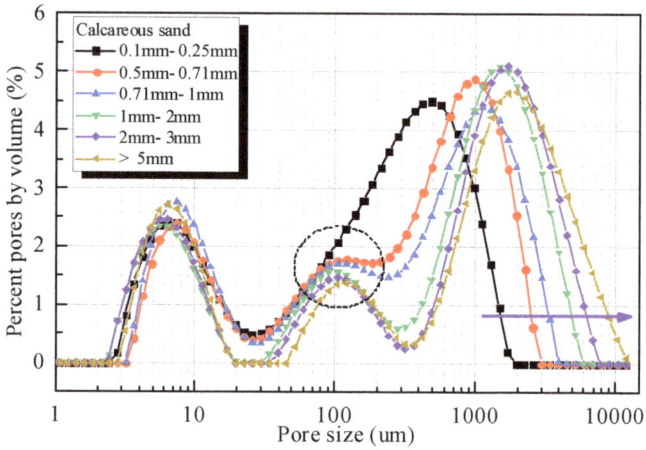

(a)

Figure 6. *Cont.*

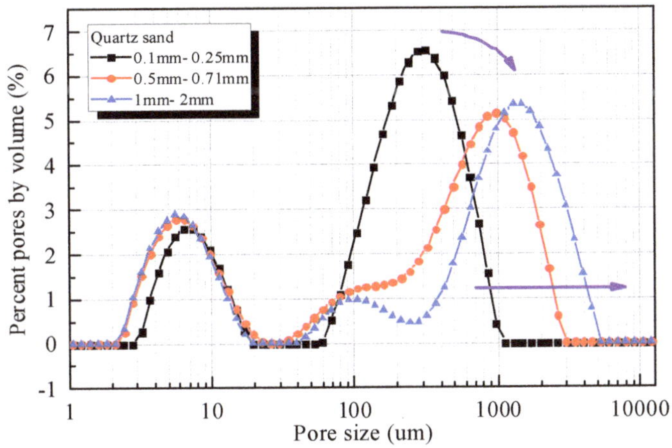

(**b**)

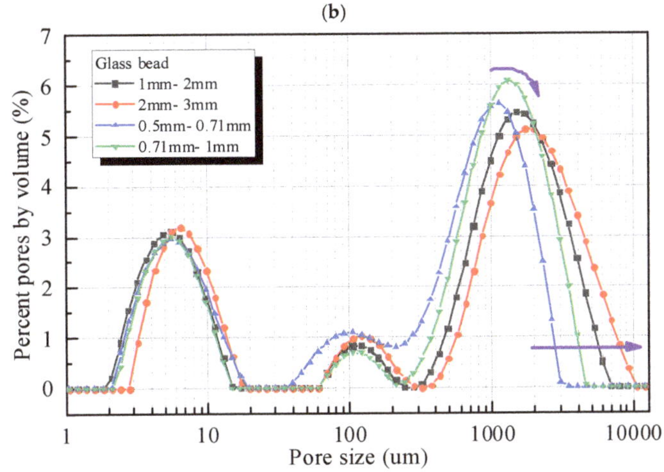

(**c**)

Figure 6. Pore size distribution of the tested materials with different grain sizes: (**a**) Calcareous sand; (**b**) Quartz sand; (**c**) Glass bead.

In order to better understand the relative change of the pore size distribution of the three tested sands with different grain sizes, the difference of the pore size distribution curve under different grain sizes is presented in Figure 7. Figure 7a is the difference between the pore size distribution of calcareous sand with different grain sizes and the pore size distribution of calcareous sand with a grain size of 0.1 mm–0.25 mm. Figure 7b is the difference between the pore size distribution of quartz sand with different grain sizes and the pore size distribution of quartz sand with a grain size of 0.1 mm–0.25 mm. Figure 7c is the difference between the pore size distribution of glass bead with different grain sizes and the pore size distribution of glass beads with a grain size of 0.5 mm–0.71 mm. It can be seen from Figure 7 that the pore size distribution of the three tested materials has a relatively consistent changing trend with the grain size. This indicates that the influence mechanism of grain size on the pore size distribution of the three materials is similar. Specifically, the increase in grain size increases the proportion of large pores with a size larger than 1000 μm, and correspondingly reduces the proportion of medium pores with a size between 100 μm–1000 μm. However, grain size has a minor effect on small pores with a size less than 100 um.

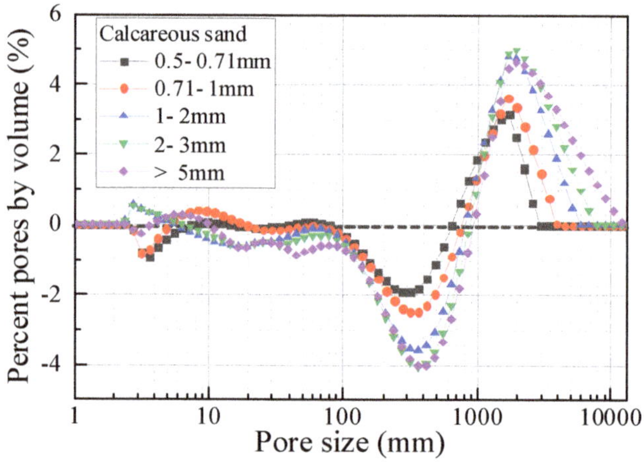

(a)

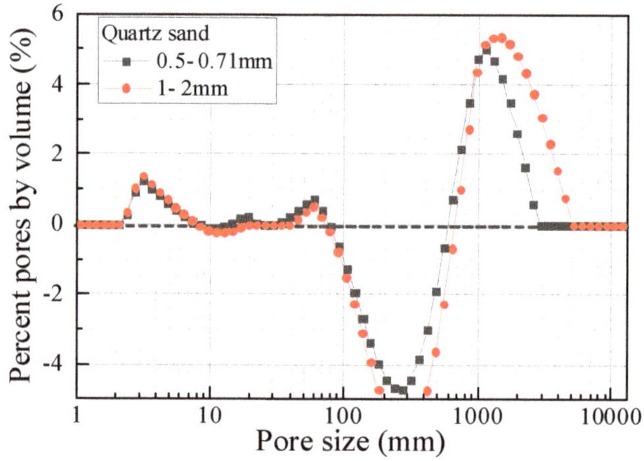

(b)

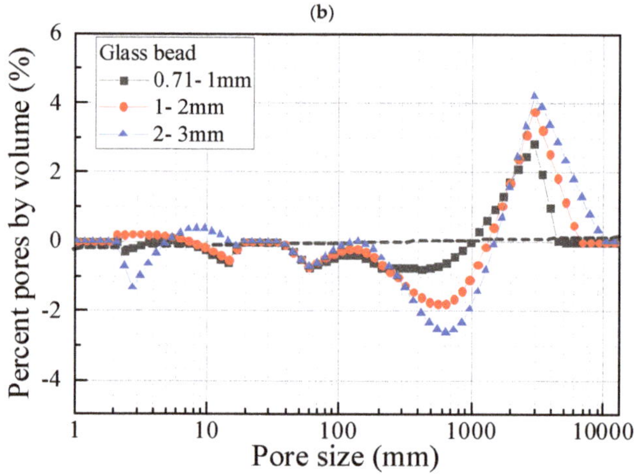

(c)

Figure 7. Differences in the pore size distribution of tested materials: (**a**) Calcareous sand; (**b**) Quartz sand; (**c**) Glass bead.

3.2. *Effect of Grain Size on the Microsrcture's Fractal Dimension*

Liang, et al. [44] reported that the lgS_v-lg$R$ curve of soil included two distinct parts and proposed that the PSD in soils showed multi-fractal characteristics. The lgS_v-lg$R$ curve of calcareous sand, quartz sand, and glass bead are shown in Figures 8–10, respectively. There is an obvious inflection point around the pore diameter of 10 um in the lgS_v-lg$R$ curve, indicating that the pore size of about 10 μm is a limitation for the change of pore properties in the carbonate-based sand and silicate-based sand. The dominant pore structure content of carbonate-based sand and silicate-based sand is distributed in the large-size range (i.e., the pore structure with a size greater than 10 μm. Furthermore, the larger pore structure of sand assembly more tends to be a skeleton pore structure with strong contact and control of the mechanical behavior. Therefore, the fractal dimension (D) in a large-size range of pores is supposed to be the dominant parameter reflecting the principal characteristics of the PSD. Based on Equation (16), the fractal dimension of the macropore (i.e., on the right side of the turning point) was obtained through the regression analysis, as shown in Figures 8–10. The fitting results of fractal dimensions demonstrate that the pore size distribution of calcareous sand, quartz sand, and glass bead all complies with the fractal law. In addition, the correlation coefficient (R^2) of calcareous sand is between 0.95–0.99; the correlation coefficient of quartz sand is between 0.87–0.93 and the correlation coefficient of glass beads is between 0.87–0.94. This indicates that the fractal characteristics of calcareous sand of various particle sizes are more obvious than that of quartz sand and glass beads.

Figure 11 shows the relationship between the fractal dimension and the mean grain size (d_{50}) of the three sands. It can be clearly observed that the fractal dimensions of calcareous sand and quartz sand all increase significantly with the increase of grain size. Compared with calcareous sand and quartz sand, the fractal dimension of glass beads increases more slowly with grain size. This may be due to the single shape (i.e., round shape) of glass beads. Previous studies pointed that the higher fractal dimension of pore size distribution means a more complex pore structure distribution and a stronger heterogeneity of the pore structure distribution [29]. The increase of grain size can lead to a significant increase of D for calcareous sand, quartz sand and glass bead. This indicates that the increase of grain size can cause a more complex pore structure distribution in the sand. Furthermore, under the same grain size, the fractal dimension of calcareous sand is the smallest, the fractal dimension of glass beads is the largest, and quartz sand is somewhere in between. In addition, the fractal dimension of calcareous sand increases more slowly with grain size than quartz sand. The larger the grain size, the greater the gap between the fractal dimensions of the two sands. This indicates that as the grain size increases, the difference in particle shape has a more obvious impact on the pore structure.

It is widely acknowledged that the determination of the fractal dimension plays an important role in predicting the physical and mechanical properties of geomaterials [45,46]. Huang et al. [45] quantified the fractal characteristics of the pore structure of shale and found a significant relationship between the fractal dimension and permeability. Kong et al. [47] established a fractal model of pore size distribution for porous geomaterials based on the fractal characteristics of pores, which could be used to study hydraulic properties. Based on the mercury intrusion test and image analysis, Dou et al. [46] analyzed the effect of different fractal models of pores on the evaluation of sandstone's permeability and highlighted the role of the fractal dimension in determining the permeability. Therefore, accurate prediction of the fractal dimension is crucial for grasping and analyzing the engineering properties (e.g., permeability) of geomaterials. Considering this, a suitable correlation (see Figure 11) was proposed to capture the fractal dimensions of carbonate-based sand and silicate-based sand with different mean grain sizes, which is as follows:

$$D = \alpha {d_{50}}^{\beta} \tag{17}$$

where α and β are model parameters, and through regression analysis, the best-fit α and β for carbonate-based sand and silicate-based are summarized in Table 2. The proposed correlation between D and d_{50} can be applied to the research of a permeability model, water-retention ability analysis, and other related aspects for granular geomaterials, which is useful for guiding the construction of geotechnical engineering.

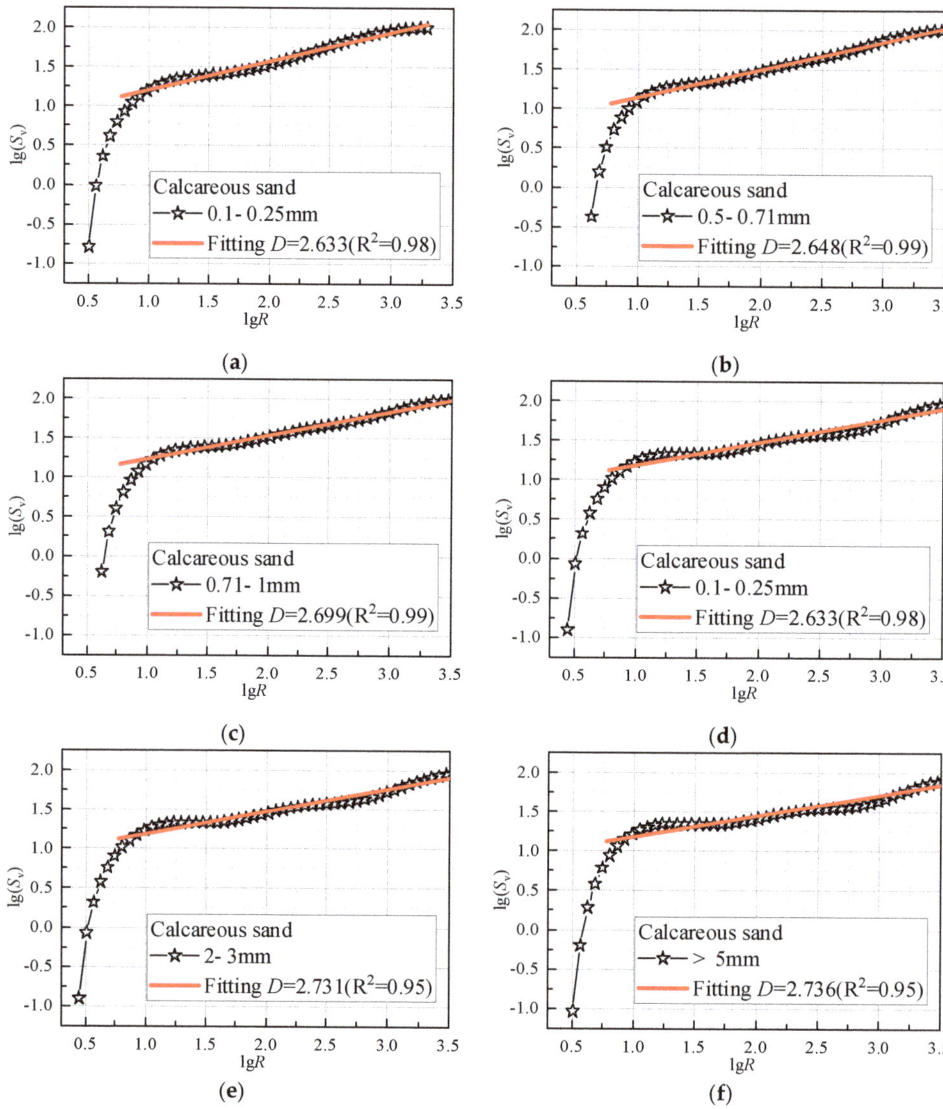

Figure 8. Determination of fractal dimension of calcareous sand with different grain sizes: (**a**) 0.1–0.25 mm; (**b**) 0.5–0.71 mm; (**c**) 0.71–1 mm; (**d**) 0.1–0.25 mm; (**e**) 2–3 mm; (**f**) >5 mm.

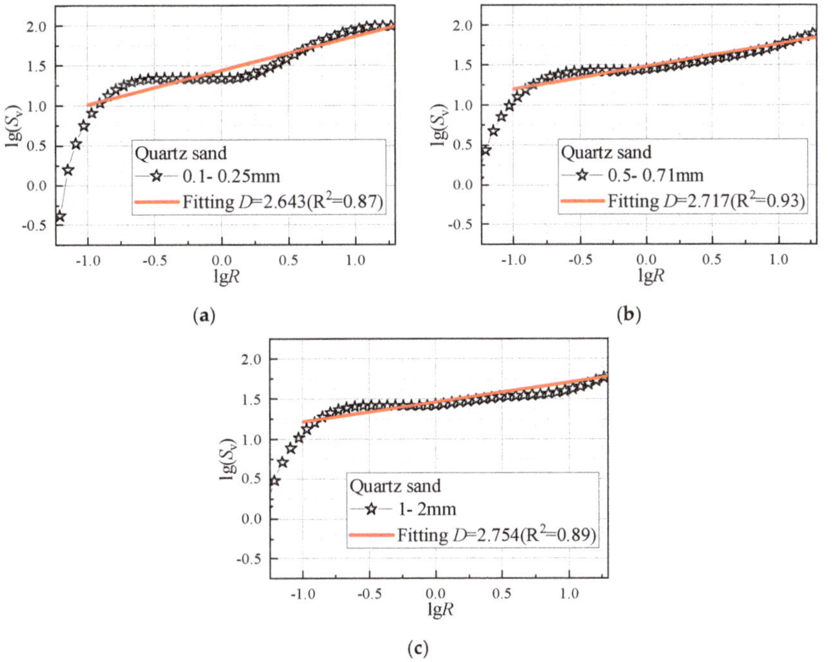

Figure 9. Determination of fractal dimension of quartz sand with different grain sizes: (**a**) 0.1–0.25 mm; (**b**) 0.5–0.71 mm; (**c**) 1–2 mm.

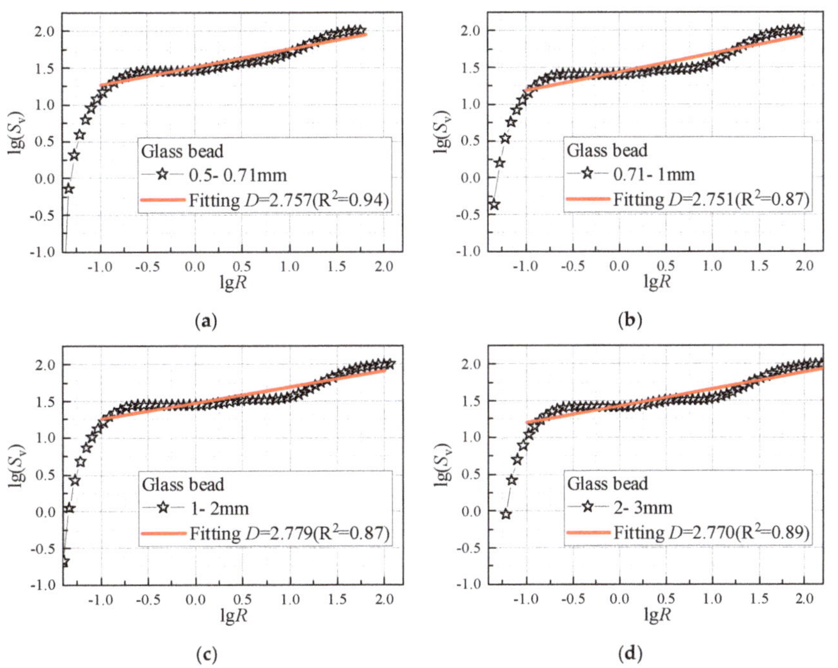

Figure 10. Determination of fractal dimension of glass bead with different grain sizes: (**a**) 0.5–0.71 mm; (**b**) 0.71–1 mm; (**c**) 1–2 mm; (**d**) 2–3 mm.

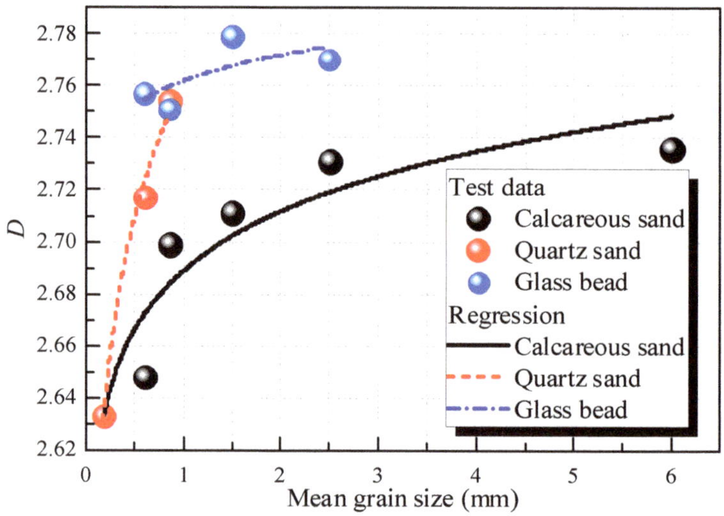

Figure 11. Fractal dimensions of the tested materials with different grain sizes.

Table 2. Summaries of model parameters for carbonate-based sand and silicate-based sand.

Parameter	Calcareous Sand	Quartz Sand	Glass Bead
α	2.689	2.761	2.762
β	0.0122	0.0276	0.00524

4. Conclusions

Carbonate-based sand and silicate-based sand are widely used in the construction of geotechnical engineering. To investigate the effect of grain size on the micro-pore structure and fractal characteristics of carbonate-based sand and silicate-based sand, a series of NMR tests were conducted on calcareous sand, quartz sand, and glass bead. The effect of grain size on the pore size distribution and fractal dimension was discussed in detail. The following main conclusions were drawn:

(1) Grain size had a significant effect on the pore size distribution of carbonate-based sand and silicate-based sand. As the grain size increased, the pore size distribution curve of sands gradually evolved from a binary structure with two peaks to a ternary structure with three peaks. The increase in the grain size could cause a remarkable increase in the maximum pore size. Moreover, the increase in grain size increased the proportion of large pores with a size larger than 1000 μm, and correspondingly reduced the proportion of medium pores with a size between 100 μm–1000 μm. However, grain size had a minor effect on small pores with a size less than 100 um.

(2) Carbonate-based sand differed significantly from silicate-based sand with respect to the pore size distribution, owing to the significant divergence in particle shapes. In the pore size range between 100 μm and 1000 μm, carbonate-based sand had a continuous pore size distribution, whereas the pore size distribution of silicate-based sand was discontinuous. This indicated that the more irregular the particle shape, the better the continuity between the large and medium pores.

(3) The pore size distribution of calcareous sand, quartz sand and glass beads all exhibited good fractal characteristics. The fractal dimension (D) of the pore size distribution of the three materials was determined by fractal theory. Compared with the quartz sand and glass beads, the calcareous sand with the most complex particle shape had the most significant fractal characteristics. Moreover, grain size had a considerable effect on the fractal dimension of the carbonate-based sand and the silicate-based sand. The

increase of grain size could lead to a significant increase of D for calcareous sand, quartz sand and glass bead, indicating that the increase of grain size would increase the heterogeneity of pore structure distribution in sands. A refined correlation was proposed to determine D for carbonate-based sand and silicate-based sand with different grain sizes.

Author Contributions: Conceptualization, Z.D.; methodology, S.-H.H. and Z.D.; formal analysis, S.-H.H. and H.-B.H.; investigation, S.-H.H. and M.G.; resources, Z.D.; writing—original draft preparation, S.-H.H. and Z.D.; visualization, H.-B.H. and M.G.; funding acquisition, S.-H.H. and Z.D. All authors have read and agreed to the published version of the manuscript.

Funding: This work was financially supported by the Chinese National Natural Science Foundation (Grant No. 51508506), Joint Fund of Zhejiang Provincial Natural Science Foundation (Grant No. LHZ20E080001), Hangzhou Science Technology Plan Project (Grant No. 20172016A06, 20180533B06, 20180533B12, 20191203B44), Scientific Research Cultivation Fund of Zhejiang University City College (J-202112), and the China Scholarship Council.

Institutional Review Board Statement: Not applicable.

Informed Consent Statement: Not applicable.

Data Availability Statement: The data are available from the corresponding author upon request.

Acknowledgments: The authors are very grateful to Yifei Sun from Ruhr University Bochum for his kind guidance and help in writing this article.

Conflicts of Interest: The authors declare no conflict of interest.

References

1. Ding, Z.; He, S.-H.; Sun, Y.; Xia, T.-D.; Zhang, Q.-F. Comparative study on cyclic behavior of marine calcareous sand and terrigenous siliceous sand for transportation infrastructure applications. *Constr. Build. Mater.* **2021**, *283*, 122740. [CrossRef]
2. He, S.H.; Ding, Z.; Xia, T.D.; Zhou, W.H.; Gan, X.L.; Chen, Y.Z.; Xia, F. Long-term behaviour and degradation of calcareous sand under cyclic loading. *Eng. Geol.* **2020**, *276*, 105756. [CrossRef]
3. Peng, Y.; Ding, X.; Xiao, Y.; Deng, X.; Deng, W. Detailed amount of particle breakage in non-uniformly graded sands under one-dimensional compression. *Can. Geotech. J.* **2019**, *57*, 1239–1246. [CrossRef]
4. Rui, S.; Wang, L.; Guo, Z.; Cheng, X.; Wu, B. Monotonic behavior of interface shear between carbonate sands and steel. *Acta Geotech.* **2021**, *16*, 167–187. [CrossRef]
5. Shahnazari, H.; Rezvani, R. Effective parameters for the particle breakage of calcareous sands: An experimental study. *Eng. Geol.* **2013**, *159*, 98–105. [CrossRef]
6. Sharma, S.S.; Ismail, M.A. Monotonic and Cyclic Behavior of Two Calcareous Soils of Different Origins. *J. Geotech. Geoenviron.* **2006**, *132*, 1581–1591. [CrossRef]
7. Wang, G.; Wang, Z.; Ye, Q.; Wei, X. Particle Breakage and Deformation Behavior of Carbonate Sand under Drained and Undrained Triaxial Compression. *Int. J. Geomech.* **2020**, *20*, 04020012. [CrossRef]
8. Wang, G.; Wang, Z.; Ye, Q.; Zha, J. Particle breakage evolution of coral sand using triaxial compression tests. *J. Rock Mech. Geotech. Eng.* **2021**, *13*, 321–334. [CrossRef]
9. Wang, H.; Sun, P.; Liu, E.; Li, R. Dynamic properties of Tianshui saturated remolded loess: A laboratory study. *Eng. Geol.* **2020**, *272*, 105570. [CrossRef]
10. Wang, J.; Li, P.; Ma, Y.; Vanapalli, S. Evolution of pore-size distribution of intact loess and remolded loess due to consolidation. *J. Soils Sediments* **2019**, *19*, 1226–1238. [CrossRef]
11. Wang, J.-D.; Li, P.; Ma, Y.; Vanapalli, S.; Wang, X. Change in pore-size distribution of collapsible loess due to loading and inundating. *Acta Geotech.* **2020**, *15*, 1081–1094. [CrossRef]
12. Wang, X.Z.; Weng, Y.L.; Wei, H.Z.; Meng, Q.S.; Hu, M.J. Particle obstruction and crushing of dredged calcareous soil in the Nansha Islands, South China Sea. *Eng. Geol.* **2019**, *261*, 105274. [CrossRef]
13. Wu, Y.; Li, N.; Wang, X.Z.; Cui, J.; Chen, Y.L.; Wu, Y.H.; Yamamoto, H. Experimental investigation on mechanical behavior and particle crushing of calcareous sand retrieved from South China Sea. *Eng. Geol.* **2021**, *280*, 105932. [CrossRef]
14. Cao, Z.; Chen, J.; Ye, X.; Gu, C.; Guo, Z.; Cai, Y. Experimental study on particle breakage of carbonate gravels under cyclic loadings through large-scale triaxial tests. *Transp. Geotech.* **2021**, *30*, 100632. [CrossRef]
15. Donohue, S.; O'sullivan, C.; Long, M. Particle breakage during cyclic triaxial loading of a carbonate sand. *Géotechnique* **2009**, *59*, 477–482. [CrossRef]
16. Gao, Q.-F.; Jrad, M.; Hattab, M.; Fleureau, J.-M.; Ighil Ameur, L. Pore Morphology, Porosity, and Pore Size Distribution in Kaolinitic Remolded Clays under Triaxial Loading. *Int. J. Geomech.* **2020**, *20*, 04020057. [CrossRef]

17. Li, Y.; Zhang, H.; Huang, M.; Yin, H.; Jiang, K.; Xiao, K.; Tang, S. Influence of Different Alkali Sulfates on the Shrinkage, Hydration, Pore Structure, Fractal Dimension and Microstructure of Low-Heat Portland Cement, Medium-Heat Portland Cement and Ordinary Portland Cement. *Fractal Fract.* **2021**, *5*, 79. [CrossRef]
18. Fujikura, Y. Estimation of Permeability for Sand and Gravel Based on Pore-Size Distribution Model. *J. Mater. Civ. Eng.* **2019**, *31*, 04019289. [CrossRef]
19. Gao, Q.-F.; Zhao, D.; Zeng, L.; Dong, H. A Pore Size Distribution-based Microscopic Model for Evaluating the Permeability of Clay. *KSCE J. Civ. Eng.* **2019**, *23*, 5002–5011. [CrossRef]
20. Romero, E.; Gens, A.; Lloret, A. Water permeability, water retention and microstructure of unsaturated compacted Boom clay. *Eng. Geol.* **1999**, *54*, 117–127. [CrossRef]
21. Tang, S.; Wang, Y.; Geng, Z.; Xu, X.; Yu, W.; Hubao, A.; Chen, J. Structure, Fractality, Mechanics and Durability of Calcium Silicate Hydrates. *Fractal Fract.* **2021**, *5*, 47. [CrossRef]
22. Ullah, A.S.; D'Addona, D.M.; Seto, Y.; Yonehara, S.; Kubo, A. Utilizing Fractals for Modeling and 3D Printing of Porous Structures. *Fractal Fract.* **2021**, *5*, 40. [CrossRef]
23. Feng, S.; Xu, Z.; Chai, J.; Li, Y. Using pore size distribution and porosity to estimate particle size distribution by nuclear magnetic resonance. *Soils Found.* **2020**, *60*, 1011–1019. [CrossRef]
24. Griffiths, F.; Joshi, R. Discussion: Change in pore size distribution due to consolidation of clays. *Geotechnique* **1990**, *40*, 303–309. [CrossRef]
25. Kong, B.; Xia, F.; Yu, B.; Xia, T.; Ding, Z. Pore Size Changes in Marine Soft Soil under Various Freezing Conditions. *J. Mar. Sci. Eng.* **2020**, *8*, 170. [CrossRef]
26. Li, X.; Zhang, L. Characterization of dual-structure pore-size distribution of soil. *Can. Geotech. J.* **2009**, *46*, 129–141. [CrossRef]
27. Mukunoki, T.; Miyata, Y.; Mikami, K.; Shiota, E. X-ray CT analysis of pore structure in sand. *Solid Earth* **2016**, *7*, 929–942. [CrossRef]
28. Cai, Y.Q.; Hao, B.B.; Gu, C.; Wang, J.; Pan, L.Y. Effect of anisotropic consolidation stress paths on the undrained shear behavior of reconstituted Wenzhou clay. *Eng. Geol.* **2018**, *242*, 23–33. [CrossRef]
29. Tang, l.; Song, Y.; Jiang, Z.; Jiang, S.; Li, Q. Pore Structure and Fractal Characteristics of Distinct Thermally Mature Shales. *Energy Fuels* **2019**, *33*, 5116–5128. [CrossRef]
30. Yao, Y.; Liu, D.; Tang, D.; Tang, S.; Huang, W. Fractal characterization of adsorption-pores of coals from North China: An investigation on CH_4 adsorption capacity of coals. *Int. J. Coal Geol.* **2008**, *73*, 27–42. [CrossRef]
31. Sun, Y.; Gao, Y.; Zhu, Q. Fractional Order Plasticity Modelling of State-Dependent Behaviour of Granular Soils without Using Plastic Potential. *Int. J. Plast.* **2018**, *102*, 53–69. [CrossRef]
32. Sun, Y.; Sumelka, W. Multiaxial stress-fractional plasticity model for anisotropically overconsolidated clay. *Int. J. Mech. Sci.* **2021**, *205*, 106598. [CrossRef]
33. Shao, X.; Pang, X.; Li, H.; Zhang, X. Fractal Analysis of Pore Network in Tight Gas Sandstones Using NMR Method: A Case Study from the Ordos Basin, China. *Energy Fuels* **2017**, *31*, 10358–10368. [CrossRef]
34. Sun, W.; Zuo, Y.; Zhonghu, W.; Liu, H.; Xi, S.; Shui, Y.; Wang, J.; Liu, R.; Lin, J. Fractal analysis of pores and the pore structure of the Lower Cambrian Niutitang shale in northern Guizhou province: Investigations using NMR, SEM and image analyses. *Mar. Pet. Geol.* **2018**, *99*, 416–428. [CrossRef]
35. Li, A.; Ding, W.; Jiu, K.; Wang, Z.; Wang, R.; He, J. Investigation of the pore structures and fractal characteristics of marine shale reservoirs using NMR experiments and image analyses: A case study of the Lower Cambrian Niutitang Formation in northern Guizhou Province, South China. *Mar. Pet. Geol.* **2017**, *89*, 530–540. [CrossRef]
36. Gao, Y.; Zhang, N.; Li, D.; Liu, H.; Cai, Y.; Wu, Y. Effects of Topographic Amplification Induced by a U-Shaped Canyon on Seismic Waves. *Bull. Seismol. Soc. Am.* **2012**, *102*, 1748–1763. [CrossRef]
37. Gao, Y.; Wu, Y.; Li, D.; Liu, H.; Zhang, N. An improved approximation for the spectral representation method in the simulation of spatially varying ground motions. *Probabilistic Eng. Mech.* **2012**, *29*, 7–15. [CrossRef]
38. Zhang, N.; Gao, Y.; Cai, Y.; Li, D.; Wu, Y. Scattering of SH waves induced by a non-symmetrical V-shaped canyon. *Geophys. J. Int.* **2012**, *191*, 243–256. [CrossRef]
39. Jafarian, Y.; Javdanian, H. Small-strain dynamic properties of siliceous-carbonate sand under stress anisotropy. *Soil Dyn. Earthq. Eng.* **2020**, *131*, 106045. [CrossRef]
40. Tian, H.; Wei, C.; Wei, H.; Yan, R.; Chen, P. An NMR-Based Analysis of Soil–Water Characteristics. *Appl. Magn. Reson.* **2013**, *45*, 49–61. [CrossRef]
41. Zhou, S.; Liu, D.; Cai, Y.; Yao, Y. Fractal characterization of pore–fracture in low-rank coals using a low-field NMR relaxation method. *Fuel* **2016**, *181*, 218–226. [CrossRef]
42. Turcotte, D.L. Fractals and fragmentation. *J. Geophys. Res. Atmos.* **1986**, *91*, 1921–1926. [CrossRef]
43. Tyler, S.; Wheatcraft, S. Fractal Scaling of Soil Particle-Size Distributions: Analysis and Limitations. *Soil Sci. Soc. Am. J.* **1992**, *56*, 362–369. [CrossRef]
44. Liang, Z.K.; Li, Z.; Jiang, Z.X.; Gao, F.L.; Zhang, Y.H.; Xiao, L.; Yang, Y.D.; Hou, Y.F.; Wang, L.W. Characteristics of pore structure and fractal dimension in continental shale based on NMR experiments and SEM image analyses—A case study of Shahezi formation shale in Changling fault depression of Songliao Basin, China. *J. Earth Sci. Environ.* **2020**, *42*, 313–328.

45. Huang, Y.; Zhang, P.; Zhang, J.; Tang, X.; Liu, C.; Yang, J. Fractal Characteristics of Pores in the Longtan Shales of Guizhou, Southwest China. *Geofluids* **2020**, *2020*, 8834758. [CrossRef]
46. Dou, W.; Liu, L.; Jia, L.; Xu, Z.; Wang, M.; Du, C. Pore structure, fractal characteristics and permeability prediction of tight sandstones: A case study from Yanchang Formation, Ordos Basin, China. *Mar. Pet. Geol.* **2021**, *123*, 104737. [CrossRef]
47. Kong, L.; Ostadhassan, M.; Liu, B.; Li, C.; Liu, K. Multifractal Characteristics of MIP-Based Pore Size Distribution of 3D-Printed Powder-Based Rocks: A Study of Post-Processing Effect. *Transp. Porous Media* **2019**, *129*, 599–618. [CrossRef]

fractal and fractional

Article

Pore Structure and Fractal Characteristics of Frozen–Thawed Soft Soil

Bowen Kong [1,2,*], Shao-Heng He [3,*], Yanli Tao [1,2] and Jianzhong Xia [1]

1 School of Civil Engineering and Architecture, Zhejiang University of Science & Technology, Hangzhou 310023, China; 115022@zust.edu.cn (Y.T.); 197002@zust.edu.cn (J.X.)
2 School of Physical and Mathematical Sciences, Nanjing Tech University, Nanjing 211816, China
3 Research Center of Coastal and Urban Geotechnical Engineering, Zhejiang University, Hangzhou 310058, China
* Correspondence: kbw@zust.edu.cn (B.K.); heshaoheng@zju.edu.cn (S.-H.H.); Tel.: +86-137-3547-6567 (B.K.)

Abstract: Freezing action always results in great damage to the pore structure and fractal characteristics of freezing–thawing soft soil. Nuclear magnetic resonance (NMR) can help collect data to describe the microstructure of frozen–thawed soft soil. This paper conducted a series of nuclear magnetic resonance (NMR) tests to interpret the effect of freezing duration, freezing–thawing pressure and freezing temperature on soil's micro-pore structure and fractal characteristics. The pore size distributions (PSDs) of the tested materials were obtained from the NMR T_2 spectra, and fractal theory was introduced to describe the fractal properties of PSD. The results demonstrate that the soil assembly with a larger pore structure tends to be a skeleton pore structure with strong fractal characteristics; the shorter the freezing duration is, the less damage caused by the freezing action; a larger pressure during freezing and thawing results in a lower fractal dimension value for thawing soil's structure, while a lower freezing temperature causes a larger fractal dimension of frozen-thawed soil.

Keywords: frozen–thawed soft clay; fractal characteristics; NMR; pore structure; artificial ground freezing

Citation: Kong, B.; He, S.-H.; Tao, Y.; Xia, J. Pore Structure and Fractal Characteristics of Frozen–Thawed Soft Soil. *Fractal Fract.* **2022**, *6*, 183. https://doi.org/10.3390/fractalfract6040183

Academic Editors: Norbert Herencsar and Zine El Abiddine Fellah

Received: 29 January 2022
Accepted: 22 March 2022
Published: 25 March 2022

1. Introduction

In recent years, with the expansion of cities, a large number of subway tunnels pass through soft soil areas. When constructing the connecting passages between two shield tunnels, due to the small working space and the difficulty in accessing the equipment, the freezing ground method is used in construction to enhance the strength of the soil [1–3]. However, the dynamic characteristics of soft soils are significantly weakened after freezing and thawing cycles [4,5]. The uneven settlement of the frozen areas caused by the difference in freezing conditions seriously threatens the safety of subway operations. The microstructure of soil has been extensively acknowledged as being related to mechanical properties [6–12]. The change in the soil microstructure caused by frost heave is the essential reason for the weakening of frozen–thawed soil mechanical performance [13]. Therefore, it is necessary to study the microstructure characteristics of frozen–thawed soft soil under different freezing conditions.

In 1975, fractal geometry theory was established by Mandelbrot to evaluate the irregularity of complex porous media and the self-similarity characteristics of graphs [14,15]. The pore structure and fractal characteristics of soil can directly reflect the looseness of the soil skeleton, which is beneficial for predicting the amount of soil volume change. At present, pore structure and fractal dimension are mainly obtained by scanning electron microscope (SEM) image analysis, mercury intrusion porosimetry (MIP), transmission electron microscopy (TEM), X-ray micro-tomography, nitrogen adsorption and nuclear magnetic resonance (NMR) for soil pore characterization [16–25]. Previous works reveal that fractal theory can be used to study the fractal characteristics of pores in soil [26–30]. It is beneficial to analyze the pore structure of materials by combining more than one method.

Sun et al. [31] investigated the relationships between fractal dimensions of micro-pore structure and mineral compositions by SEM, X-ray diffraction observations and nuclear magnetic resonance (NMR). Yang et al. [32] used SEM, low pressure gas adsorption, helium pycnometry, X-ray diffraction and geochemical analyses to investigate the pore structure and fractal dimension of the pores in shale formation. According to fractal theory, Shao et al. [33] brought to light that the fractal dimension can reflect its physical properties in tight sandstone, which consists of quartz and clay minerals. Through a series of NMR tests, Li et al. [34] found that shale with a larger surface fractal dimension has a higher methane adsorption capacity. Furthermore, Liang et al. [35] reported that the lg*S*v-lg*R* curve of soil included two distinct parts and proposed that the PSD in soil showed multi-fractal characteristics. Technological advances provide new ideas for the study of the microstructure of frozen–thawed soft soils.

Due to the high water content, the freezing action results in irreversible damage to the soil's structure. Water freezing causes expansion in saturated soft clay, which damages the soft clay structure [36]. Frozen soft clay has many surface fissures, and the number of fissures multiplies after thawing [13]. Ding et al. [37] studied the microstructure of frozen–thawed soil by SEM and proposed that after freezing and thawing, the soil's microstructure appears flocculated, and the lower the freezing temperature, the more fragmented the microstructure. However, previous research does not provide a comprehensive guide for describing structural changes of frozen–thawed soil under various freezing conditions, and further research is needed. In actual construction, when moving away from the freezing pipe, the freezing temperature of the soil rises. Furthermore, the confining pressure near the freezing pipe is greater than that near the ground. Therefore, there are various combinations of freezing duration, temperatures and confining pressures in frozen areas, resulting in different soil damage over the range of the freezing ground. In soil samples, the relaxation time obtained by NMR is closely related to the properties of fluid, so accurate data on pore structure and fractal characteristics can be obtained [13].

In this work, a series of nuclear magnetic resonance (NMR) tests were conducted on soil samples frozen under controllable freezing duration, temperature and confining pressure. The characteristic curves of the pore size distribution and its change rule of the frozen-thawed soil at different stages were explored. Then, fractal theory was introduced to describe the fractal properties of frozen–thawed soil. The research results can be used as a reliable basis for predicting the degree of structural damage of frozen–thawed soil under construction conditions. A better interpretation could then be obtained from tests for uneven settlement of connecting passages in subway systems.

2. Materials and Methods

2.1. Soft Clay Samples

The soft clay was taken from 15~20 m below ground in Hangzhou. The soil samples were mainly composed of illite and illite-montmorillonite minerals. The physical and mechanical properties of the soil sample were characterized by low strength, low permeability, high compressibility and high sensitivity. For the saturated soft clay, its natural weight is 17.7 kN/m^3, density ρ = 1.82 g/cm^2, relative density Gs = 2.75, water content ω = 45%, plastic limit ω_P = 33.4%, liquid limit ω_L = 51.6%, plasticity index I_p = 18.2 and reference thermal conductivity 0.8^{-1} $W{\cdot}m^{-1}{\cdot}K^{-1}$. The freezing soil sample is shown in Figure 1.

The remolded soft clay was prepared in a cylinder 38 mm in diameter and 76 mm in height. Then, the cylinder was placed in a vacuum saturation barrel at −100 kPa for 3 h. Finally, the soft clay was soaked in the cylinder at atmospheric pressure for 12 h to obtain saturated soft clay samples, which were full of water. In order to avoid error caused by the poor uniformity of undisturbed soil, the undisturbed soil was remolded by the vacuum preloading method, and remolded soil samples with good uniformity were obtained [23].

Figure 1. Photograph of the freezing specimen.

Figure 2 shows representative SEM images of soil microstructure before and after freezing and thawing at −20 °C. From Figure 2a, it can be observed that the micro-structural characteristics of the remodeled soil before freezing mostly are in the form of edge–edge and edge-surface contact between the structures. The clumps are inlaid with each other in outline. There are obvious fissures between the clumps, and the fissures are filled with agglomerated fragments of sheet-like soil particles. After freezing, it can be clearly observed that the soil structure became looser; a honeycomb flocculation structure and feather-like flake structure appears around the soil particle aggregates, as shown in Figure 3b. It also can be seen that the agglomeration structure between soil particles was destroyed, which increased the complexity level of the pore morphology.

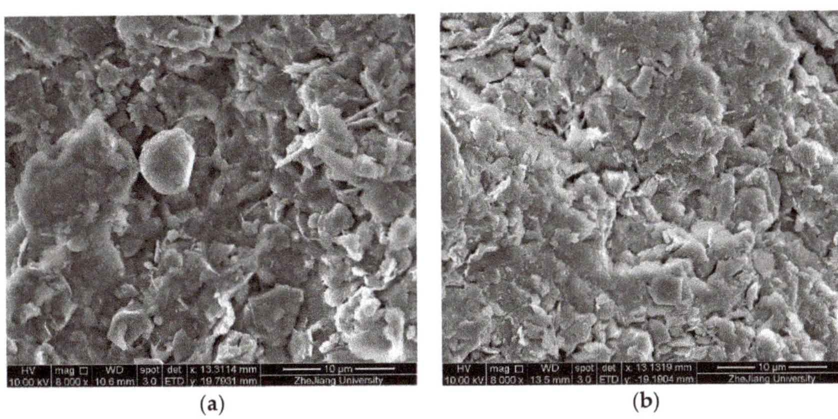

Figure 2. Scanning electron microscope (SEM) images of soft clay: (**a**) unfrozen soil; (**b**) frozen-thawed soil.

2.2. Freezing Plan for Soft Clay

Water freezing causes expansion in saturated soft clay, which damages the soft clay structure. Variable settings were in this paper based on different construction situations, including different freezing durations, freezing pressures and freezing temperatures. Variables were set based on the above conditions. All the prepared samples were frozen under specific temperatures, pressures and durations following the experimental program described in Table 1.

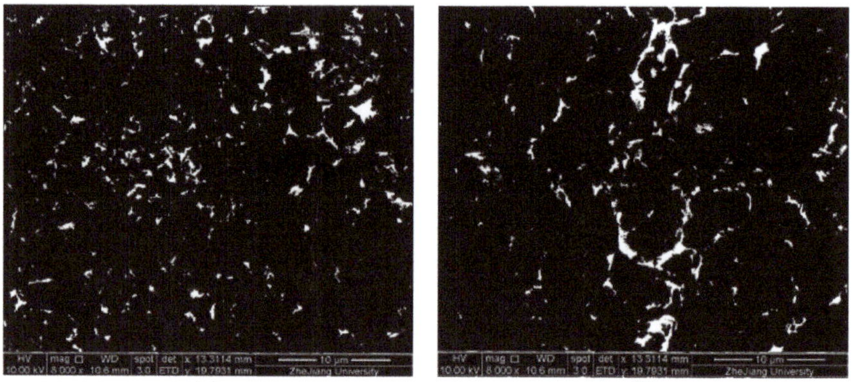

(a) (b)

Figure 3. Pore morphology of soft clay: (**a**) unfrozen soil; (**b**) frozen-thawed soil.

Table 1. Experimental program.

Sample No.	Sample Size (mm)	Freezing Duration (h)	Freezing Pressure (kPa)	Freezing Temperature (°C)
A1	38 × 38	1	-	−20
A2	38 × 38	2	-	−20
A4	38 × 38	4	-	−20
A5	38 × 38	6	-	−20
A6	38 × 38	12	-	−20
A7	38 × 38	18	-	−20
B1	38 × 38	24	50	−20
B2	38 × 38	24	100	−20
B3	38 × 38	24	200	−20
B4	38 × 38	24	400	−20
C1	38 × 38	24	-	−3
C2	38 × 38	24	-	−8
C3	38 × 38	24	-	−20

Three groups of variables were set to analyze pore structure and fractal characteristic change under different freezing conditions. The A1–A7 samples were frozen under freezing durations from 1 to 24 h at −20 °C. The second set of frozen soft clay samples, with freezing–thawing pressure from 50 to 400 kPa, were labelled as B1–B4. Three different thawing temperatures, from −3 °C to −20 °C, were used frozen soft clay samples, set as C1–C3. The testers tried to avoid any possible accidents to obtain highly reproducible and accurate results. For every serial number in Table 1, there were two samples for parallel trials; the average of the two experiments were taken into account if the trends were consistent. The test was repeated if the parallel trials differed widely, to make sure the test results were reliable.

2.3. Calculation Method

2.3.1. Principle of NMR Spectrometry

The MesoMR23-060H-I low-field NMR (Suzhou Niumag Analytical Instrument Corporation, Suzhou, China) equipment used is depicted in Figure 4. With the help of this equipment, the pore structure and fractal characteristics could be studied. The central principle of NMR is that a nucleus always spins around its own axis at a constant frequency, generating a magnetic field when the number of protons is odd and the number of neutrons is even. Because the H element in water follows this law, it can be used to calculate water content. Furthermore, the bound water and flow water can be distinguished by NMR spectra [38].

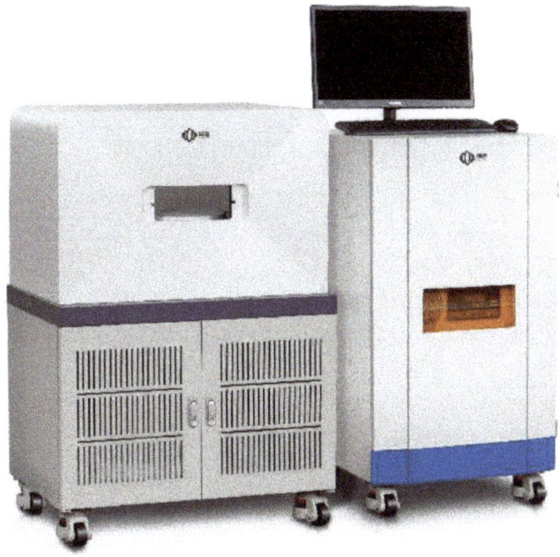

Figure 4. Nuclear magnetic resonance (NMR) equipment.

The parameter T_2, which is called transverse relaxation time, obtained from the NMR signal, was introduced to represent the time required for the transverse magnetization vector decaying to 37% of its maximum value. The T_2 value conforms to the following formula [22]:

$$M_{xy} = M_0 \exp\left(-\frac{t}{T_2}\right) \tag{1}$$

NMR operates on the principle of nuclear spin in a magnetic field. Hydrogen, which has an odd number of protons and an even number of neutrons, can be used to calculate pore water distribution. The relationship of pore radius R and T_2 is given by Equation (2), in which ρ_2 is the surface relativity, S represents the surface area of the pore, and V represents the volume of the water. Then, the size of the pores in the soil can be obtained by the T_2 spectra [23]:

$$\frac{1}{T_2} \approx \rho_2\left(\frac{S}{V}\right)_{pore} = \rho_2\frac{2}{R} \tag{2}$$

2.3.2. Principle of Fractal Dimension

Fractal dimension reflects the validity of space occupied by complex shapes, and it is a measure of the irregularity of complex shapes. Current studies show that the more complex the pore structure is, the stronger the heterogeneity and the higher the fractal dimension [36–38]. NMR tests can measure pore size distribution characteristics and characterize the heterogeneity of soil samples.

According to previous works, the pore structures of porous media show fractal characteristics and can be studied by fractal theory [39]. At present, fractal theory has been widely used in quantitatively describing and studying the geometric characteristics of the PSD [40]. The pore size distribution in a finer percentage by weight is as follows:

$$M(R > R_a)/\rho_P C_m = (R/R_{max})^{3-D} \tag{3}$$

where R is the pore size and R_{max} is the pore size accounting for the maximum percentage; M is the mass of pores whose size R is greater than the specific measurement scale R_a; C_m are constants related to pore shape and size; and ρ_P is the pore density under the assumed conditions. For the fractal nature of soil pore size distribution, the fractal dimension D is strictly limited to the range of $0 < D < 3$ [35].

The fractal dimension D can be obtained when the percentage of cumulative pore volume less than a certain value is known. Because the T_2 value has a linear relationship with the pore size R, based on the equation to calculate moveable fluid space fractal dimension established by Zhou et al. [39], the fractal geometric approximation equation corresponding to pore size R can be expressed as

$$S_V = (R_{max} \backslash R)^{D-3} \quad (4)$$

where S_V (%) is the percentage of the pore volume, which is calculated by the pore size smaller than the specific pore diameter R to the total pore volume; D is the fractal dimension; and R_{max} is the maximum pore size. After taking the logarithms, the model of fractal dimension D in PSD corresponding to the pore size R obtained by NMR is as follows:

$$lg\, S_v = (3 - D)lgR + (D - 3)lgR_{max} \quad (5)$$

3. Results and Discussion

3.1. Effect of Freezing Duration on the Pore Size Distribution

Figure 5a shows the pore distribution of frozen soil after thawing for 24 h at different freezing times. At this time, all the ice in the soil melts into water, and all signals can be detected. Between 1 h of freezing and 2 h of freezing, the pore distribution in the soil changed greatly. After 2 h of freezing, the peak proportion of small pores decreased by nearly 0.5%, and the peak proportion of mesopores increased by nearly 0.5%. The changes in soil pores within 2–18 h of freezing are the same as those in the two hours before freezing. This indicates that during the whole freezing process, the freezing speed of water decreased sharply with a time increase.

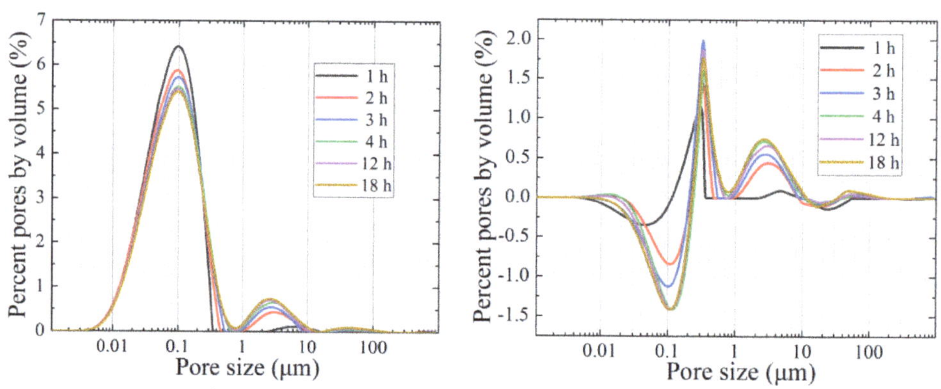

Figure 5. Effect of freezing duration: (**a**) pore size distribution; (**b**) differences in the pore size distribution.

It is known that unfrozen water is a key part in determining the physical properties of frozen soils [41]; however, for frozen–thawed soil, the water in the frozen part is the key cause of soil microstructure damage. In order to better explain the principle of freezing failure, Figure 5b plots the change in porosity proportion under different freezing durations. There is little difference in pore changes between freezing for 6 h and freezing for 18 h, indicating that for the sample size selected in this paper at −20 °C, the freezing process is mostly completed after freezing for 6 h, and the freezing process is basically completed after freezing for 18 h. After six hours of freezing, the freezing time was no longer the main cause of soil microstructure damage [13].

3.2. Effect of Freezing Pressure on the Pore Size Distribution

From the previous section, the proportion of smaller sized pores decreases while the proportion of pores of a larger size increases after the freeze–thaw cycle. Overburden earth pressure during the actual construction can influence the degree of pore size change. As pressure is applied to the soil samples during freezing and thawing, the effect of pressure on the proportion of pore sizes can be obtained by comparing the changes of soil pore distribution under different pressures. Figure 6a,b shows that the proportion of pore sizes of 0.01–0.2 μm decreases, and the pore size of 0.2–0.4 μm increases significantly. The proportion of 1–10 μm pore size increases, and the 10–100 μm pore size decreases to 0. In the selected pressure range, after the freezing confining pressure exceeds 100 kPa, the effect of freezing on the change of soil pores of 0.2–1 μm becomes small. It can be seen from the change of pore size distribution that, when freezing under a certain confining pressure, the water in small pores is difficult to freeze, and the freezing ratio of water in the mesopores decreases as the pressure increases. During the melting process, the pore volume of the mesopores is greatly reduced under the dual action of confining pressure and self-weight. In terms of pore size, the greater the pressure applied to the soil sample during the freezing and thawing process, the lower the percentage the middle pores occupied, and the larger the percentage the tiny pores occupied.

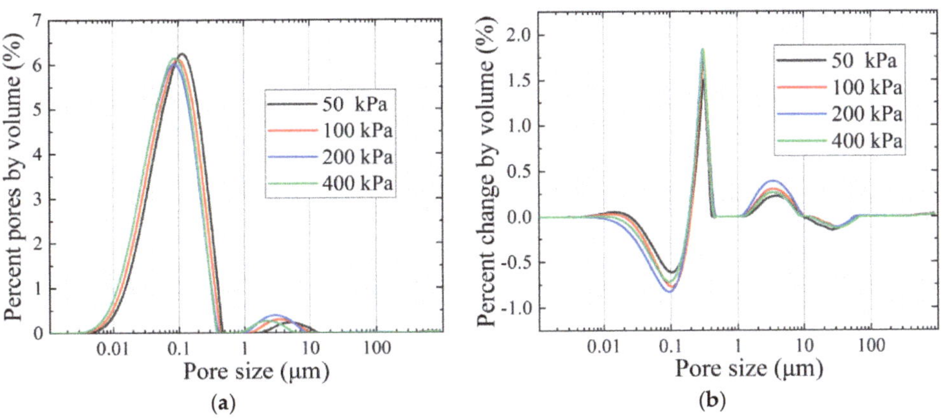

Figure 6. Effect of freezing pressure: (**a**) pore size distribution; (**b**) differences in the pore size distribution.

3.3. Effect of Freezing Temperature on the Pore Size Distribution

Figure 7a shows the pore distribution of frozen–thawed soil at different freezing temperatures. With the decrease of the freezing temperature, the peak value of the proportion of small pores decreases, while the proportion of medium pores increases. In Figure 7b, as the temperature dropped, the percentage change of pore size from 0.01 μm to 0.2 μm decreased and formed a main trough. At the same time, the percentage change of pore size between 0.2 μm and 1 μm increased to form a main peak. It is quite clear that the increase of the main peak was a result of the reduction of pores from 0.01 μm to 0.2 μm in diameter.

From the percentage change in different freezing temperatures, the lower the temperature applied to the soil sample during the freezing process, the greater the peak value of the sub-peak. This is because the proportion of frozen water increases with the temperature decrease. The bound fluid in soil cannot be frozen easily under the action of hydrogen bonds, while the movable fluid can be frozen totally in a short time. Thus, the freezing effect is more obvious in larger pores, which occupy the vast majority of movable fluid.

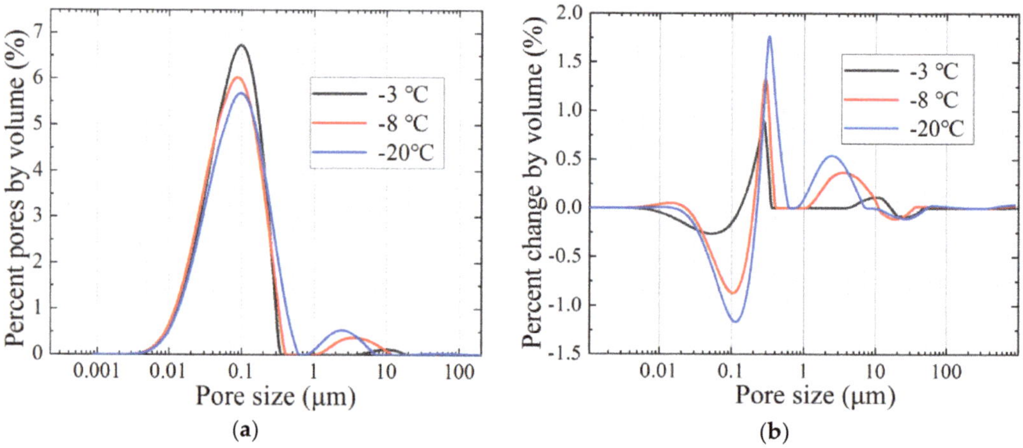

Figure 7. Effect of freezing temperature: (**a**) pore size distribution; (**b**) differences in the pore size distribution.

3.4. Fractal Dimension of Frozen–Thawed Soil

From Section 3.3, it can be seen that the freezing action affects the distribution of pores in soil. The variation of pore distribution in soil results in the change of fractal dimensions of the soil pores. The pore distribution has self-similarity, so it can be characterized by fractal theory. Fractal dimension results from the NMR test can be calculated by linear fitting of two fluid types, which can be distinguished by 2.8 ms for T_2 value [13]. In this paper, D_b is the fractal dimension of the bound fluid pore, while D_m is the fractal dimension of the pore of movable fluid. Based on Formula (5), it can be concluded that

$$lg\ S_v = (3 - D_b)lgR + (D_b - 3)lgR_{max} \text{ (for bound water)} \quad (6)$$

$$lg\ S_v = (3 - D_m)lgR + (D_m - 3)lgR_{max} \text{ (for movable water)} \quad (7)$$

If the pores have a self-similar pore structure and fractal characteristics, then Equations (6) and (7) present linear relationships between lgS_v and lgR. The calculated fractal dimension values for the NMR test are listed in Table 2.

Table 2. Statistical results of nuclear magnetic resonance (NMR) fractal dimensions.

Sample No.	Variable	D_b	R^2	D_m	R^2
A1	1 h FD	0.245	0.51	2.647	0.92
A2	2 h FD	0.281	0.55	2.681	0.93
A4	4 h FD	0.342	0.61	2.685	0.87
A5	6 h FD	0.326	0.59	2.679	0.94
A6	12 h FD	0.344	0.66	2.757	0.92
A7	18 h FD	0.345	0.63	2.761	0.89
B1	50 kPa FP	0.331	0.78	2.704	0.92
B2	100 kPa FP	0.337	0.62	2.691	0.88
B3	200 kPa FP	0.335	0.64	2.678	0.87
B4	400 kPa FP	0.336	0.77	2.671	0.92
C1	−3 °C FT	0.233	0.56	2.673	0.90
C2	−8 °C FT	0.315	0.54	2.709	0.86
C3	−20 °C FT	0.347	0.49	2.776	0.92

1. FD represents freezing duration; 2. FP represents freezing pressure; 3. FT represents freezing temperature.

The lgSv-lg$R$ curves of soil samples under various freezing durations, freezing pressures, and freezing temperatures are shown in Figure 8a–f, Figures 9a–d and 10a–c, respectively. Based on Equation (8), the fractal dimension of the bound fluid and movable fluid is obtained through the regression analysis. The red lines in the figures represent the fitting line of the bound fluid, while the green lines represent the fitting line of movable fluid. The fitting results of fractal dimensions of freezing duration, freezing pressure, and freezing temperature all comply with the fractal law. The correlation coefficient (R^2) of all specimens is 0.49–0.78 for D_b and 0.86–0.94 for D_m. In these figures, the fractal characteristics of movable fluid are more obvious than those of bound fluid, indicating that frost heave and thawing settlement mainly affect macropores in the specimens. Further explanation is given in the next section.

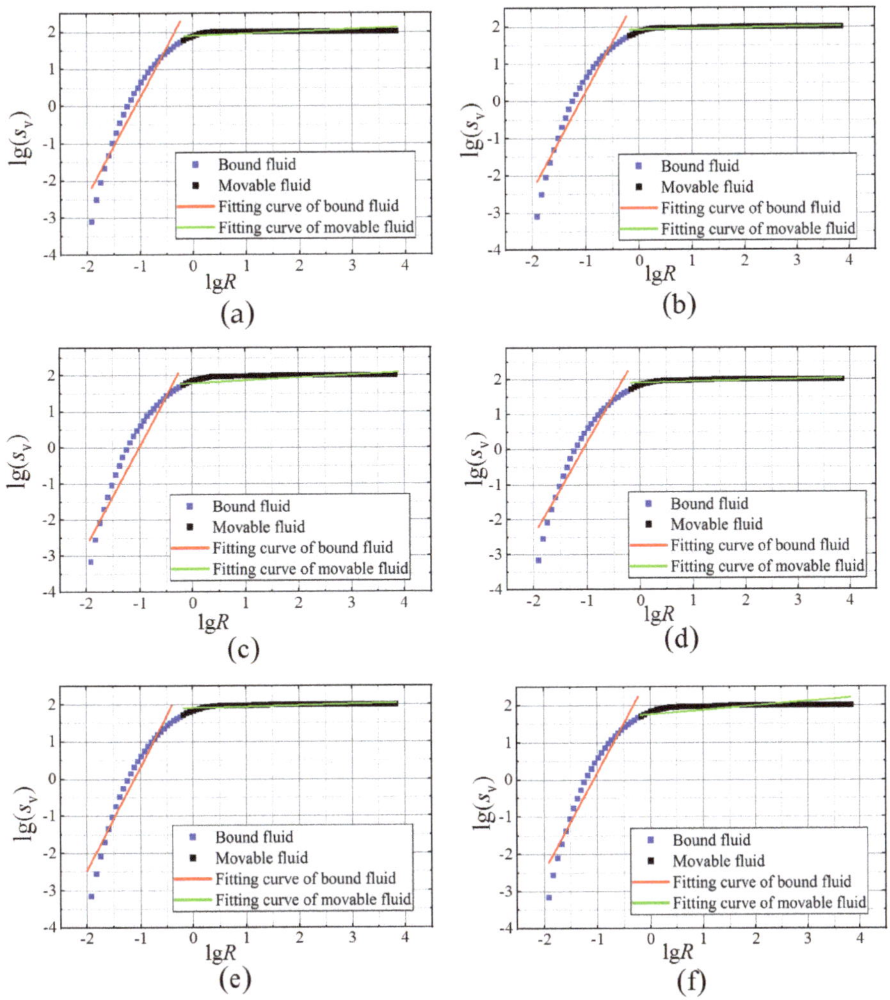

Figure 8. The lgSv-lg$R$ curves under various freezing durations: **(a–f)** represent 1 h, 2 h, 4 h, 6 h, 12 h and 18 h, respectively.

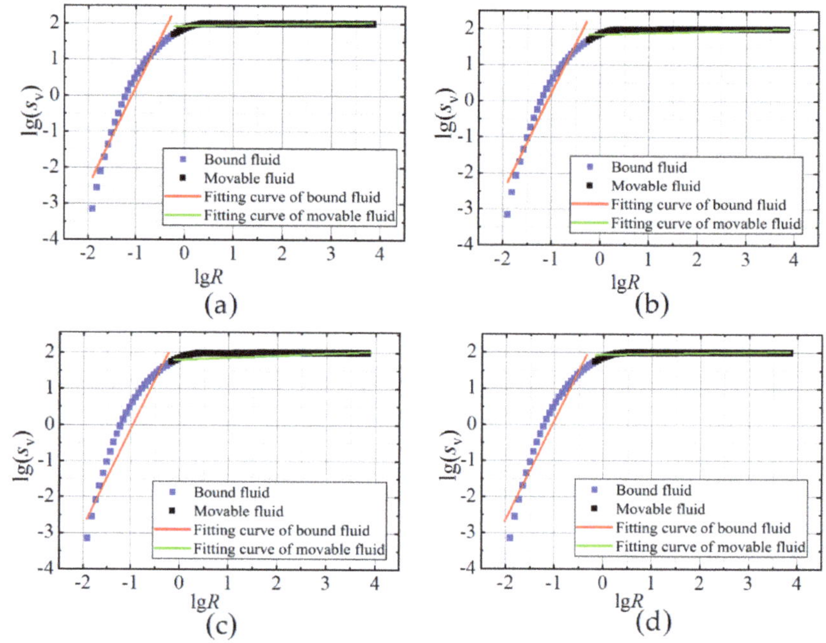

Figure 9. The lg$S$v-lg$R$ curves under various freezing pressures: (**a–d**) represent 50 kPa, 100 kPa, 200 kPa and 400 kPa, respectively.

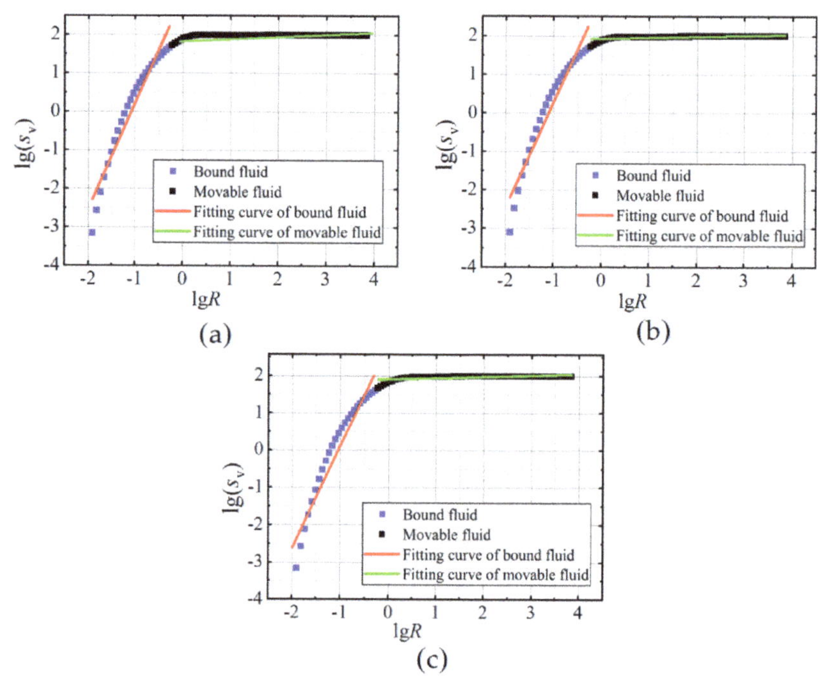

Figure 10. The lg$S$v-lg$R$ curves under various freezing temperatures: (**a–c**) represent −3 °C, −8 °C and −20 °C, respectively.

4. Discussion

The freezing process of water in soil is fast at first and then slow, as specified in reference [13]. It can be seen from Figure 11 that the fractal dimension of pores in soil can be reflected by associating the D_b and D_m values measured by NMR with freezing duration, as follows:

$$D = at^b \tag{8}$$

where D is the fractal dimension values, t is the freezing duration, a is the parameter related to soil structure properties, and b is related to the degree of freezing failure. The value of a and b is shown in Table 3. The fractal dimension increases with the freezing duration and tends to a fixed value a. A large proportion of the freezing process was concentrated in the first 6 h. Considering the energy required to freeze water, the movable fluid will be frozen quickly, and the bound fluid could be partly frozen, with the freezing time growing during the freezing process. Water freezing can break the structure of the soil's skeleton, so the fractal dimension grows as the freezing progresses.

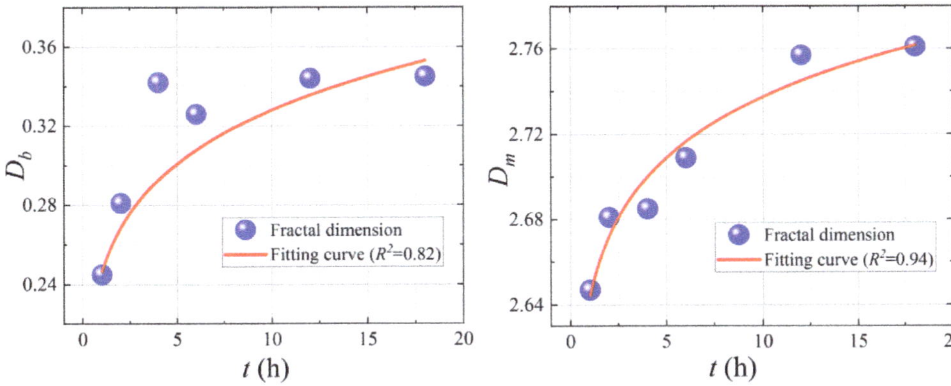

Figure 11. Fractal dimension under various freezing durations: (**a**) fractal dimension for bound fluid; (**b**) fractal dimension for movable fluid.

Table 3. Summaries of model parameters for fractal dimension under various freezing durations.

	a	b	R^2
D_b	2.644	0.014	0.82
D_m	0.247	0.124	0.94

Figure 12 shows that the value of D_b has no apparent linear relationship with the freezing pressure, while D_m is negatively correlated with the value of freezing pressure. The reason is that, for frozen-thawed soil, fractal dimension change is the result of frost heave action during freezing progress. When water freezes under a certain confining pressure, the freezing proportion of the water in the mesopores decreases. When the ice melts under a certain confining pressure, the pores' structure is compressed greatly under the dual action of confining pressure and dead weight. In conclusion, the greater the pressure applied to the soil samples during freezing and thawing, the less the effect from frost heaving, and the lower the fractal dimension value of the thawing soil's structure.

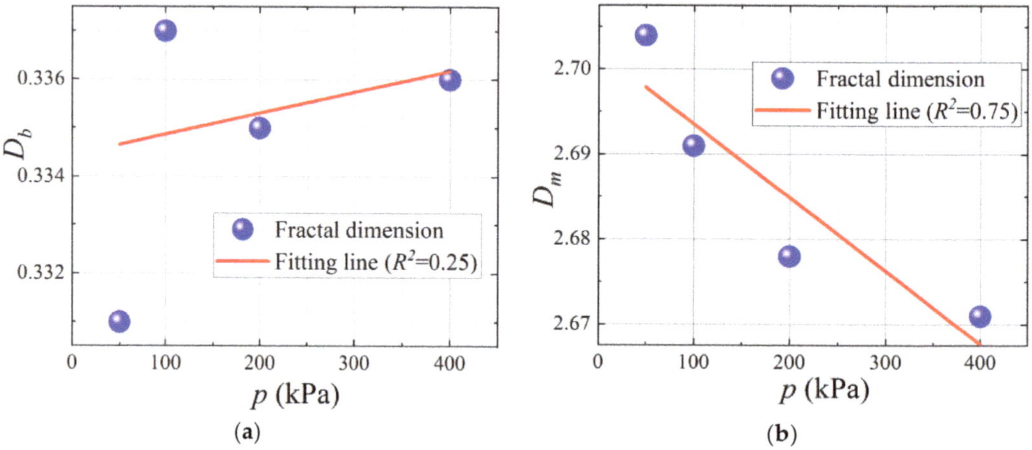

Figure 12. Fractal dimension under various freezing pressures: (**a**) fractal dimension for bound fluid; (**b**) fractal dimension for movable fluid.

Figure 13a,b shows the relationship between fractal dimension and freezing temperature, and the degree of damage to the pore structure caused by freezing temperature can be clearly obtained. There is an obvious correlation between the fractal dimensions of the movable fluid (Figure 13b). Freezing progress is an energy exchange process; a lower freezing temperature has greater energy to break the chemical bonds and physical bonds between the pore water hydron. It can be seen that the lower the freezing temperature, the more loose the skeleton structure, and the larger the fractal dimension. The water in the macropores is more easily frozen than the water in the micropores; therefore, the relativity between the fractal dimension of movable fluid to the freezing temperature is larger than that of the fractal dimension of the bound fluid.

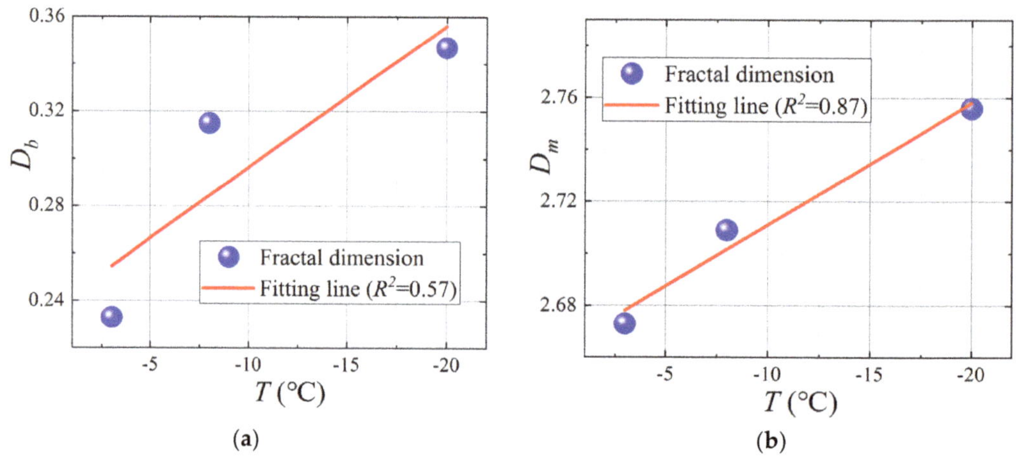

Figure 13. Fractal dimension under various freezing temperatures: (**a**) fractal dimension for bound fluid; (**b**) fractal dimension for movable fluid.

Overall, NMR technology provides an accurate and reliable method to obtain fractal dimension information from frozen–thawed soil, which is helpful to predict the physical and mechanical properties [42]. The test result in this paper provides further research on the frozen–thawed soil's micro-structure to that in reference [37]. The impact of various freezing conditions, including freezing duration, temperature and confining pressure, was obtained from the analysis of the test results, which describe the pore structure and fractal characteristics of frozen–thawed soft soil. The mathematical relationship between the fractal characteristics and thawing settlement characteristics deserves further investigation. In addition, some scientists found a relationship between hydraulic properties, grain size and fractal dimension [40,43–45], which can be applied in the field of frozen–thawed soil in future research.

5. Conclusions

The artificial ground freezing method is widely used in the construction of subway sections connecting passages in soft soil areas. The performance of frozen–thawed soil is strongly associated with the freezing conditions. According to the mechanism of water freezing, crystallization is the main cause of fractal dimension change for frozen–thawed soil. Various freezing conditions bring different types of damage to the soil structure. Accounting for the pore size distribution curves, the water in larger pores contains more movable fluid than that of smaller pores. Using NMR technology, the micro-structural change in soil could be observed, and the fractal dimension of pores can be calculated using pore size distribution curves. The following main conclusions were drawn:

(1) The shorter the freezing duration, the less damage was caused by freezing action. Movable fluid was more easily frozen than bound fluid, which caused the freezing rate to be rather high at the initial stage and then to slow down.

(2) The larger the confining pressure during freezing–thawing, the smaller the proportion of pores in the thawed soil, and the tighter the skeleton structure. The freezing confining pressure had little effect on the bound fluid freezing.

(3) The degree of damage to the soil microstructure varied with the freezing temperature, and the effect of freezing on water frost heaving was more obvious with the decrease of the freezing temperature.

(4) The larger pore structure of the soil assembly tended to be a skeleton pore structure with strong fractal fract. Therefore, the fractal dimension D_m for movable fluid was the dominant parameter reflecting the principal characteristics of the pore. This is because the frost heave rate of bound fluid is much higher than that of movable fluid.

(5) Water freezing could break the structure of the soil's skeleton, so the fractal dimension grew as the freezing process progressed. The larger pressure applied to soil samples during freezing and thawing, the lower the fractal dimension value for the thawing soil's structure, and the lower the freezing temperature, the larger the fractal dimension of the frozen–thawed soil.

Author Contributions: Conceptualization, B.K. and S.-H.H.; methodology, B.K.; validation, Y.T.; formal analysis, B.K.; investigation, Y.T.; resources, S.-H.H.; data curation, S.-H.H.; writing—original draft preparation, B.K.; writing—review and editing, B.K.; visualization, S.-H.H.; supervision, J.X.; project administration, J.X.; funding acquisition B.K. and J.X. All authors have read and agreed to the published version of the manuscript.

Funding: This research was funded by the Commonweal Project of Zhejiang Province, grant number LGG22E090002.

Institutional Review Board Statement: Not applicable.

Informed Consent Statement: Not applicable.

Data Availability Statement: The data are available from the corresponding author upon request.

Acknowledgments: The authors are very grateful to Yifei Sun from Ruhr University Bchum for his kind guidance in writing this article.

Conflicts of Interest: The authors declare no conflict of interest.

References

1. Liu, J.; Ma, B.; Cheng, Y. Design of the gongbei tunnel using a very large cross-section pipe-roof and soil freezing method. *Tunn. Undergr. Sp. Technol.* **2018**, *72*, 28–40.
2. Leung, C.K.W.; Leung, R.K.Y.; Cheung, A.K.K.; Chan, W.L. Application of artificial ground freezing method for tunnel. In Proceedings of the HKIE Civil Division International Conference, Hong Kong, China, 26–28 April 2012.
3. Chang, D.K.; Lacy, H.S. Artificial ground freezing in geotechnical engineering. In Proceedings of the International Conference on Case Histories in Geotechnical Engineering, Arlington, VA, USA, 14 August 2008; pp. 1–11.
4. Cui, Z.D.; Tang, Y.Q.; Yan, X.X. Centrifuge modeling of land subsidence caused by the high-rise building group in the soft soil area. *Environ. Earth Sci.* **2009**, *59*, 1819–1826.
5. Casini, F.; Olivella, S.; Gens, A.; Viggiani, G.M.B. Artificial ground freezing of a volcanic ash: Laboratory tests and modelling. *Environ. Geotech.* **2016**, *3*, 141–154.
6. Gao, Q.-F.; Jrad, M.; Hattab, M.; Fleureau, J.-M.; Ameur, L.I. Pore Morphology, Porosity, and Pore Size Distribution in Kaolinitic Remolded Clays under Triaxial Loading. *Int. J. Geomech.* **2020**, *20*, 04020057. [CrossRef]
7. Li, Y.; Zhang, H.; Huang, M.; Yin, H.; Jiang, K.; Xiao, K.; Tang, S. Influence of Different Alkali Sulfates on the Shrinkage, Hydration, Pore Structure, Fractal Dimension and Microstructure of Low-Heat Portland Cement, Medium-Heat Portland Cement and Ordinary Portland Cement. *Fractal Fract.* **2021**, *5*, 79. [CrossRef]
8. Fujikura, Y. Estimation of Permeability for Sand and Gravel Based on Pore-Size Distribution Model. *J. Mater. Civ. Eng.* **2019**, *31*, 04019289. [CrossRef]
9. Gao, Q.-F.; Zhao, D.; Zeng, L.; Dong, H. A Pore Size Distribution-based Microscopic Model for Evaluating the Permeability of Clay. *KSCE J. Civ. Eng.* **2019**, *23*, 5002–5011. [CrossRef]
10. Romero, E.; Gens, A.; Lloret, A. Water permeability, water retention and microstructure of unsaturated compacted Boom clay. *Eng. Geol.* **1999**, *54*, 117–127. [CrossRef]
11. Tang, S.; Wang, Y.; Geng, Z.; Xu, X.; Yu, W.; Hubao, A.; Chen, J. Structure, Fractality, Mechanics and Durability of Calcium Silicate Hydrates. *Fractal Fract.* **2021**, *5*, 47.
12. Ullah, A.S.; D'Addona, D.M.; Seto, Y.; Yonehara, S.; Kubo, A. Utilizing Fractals for Modeling and 3D Printing of Porous Structures. *Fractal Fract.* **2021**, *5*, 40.
13. Kong, B.; He, S.; Xia, T.; Ding, Z. Research on Microstructure of Soft Clay under Various Artificial Ground Freezing Conditions Based on NMR. *Appl. Sci.* **2021**, *11*, 1810. [CrossRef]
14. Pfeifer, P.; Avnir, D. Chemistry in noninteger dimensions between two and three. I. Fractal theory of heterogeneous surfaces. *J. Chem. Phys.* **1983**, *79*, 7.
15. Avnir, D.; Jaroniec, M. An isotherm equation for adsorption on fractal surfaces of heterogeneous porous materials. *Langmuir* **1989**, *5*, 1431–1433.
16. Dathe, A.; Eins, S.; Niemeyer, J.; Gerold, G. The surface fractal dimension of the soil-pore interface as measured by image analysis. *Geoderma* **2001**, *103*, 203–229.
17. Wang, P.; Jiang, Z.; Ji, W.; Zhang, C.; Yuan, Y.; Chen, L. Heterogeneity of intergranular, intraparticle and organic pores in longmaxi shale in Sichuan basin, south China: Evidence from SEM digital images and fractal and multifractal geometries. *Mar. Pet. Geol.* **2016**, *72*, 122–138.
18. Song, Z.; Liu, G.; Yang, W.; Zou, H.; Sun, M.; Wang, X. Multi-fractal distribution analysis for pore structure characterization of tight sandstone: A case study of the upper paleozoic tight formations in the longdong district. *Ordos Basin Mar. Pet. Geol.* **2018**, *92*, 842–854.
19. Wang, J.; Li, P.; Ma, Y.; Vanapalli, S. Evolution of pore-size distribution of intact loess and remolded loess due to consolidation. *J. Soils Sediments* **2019**, *19*, 1226–1238. [CrossRef]
20. Wang, J.-D.; Li, P.; Ma, Y.; Vanapalli, S.; Wang, X. Change in pore-size distribution of collapsible loess due to loading and inundating. *Acta Geotech.* **2020**, *15*, 1081–1094. [CrossRef]
21. Feng, S.; Xu, Z.; Chai, J.; Li, Y. Using pore size distribution and porosity to estimate particle size distribution by nuclear magnetic resonance. *Soils Found.* **2020**, *60*, 1011–1019. [CrossRef]
22. Griffiths, F.; Joshi, R. Discussion: Change in pore size distribution due to consolidation of clays. *Geotechnique* **1990**, *40*, 303–309. [CrossRef]
23. Kong, B.; Xia, F.; Yu, B.; Xia, T.; Ding, Z. Pore Size Changes in Marine Soft Soil under Various Freezing Conditions. *J. Mar. Sci. Eng.* **2020**, *8*, 170. [CrossRef]
24. Li, X.; Zhang, L. Characterization of dual-structure pore-size distribution of soil. *Can. Geotech. J.* **2009**, *46*, 129–141. [CrossRef]
25. Mukunoki, T.; Miyata, Y.; Mikami, K.; Shiota, E. X-ray CT analysis of pore structure in sand. *Solid Earth* **2016**, *7*, 929–942. [CrossRef]
26. Cai, Y.Q.; Hao, B.B.; Gu, C.; Wang, J.; Pan, L.Y. Effect of anisotropic consolidation stress paths on the undrained shear behavior of reconstituted Wenzhou clay. *Eng. Geol.* **2018**, *242*, 23–33. [CrossRef]
27. Tang, l.; Song, Y.; Jiang, Z.; Jiang, S.; Li, Q. Pore Structure and Fractal Characteristics of Distinct Thermally Mature Shales. *Energ. Fuel.* **2019**, *33*, 5116–5128. [CrossRef]
28. Yao, Y.; Liu, D.; Tang, D.; Tang, S.; Huang, W. Fractal characterization of adsorption-pores of coals from North China: An investigation on CH 4 adsorption capacity of coals. *Int. J. Coal. Geol.* **2008**, *73*, 27–42. [CrossRef]

29. Sun, Y.; Gao, Y.; Zhu, Q. Fractional Order Plasticity Modelling of State-Dependent Behaviour of Granular Soils without Using Plastic Potential. *Int. J. Plast.* **2018**, *102*, 53–69. [CrossRef]
30. Sun, Y.; Sumelka, W. Multiaxial stress-fractional plasticity model for anisotropically overconsolidated clay. *Int. J. Mech. Sci.* **2021**, *205*, 106598. [CrossRef]
31. Sun, W.; Zuo, Y.; Zhonghu, W.; Liu, H.; Xi, S.; Shui, Y.; Wang, J.; Liu, R.; Lin, J. Fractal analysis of pores and the pore structure of the Lower Cambrian Niutitang shale in northern Guizhou province: Investigations using NMR, SEM and image analyses. *Mar. Pet. Geol.* **2018**, *99*, 416–428. [CrossRef]
32. Yang, R.; He, S.; Yi, J.; Hu, Q. Nano-scale pore structure and fractal dimension of organic-rich Wufeng-Longmaxi shale from Jiaoshiba area, Sichuan Basin: Investigations using FE-SEM, gas adsorption and helium pycnometry. *Mar. Pet. Geol.* **2015**, *7*, 27–45. [CrossRef]
33. Shao, X.; Pang, X.; Li, H.; Zhang, X. Fractal Analysis of Pore Network in Tight Gas Sandstones Using NMR Method: A Case Study from the Ordos Basin, China. *Energy Fuels* **2017**, *31*, 10358–10368. [CrossRef]
34. Li, A.; Ding, W.; Jiu, K.; Wang, Z.; Wang, R.; He, J. Investigation of the pore structures and fractal characteristics of marine shale reservoirs using NMR experiments and image analyses: A case study of the Lower Cambrian Niutitang Formation in northern Guizhou Province, South China. *Mar. Pet. Geol.* **2017**, *89*, 530–540. [CrossRef]
35. Liang, Z.K.; Li, Z.; Jiang, Z.X.; Gao, F.L.; Zhang, Y.H.; Xiao, L.; Yang, Y.D.; Hou, Y.F.; Wang, L.W. Characteristics of pore structure and fractal dimension in continental shale based on NMR experiments and SEM image analyses:A case study of Shahezi formation shale in Changling fault depression of Songliao Basin, China. *J. Earth Sci. Environ.* **2020**, *42*, 313–328.
36. Cetin, H.; Fener, M.; Söylemez, M.; Günaydin, O. Soil structure changes during compaction of a cohesive soil. *Eng. Geol.* **2007**, *92*, 38–48.
37. Ding, Z.; Kong, B.; Wei, X.; Zhang, M.; Xu, B.; Zhao, F. Laboratory Testing to Research the Micro-Structure and Dynamic Characteristics of Frozen–Thawed Marine Soft Soil. *J. Mar. Sci. Eng.* **2019**, *7*, 85. [CrossRef]
38. Jia, B.; Chen, Z.; Xian, C. Investigations of CO_2 storage capacity and flow behavior in shale formation. *J. Petrol. Sci. Eng.* **2022**, *208*, 109659. [CrossRef]
39. Zhou, S.; Liu, D.; Cai, Y.; Yao, Y. Fractal characterization of pore–fracture in low-rank coals using a low-field NMR relaxation method. *Fuel* **2016**, *181*, 218–226.
40. He, S.-H.; Ding, Z.; Hu, H.-B.; Gao, M. Effect of Grain Size on Microscopic Pore Structure and Fractal Characteristics of Carbonate-Based Sand and Silicate-Based Sand. *Fractal Fract.* **2021**, *5*, 152. [CrossRef]
41. Kruse, A.M.; Darrow, M. Adsorbed cation effects on unfrozen water in fine-grained frozen soil measured using pulsed nuclear magnetic resonance. *Cold Reg. Sci. Technol.* **2017**, *142*, 42–54. [CrossRef]
42. Huang, Y.; Zhang, P.; Zhang, J.; Tang, X.; Liu, C.; Yang, J. Fractal Characteristics of Pores in the Longtan Shales of Guizhou, Southwest China. *Geofluids* **2020**, *2020*, 1–16. [CrossRef]
43. Dou, W.; Liu, L.; Jia, L.; Xu, Z.; Wang, M.; Du, C. Pore structure, fractal characteristics and permeability prediction of tight sandstones: A case study from Yanchang Formation, Ordos Basin, China. *Mar. Pet. Geol.* **2021**, *123*, 104737. [CrossRef]
44. Kong, L.; Ostadhassan, M.; Liu, B.; Li, C.; Liu, K. Multifractal Characteristics of MIP-Based Pore Size Distribution of 3D-Printed Powder-Based Rocks: A Study of Post-Processing Effect. *Transp. Porous Media* **2019**, *129*, 599–618. [CrossRef]
45. Fu, X.; Ding, H.; Sheng, Q.; Zhang, Z.; Yin, D.; Chen, F. Fractal Analysis of Particle Distribution and Scale Effect in a Soil–Rock Mixture. *Fractal Fract.* **2022**, *6*, 120. [CrossRef]

fractal and fractional

MDPI

Article

Discrete Element Modelling of Fractal Behavior of Particle Size Distribution and Breakage of Ballast under Monotonic Loading

Cheng Chen [1], Xin Zhang [1], Yifei Sun [2], Lei Zhang [1], Rui Rui [1] and Zhide Wang [1,*]

[1] School of Civil Engineering and Architecture, Wuhan University of Technology, Wuhan 430070, China; chengchen87@whut.edu.cn (C.C.); xvtm4eotz@163.com (X.Z.); zhangleigeo@whut.edu.cn (L.Z.); r.rui@whut.edu.cn (R.R.)

[2] Faculty of Civil and Environmental Engineering, Ruhr University Bochum, 44801 Bochum, Germany; yifei.sun@ruhr-uni-bochum.de

* Correspondence: wangzhide-wuhan@whut.edu.cn

Abstract: Breakage of particles has a great influence on the particle size distribution (PSD) and the associated mechanical behavior of ballast under train loads. A discrete element method (DEM) simulation of triaxial testing under monotonic loading was carried out using FRM (fragment replacement method) breakable particles as ballast and a flexible shell model as membrane. The coupled model was validated by comparing the load-deformation responses with those measured in previous experiments and was then used to analyze the contact orientations and the distribution of particle breakage from a micromechanical perspective. The simulation results show that higher confining pressure and larger axial strain may increase the grain breakage (B_g) and the fractal dimension (D) of ballast. It was observed that most breakage was first-generation breakage, and that the proportions of the second- to fifth-generation breakage decreased successively. Moreover, as the axial strain or confining pressure increased, the percentage of small particle fragments increased in correspondence with the PSD curves that remained concave upwards, as the fractal dimension D of PSD increased. In addition, the evolution of D exhibited a linear correlation with grain breakage B_g. Contrarily, a quadratic curve relation between D and volumetric strain was exhibited under different axial strain stages. Therefore, D has the potential to be a key indicator to evaluate the degree of ballast crushing and PSD degradation, which may contribute to better decision making concerning track bed maintenance.

Keywords: discrete element method; particle breakage; ballast; fractal dimension; PSD evolution

Citation: Chen, C.; Zhang, X.; Sun, Y.; Zhang, L.; Rui, R.; Wang, Z. Discrete Element Modelling of Fractal Behavior of Particle Size Distribution and Breakage of Ballast under Monotonic Loading. *Fractal Fract.* **2022**, *6*, 382. https://doi.org/10.3390/fractalfract6070382

Academic Editor: Zine El Abiddine Fellah

Received: 25 May 2022
Accepted: 3 July 2022
Published: 6 July 2022

1. Introduction

Ballast is a granular material used for construction of railway beds. It can be characterized as having an extremely irregular particle shape with high porosity and strength. Ballast helps to transmit and distribute axle loads from the sleepers to the sub-ballast layer [1]. The dynamic stress of a passing train degrades and fouls the ballast layer which usually experiences continuous breakage, directly resulting in track settlement, densification, and the need for frequent maintenance. Ballast breakage may lead to a sharp increase in the number of fragments, weakening interlocking and impacting mechanical performances [2]. In practice, ballast needs to be maintained and replaced when its quality deteriorates to a certain extent.

Recently, fractal theory has been widely used in describing and studying the particle size distribution (PSD) of granular material [3]. Compared with traditional methods based on Euclidean geometry, fractal theory describes PSD well because it can characterize the local shape, size, and structure of particles, and any similarities among them [4]. It is beneficial to analyze the pore structure of materials by combining more than one method, considering such factors as the relationship between fractal characteristics, or the physical or chemical attributes of materials [5]. Early studies revealed that fractal theory could be used to study the fractal characteristics of soil [4–6], sand [7], rockfill [8–10] and other material [11,12]. Compared with natural soil or quartz sand of smaller size and wider grading distribution, layers of ballast are usually composed of medium-to-coarse gravel-sized particles (10–60 mm) whose uniformity coefficients vary from 1.5 to 3.0 according to the ballast specifications [13] which can be regarded as uniformly sized granular material. As more and more small fragments were produced by ballast breakage under train loads, the PSD curve of ballast also changed, with a fractal dimension which was directly proportional to the fragment content. Therefore, the fractal theory could be used to study PSD evolution and ballast breakage. However, studies on the effect of the fractal distribution of particle size on the characteristics of ballast are still not enough in themselves.

The discrete element method (DEM) has been extensively used in railway engineering and also pavement engineering [14–16]. In order to analyze the evolution of fractal PSD using the fragment replacement method (FRM), the particle is represented by a group of smaller ones without bonding as implemented in [16–21]. McDowell et al. [16] adopted octahedral shear stress as the fracture criterion and replaced broken spheres with smaller ones with overlap in their successful investigation of the evolution of fractal PSD for silica sand. Ciantia et al. [17] used a multigenerational DEM approach to simulate soil breakage in a one-dimensional compression test, and obtained a limit fractal dimension of 2.47. Zhou et al. [18] found that both particle size and particle contact force were distributed in a fractal manner by one-dimensional compression test simulation, which is consistent with the findings of Vallejo and Lobo-Guerrero [19]. De Bono et al. [20] conducted one-dimensional normal compression test simulation with different fracture criteria, and pointed out that both octahedral shear stress and the largest normal contact stress can lead to the correct normal compression lines and fractal PSDs. Li et al. [21] quantified the evolution of microstructures of carbonate sands under mechanical loading and found that the fractal dimension of crack networks increased with external loading due to crack branching via cleavage.

Although the FRMs used in the above studies can accurately simulate particle breakage based on DEM, most of the parent breakable particles involved were spherical and so could hardly simulate the interlocking between particles [16–21]. However, particle shape plays a key role in the fractal behavior of ballast with irregular shapes [22,23]. Therefore, in order to reveal the fractal characteristics of ballast, a DEM–FDM (finite differential method) coupled simulation of triaxial testing under monotonic loading using an irregular-shaped breakable ballast model is carried out in this study. The effects of axial strain and confining pressure on ballast breakage and the fractal behavior of ballast PSD are investigated. The relationships between fractal dimensions, grain breakage and the stress states of ballast are then established.

2. Ballast Breakage Model Using FRM

Firstly, the DEM study using FRM to simulate particle breakage should include a reasonable fracture criterion and a fragment replacement mode. They should be defined to check (1) whether a particle should break and (2) how any individual particle does break.

It is widely accepted that the failure of spherical particles under compression is in fact a tensile failure. Jaeger [24] suggested that the 'tensile strength' of rock grains can be measured by diametral compression between flat platens. For a grain of diameter d under a diametral force F, a characteristic tensile stress σ induced within it can be defined as

$$\sigma = F/d^2 \tag{1}$$

For railway ballast with typical sizes ranging from 10 to 60 mm, Lim et al. [25] conducted indirect diametral compression and measured the tensile strengths of ballast between flat platens pointing out that Weibull strength σ_0 is a function of the particle size d. This is usually expressed in the form

$$\sigma_0 \propto d^{\,b} \tag{2}$$

where σ_0 is the value of σ such that 37% of the blocks survive the test intact; while b is a constant and a function of the statistical distribution of flaws in the material, describing the size-hardening law. Typical values of b range from -0.69 to -0.40 for different types of ballasts. Following the single-particle crushing test results [25], the characteristic tensile strength σ_0 of ballast size d_0 (19.7 mm) is 14.7 MPa, and the value of b is -0.56, which is used to attribute random strengths to new fragments and other initial particles.

In order to determine the correct behavior of ballast under normal compression, De Bono and McDowell [20] investigated various possibilities for the fracture criterion, including octahedral shear stress, major principal stress within a particle, mean stress, and stress calculated from the maximum contact force on a particle. The criterion based on the maximum contact force (F_{max}) was shown to better match the experiments and simulations, such that

$$\sigma_{Fm} = F_{max}/d^2 \quad (3)$$

According to Raymond et al. [1], the process of ballast degradation due to loading can occur in three ways: by the grinding of small-scale asperities, by breakage of angular projections, and by the splitting of particles into approximately equal parts. In addition, an irregular clump produces more contact compared to a spherical particle, leading to a lower maximum contact force. Hence, these particles will be replaced by new fragments when the F_{max} stress exceeds than half of the characteristic strength.

From the perspective of computational efficiency, in this study, each particle was allowed to split into several smaller particles, maintaining conservation of mass when the value of σ_{Fm} was greater than or equal to its Weibull strength. Previous studies have used spheres or disks to simulate particle breakage within the bonded parent sphere, where new sphere fragments to be contained overlap sufficiently [16,18,20] or tangentially to each other [17,19]. As shown in Figure 1a, new sphere fragments overlap sufficiently to be contained within the bounding parent sphere, with the axes joining their centers aligned along the direction of the minor principal stress. This process produces undesirable local pressure spikes during breakage and continues until internal forces are completely released and the system reaches equilibrium [16].

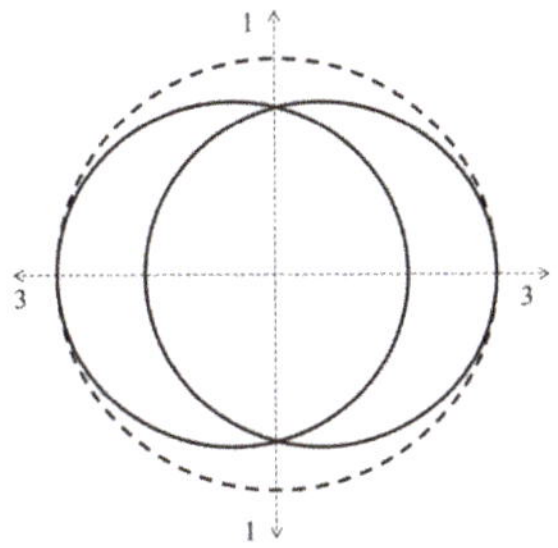

(a)

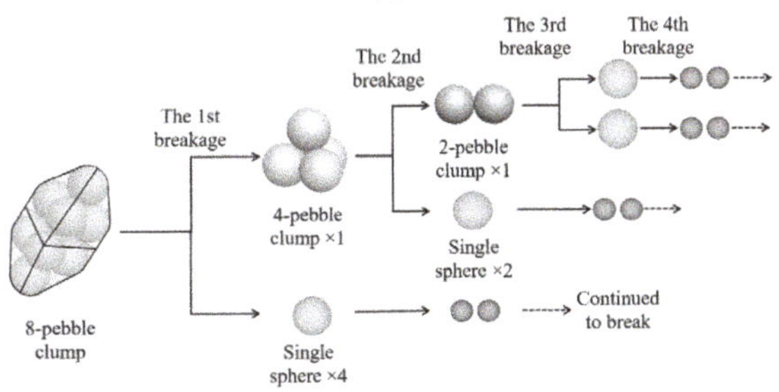

(b)

Figure 1. Fragment replacement model (FRM): (**a**) single spheres [16]; (**b**) irregular-shaped particles.

It is therefore acknowledged that it is impossible to simulate a perfectly realistic ballast fracture mechanism using disks or spheres [16]; therefore, eight-pebble clumps of irregular shape were adopted for this study. The fragment replacement mode is depicted in Figure 1b, which shows that irregular particles were replaced by smaller fragments within the bounding parent particles without overlaps. During the first-generation breakage, the eight-pebble clump might split into a four-pebble clump and four single spheres. All the fragments could continue to break and generate smaller particles. Except for single spheres, most of the particles could break and generate fragments without overlap, which means internal forces could be releases and the system could reach equilibrium immediately. Hence this fragment replacement mode can greatly improve calculation efficiency, and is therefore used in this study.

3. Simulations of Monotonic Triaxial Test

Based on the experimental monotonic triaxial tests conducted by Indraratna et al. [26], DEM–FDM coupled simulations of monotonic triaxial tests were carried out under different confining pressures ranging from 60 kPa to 240 kPa. Figure 2 shows the DEM–FDM model of a triaxial test with dimensions of 300 mm diameter and 600 mm height. It consists of 971 eight-pebble clumps as breakable ballast using FRM in PFC3D (particle flow code in three dimensions) with a flexible boundary as membrane using shell model in FLAC3D (fast Lagrangian analysis of continua in three dimensions). The top and bottom plates using disk walls have the same dimensions as the triaxial test, to maintain a consistent area of contact with the ballast sample. The eight-pebble clump has an irregular shape with an aspect ratio of 1.83, as expressed by the ratio of the particle length to the particle width. In accordance with the PSD of ballast in triaxial tests, the PSD used in these simulations is shown in Table 1.

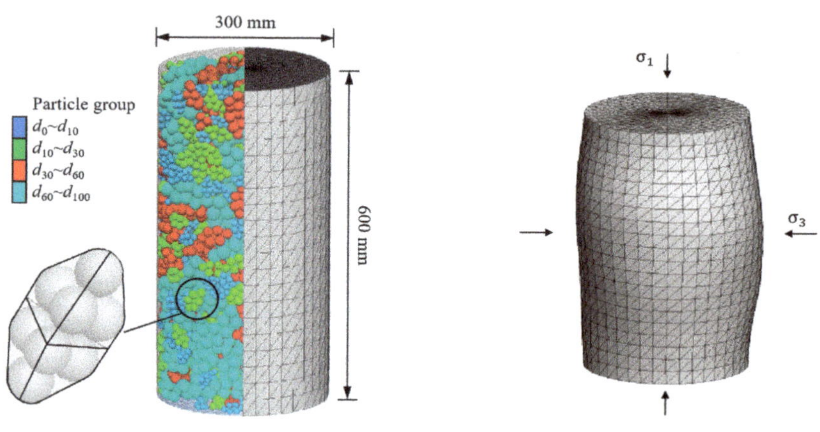

Figure 2. Monotonic triaxial test model: (**a**) test sample; (**b**) FLAC3D model of membrane.

Table 1. Grain size characteristics of ballast.

Parameters	d_{10}/mm	d_{30}/mm	d_{50}/mm	d_{60}/mm	d_{max}/mm	C_u	C_c	Size Ratio
Value	27.1	32.6	38.9	41.3	53	1.5	0.9	5.7

During sample generation, the particles generated at random orientation by the Ball Distribute command. Firstly, the sample was divided into five equal layers, each layer was generated in turn and allowed to fall under gravity, then compacted as performed in the lab. Once the sample attained equilibrium, a flexible membrane model was generated to replace the cylinder wall. The membrane model was linked with the top and bottom walls by each node. The shell model as a flexible membrane consisted of 1440 facet units which were assumed to be triangles of uniform thickness lying between three nodal points. Each shell structural element was defined by its geometric and material properties. The parameters of FRM ballast particles and shell elements are listed in Table 2.

Table 2. Simulation parameters in PFC3D and FLAC3D.

Input Parameters	Value
Normal and shear stiffness of particles: N/m	2×10^6
Friction coefficient of clumps	0.5
Density: kg/m^3	2650
Friction coefficient of walls and membrane	0.1
Normal stiffness of walls: N/m	2×10^7
Shear stiffness of walls: N/m	0
Normal stiffness of membrane: N/m	2×10^5
Shear stiffness of membrane: N/m	0
Young's modulus of membrane: kPa	400
Sample porosity	0.44

In order to reduce the influence of the initial arrangement of particles on the simulation results, the same sample was used for different loading conditions. Once the assembly generation procedure was completed, artificial isotropic compression was applied to the plates and membrane to achieve the required stress state. The assigned uniform pressure, ranging from 60 to 240 kPa, was applied to all shell elements of the membrane model, so that the required confining pressure was obtained. Monotonic loading was applied to both the top and bottom disk walls with a constant speed of 0.02 m/s, and the monotonic load test continued until the axial strain reached 20%. During the loading process, data relating to porosity, the stress–strain curve, volumetric strain and particle breakage were all recorded.

4. Results and Discussion

4.1. Stress–Strain Response

Figure 3 shows the simulations and experimental results, in terms of deviator stress and of volumetric strain against axial strain under different confining pressures. Comparing the experimental results with the simulation results, a trend of stress–strain relationship is seen to be well simulated. With the development of the axial strain, the deviator stress gradually increases and attenuates after reaching a relatively stable value, showing the strain softening behavior and the maximum deviator stress increases with the increase in confining pressure. However, the shear strength of the simulated assembly is overestimated in comparison with the experimental one. This could be caused by the irregular shape of particles with an aspect ratio of 1.83, which were recognized as flaky particles. Similarly, after having conducted a set of triaxial ballast tests to investigate the effects of ballast shape on ballast performance, Roner [27] found that randomly placed flaky material exhibited higher deviator stress and higher angles of internal friction than non-flaky material at the same void ratio. This offers a possible explanation as to why the deviator stress was higher than expected. In addition, an orientation parallel to the failure plane will cause a substantial strength reduction when a significant proportion of the particles are flaky.

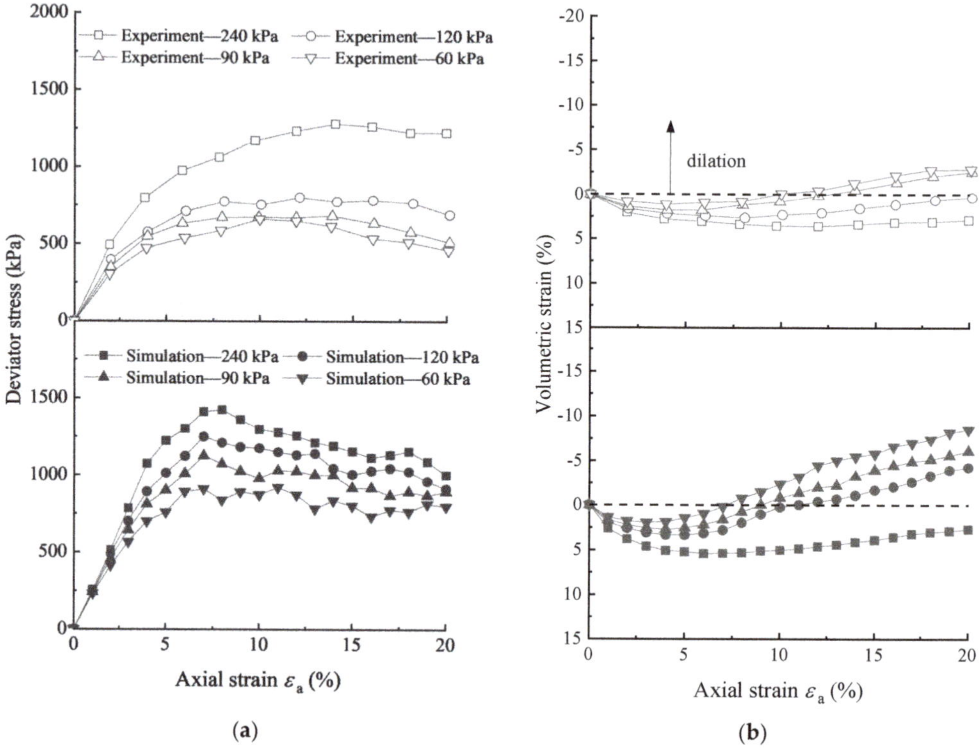

Figure 3. Simulation results under monotonic loading: (**a**) deviator stress; (**b**) volumetric strain.

However, the purpose of this study is not to resolve the particle shape problem, but rather to adopt the best approach to investigate the fragment replacement mode and particle crushing under shear. Therefore, this study applies simplified and irregular shaped particles, and focus on the fractal fragmentation and gradation evolution of particles.

Figure 3b shows that the sample exhibits shear shrinkage in the initial stage, but with the increase in axial deformation, an obvious shear expansion is observed, especially under low confining pressure. As the confining pressure decreases from 240 kPa to 60 kPa, the dilatancy effect becomes more obvious. Volumetric strain agrees reasonably well with the experimental results under a high confining pressure of 240 kPa where breakage is more pronounced. This is probably attributable to the generation of smaller size fragments from the split particles. Using more and smaller fragments to form the origin clumps may provide more realistic volumetric strains at lower confining pressures. Therefore, further investigation of the effects of fragment replacement mode is needed.

Figure 4 shows the effect of different confining pressures on the peak friction angle φ_p of the ballast sample, and both the simulation and experimental results show that the peak friction angle decreases with the increase in confining pressure. Moreover, the simulation using uniform flat shape particles shows a higher angle of internal friction than the experimental results. This is due to the fact that the FRM particle only undergoes particle splitting at higher stress thresholds and is unable to undergo corner breakage and surface abrasion to release internal forces at the initial strain stage.

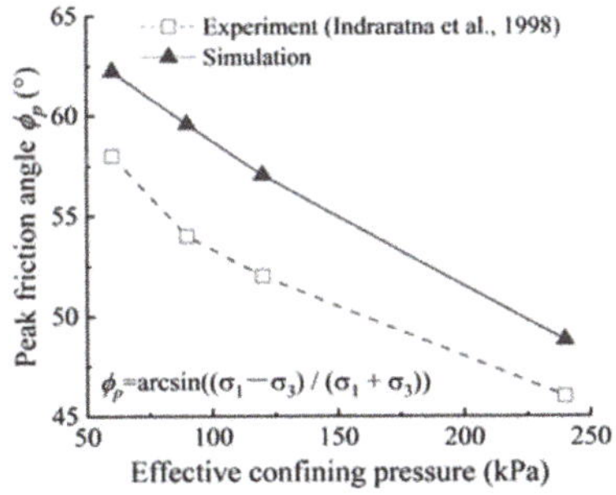

Figure 4. Influence of confining pressure on peak friction angle [26].

It can be seen from Figure 5a that the deformation of the sample is allowed to be uneven, and the expansion of the middle part is the most obvious after using the flexible membrane model. This verifies that the flexible membrane model is better than the servo-controlled cylinder walls for simulating the triaxial test. In order to study the development of porosity during loading, three measurement spheres with radii of 85, 120, and 85 mm were arranged in the sample, as shown in Figure 5b. Figure 5b shows the evolution of porosity with axial strain in the upper, middle and lower regions under 240 kPa confining pressure. As the axial strain increases to 12%, the porosity of each part exhibits the characteristics of first decreasing and then increasing. Subsequently, the porosities in the upper and lower parts are stable, but the porosity in the middle region continues to increase.

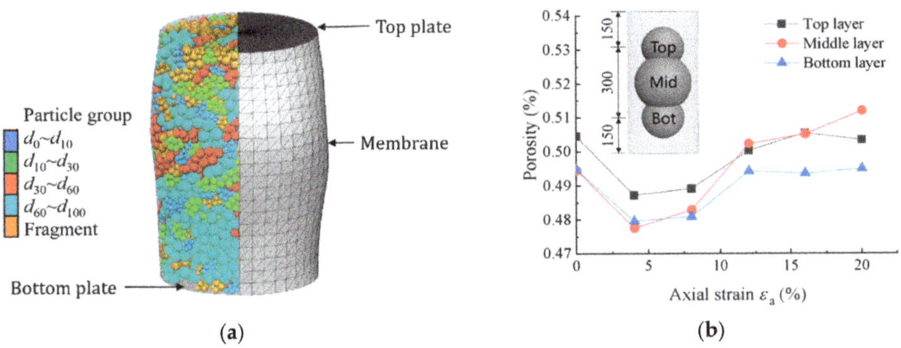

Figure 5. Simulation results under confining pressure of 240 kPa: (**a**) sample after loading; (**b**) evolution of porosity.

The evolution process of the particle displacement vector of the sample under the confining pressure of 240 kPa is shown in Figure 6. During the loading process, the particles within the top and bottom half move toward the horizontal plane in the middle. When axial strain reaches 4%, due to mutual resistance, the intermediate particles start to move outward horizontally to produce a shear sliding surface, and the angle between the sliding surface and the horizontal direction gradually increases.

Under a confining pressure of 240 kPa, the polar coordinate distribution of contact numbers and average contact forces within the sample is plotted as shown in Figure 7. With the increase in axial strain, the average contact force in the sample first increases and then decreases between 12% and 20% axial strain, which is similar to the substantial strength reduction in the deviator stress after peaking shown in Figure 3b. Most of the internal contacts of the sample are distributed in the range of ±30° from the vertical plane in the initial stage; the contact number in the vertical direction then decreases with the loading, while the contact number in the horizontal direction increases. At the end of loading, the average contact force and contact number in the vertical direction are almost 3.36 and 2.49 times larger than those in the horizontal direction.

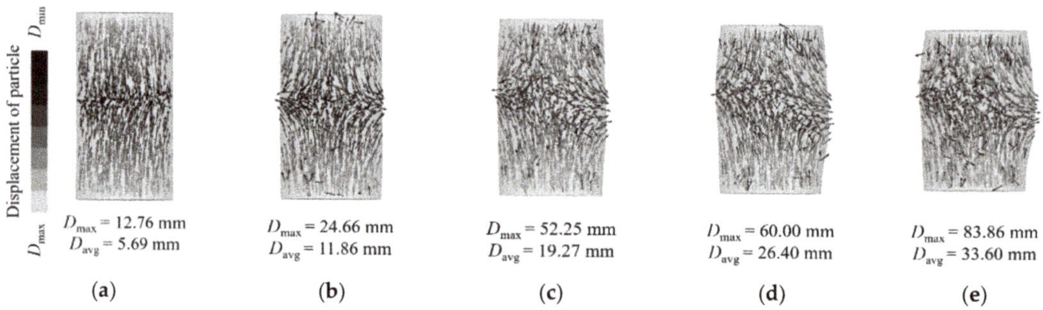

Figure 6. Evolution of particle displacement under 240 kPa confining pressure: (**a**) ε_a = 4%; (**b**) ε_a = 8%; (**c**) ε_a = 12%; (**d**) ε_a = 16%; (**e**) ε_a = 20%. (D_{max}: the maximum displacement; D_{avg}: the average displacement).

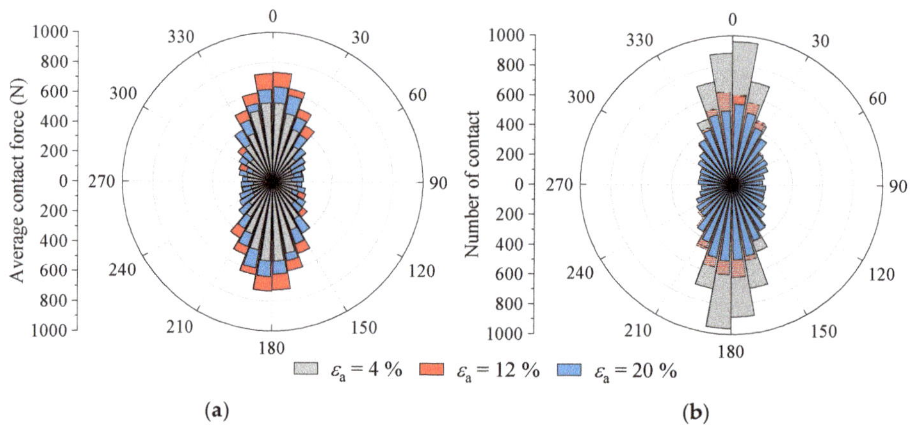

Figure 7. The orientation of contact within the sample under 240 kPa confining pressure: (**a**) average contact force; (**b**) contact number.

4.2. Particle Breakage

As shown in Figure 8, the number of fragments could be ignored at the initial compaction stage (4% axial strain) before increasing linearly above 8% axial strain. This indicates that the sample becomes denser in the compaction stage due to particle rearrangement and a reduction in the number of pores. When the porosity reaches the minimum value, the particle interlocking phenomenon is further exacerbated and a large number of particles are broken off by shearing. In addition, more particle breakage occurs under higher confining pressure.

In accordance with experiment [26], the differences in the percentage by weight of each particle size fraction before and after simulation (ΔW_k) are plotted against the aperture of the lower sieve corresponding to that fraction, as shown in Figure 9. The equivalent diameters and weights of each particle size are calculated by tracing all particles and extracting the volumes in simulation. For aperture sizes between 18.9 mm and 29 mm, the percentage change in particle size decreased and formed a main trough. Meanwhile, For aperture sizes between 18.9 mm and 53 mm, the percentage change of particle size increased and formed a main peak.

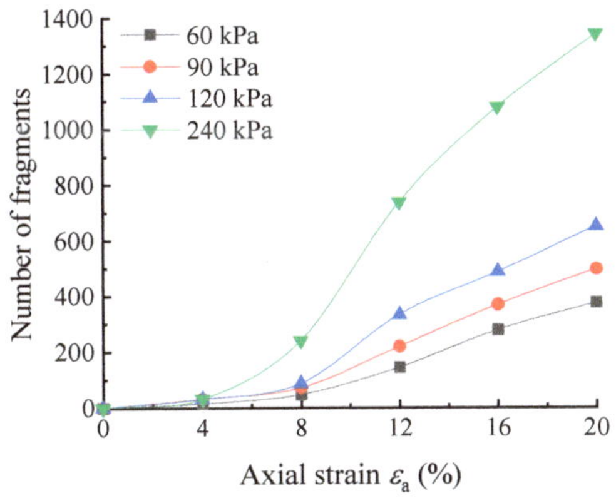

Figure 8. Number of fragments against axial strain under different confining pressures.

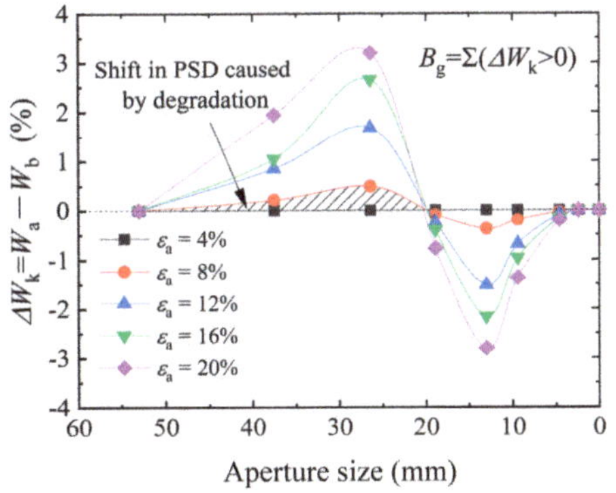

Figure 9. Variation of particle distribution with grain size.

A measure of grain breakage (B_g) equal to the sum of positive values of ΔW_k is now plotted and expressed as a percentage, as introduced by Marsal [28] for granular materials. The B_g under 240 kPa confining pressure increases with increasing axial strain. In particular, the largest increment of B_g can be found the stage from 8% to 12% axial strain, which also corresponds to the peak in the growth rate of crushing, as shown in Figure 8. Moreover, with a large number of particles broken in this phase, the peak deviator stress is gradually deceased when the axial strain reaches 8% as shown in Figure 3a.

Figure 10b shows levels of grain breakage B_g after loading under different confining pressures. It can be observed that more particles are broken off with increasing confining pressure in both the experimental test and the simulations, which proves the acceptability of the fracture criterion and fragment replacement mode used in the simulation study. It should be noted that the breakage in simulation is underestimated compared with the experimental test. This could be explained by the fact that the simple splitting model does not take particle abrasion and small corner breakage into consideration.

For increasing levels of axial strain, the quantities and positions of fragments under 240 kPa confining pressure are plotted in Figure 11. The generations of breakage are defined from first generation to fifth generation as shown in Figure 1. At the initial stage of 8% axial strain, almost all of the fragments belong to the first generation, and the lateral bulging of the sample is not obvious. When ε_a = 12%, the fragments coming from the second and third generations begin to generate at locations near the top and bottom plates. As ε_a increases to 16%, and to 20%, new fragments coming from the fourth and fifth generations, respectively, start to appear in the sample. Apart from the area near the top and bottom plates, most of these fragments (second–third generation) are distributed along the shear surface as shown in Figure 11e, which is consistent with the distribution of particle displacement in Figure 6e. In summary, 160 of 971 eight-ball clumps were split and a total number of 1341 fragments were generated. Among these, 1237 fragments were of the first generation, accounting for 92.2% of all breakage and located all over the sample, while 104 fragments of the second to fourth generation accounted for 7.8% of all breakage and are more concentrated in the upper half of the sample. Therefore, the first-generation breakage is dominant, and the proportions of breakage decrease from the second to the fifth generations, successively

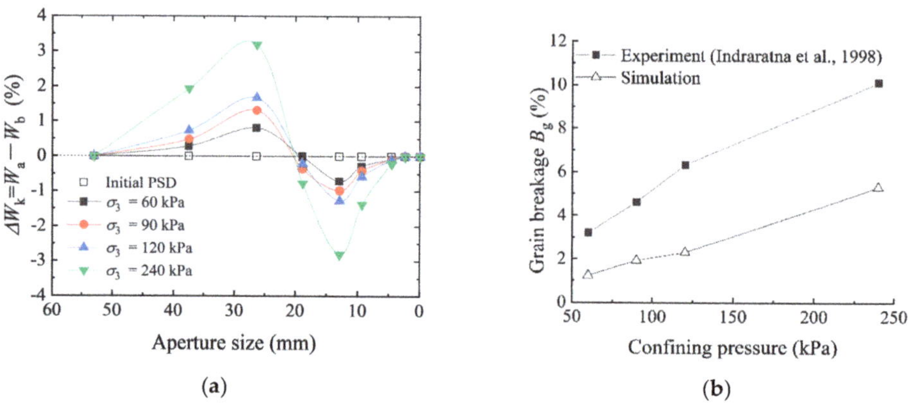

Figure 10. Effect of confining pressure on: (**a**) ΔW_k; (**b**) grain breakage B_g [26].

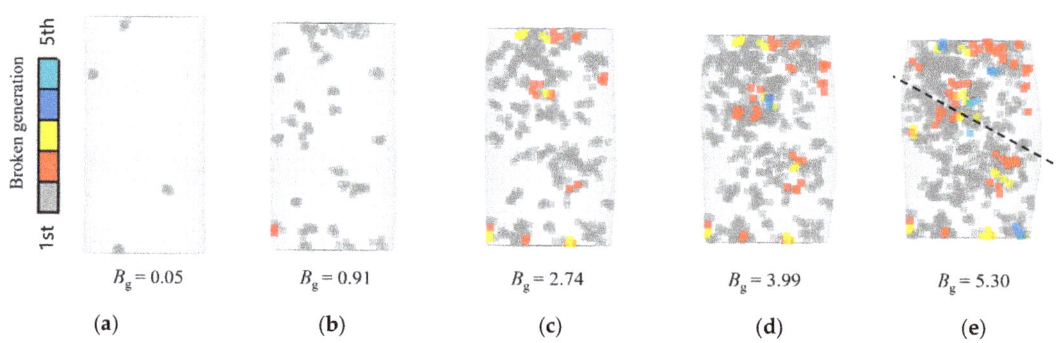

Figure 11. The development of fragment distribution under confining pressure of 240 kPa: (**a**) ε_a = 4%; (**b**) ε_a = 8%; (**c**) ε_a = 12%; (**d**) ε_a = 16%; (**e**) ε_a = 20%.

4.3. Gradation Evolution and Fractal Distribution

Turcotte et al. [3] demonstrated that any initial distribution of soil particles would tend towards a self-similar fractal distribution with increasing particle fragmentation. Railway track bed typically consists of ballast of size 10–60 mm in different ballast grading specifications, generally larger than soil or calcareous sand. Considering the ballast breakage results in the previous section, the fractal evolution of the PSD against the axial strain and confining pressure were also studied.

Figure 12a shows the evolution of PSD under monotonic loading and a confining pressure of 240 kPa. It can be seen that, as the axial strain increases, the percentage of small particle fragments ranging from 10 mm to 35 mm increases, with the grading curves remaining concave upwards. This can be attributed to the supplement of aggregates induced by particle breakage from the upper size ranges. Figure 12b shows the comparison of PSDs under different confining pressures, ranging from 60 kPa to 240 kPa. It can be seen that all the sample PSDs after test are shifted left where smaller aggregates were generated, when compared with the initial grading. With increases in confining pressure, the sample grading shifts more towards the left, especially under the highest confining pressure of 240 kPa.

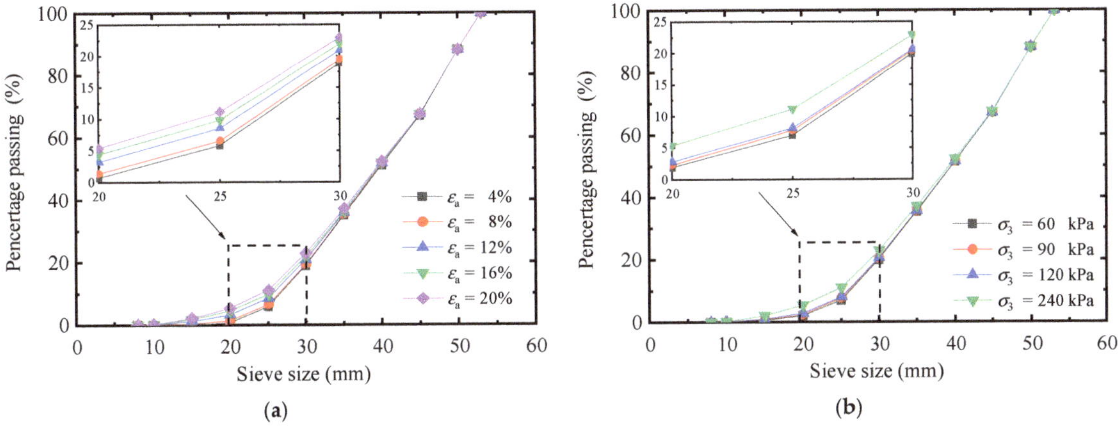

Figure 12. Evolution of PSDs: (**a**) effect of axial strain; (**b**) effect of confining pressure.

According to the study by Tyler et al. [11], the fractal grading can be described as

$$F_u(d) = M(\delta < d_i)/M_T = (d_i/d_{max})^{3-D} \quad (4)$$

where $F_u(d)$ is the percentage finer, d_{max} is the maximum diameter for the given grain assembly, M_T is the total mass of particles, δ is the particle size, and $M(\delta < d_i)$ is the cumulative mass of the particles whose grain size is smaller than d_i. The fractal dimension D can be determined by:

$$\lg[M(\delta < d_i)/M_T] = (3 - D)\lg(d_i) + (D - 3)\lg(d_{max}) \quad (5)$$

The PSD curve presents the correlations between the percentages passing and normalized particle diameters which gradually close to a linear line with the log–log scales. Figure 13a shows that the fitting lines shift upward, and the grading curve gradually becomes linear as the axial strain increases. Meanwhile, the linear portion of the curve from which the fractal dimension can be obtained increases in length, suggesting a more reliable value. Moreover, the slope becomes steeper (i.e., the fractal dimension D increases) with increasing confining pressure, with the slope appearing to become constant as shown in Figure 13b. The size ratio of the largest and smallest particle is much smaller than that of soil or quartz sand, with values of C_u ranging from 1.5 to 3.0, which can be regarded as granular material of uniform size. This is quite different from granular materials in natural state because the grading ballast used in railway is qualified by the ballast standard, leading to a much smaller initial value of D (0.2–0.3). Predictably, a continued compression will cause a further crushing and increasing of fine particles, leading to a further fractal.

With the faster and heavier of the passing train, the ballast layer are subjetced to settle down and crushing under the long-term cyclic loading, and the fractal dimension gradually increases. Ballast will need to be maintained or replaced once the fractal dimension reaches a constant value. Further experimental research is required to predict the critical value of the fractal dimension which indicates when ballast needs to be maintained.

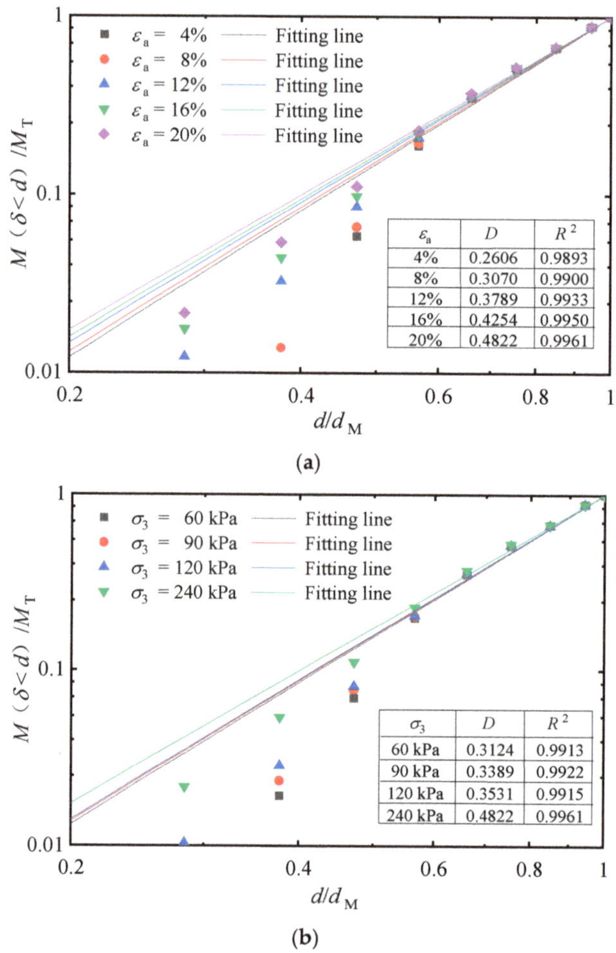

Figure 13. Linear fitting for $\lg[M(\delta < d)/M_T]$ and $\lg(d/d_M)$ for simulation sample: (**a**) effect of axial strain; (**b**) effect of confining pressure.

Figure 14 shows the evolution of fractal dimension against grain breakage under different confining pressures. It shows that the fractal dimension of the sample increases with rising confining pressure and also axial strain. The relationship between D and B_g is almost linear. Therefore, the fractal dimension can be well used to evaluate the degree of crushing and deterioration of ballast.

The relationship between D and the volumetric strain ε_v for these four samples under different confining pressures can be plotted in Figure 15. At the initial stage (4% axial strain), the samples are compressed with an almost constant value of D; a linear relationship between volumetric strain and D can then be observed above 8% axial strain under different confining pressures, which corresponds with the maximum sample dilation angle as shown in Figure 3b. A quadratic curve relation between the volumetric strain and fractal dimension can be obtained for different levels of axial strain, as shown in Figure 15. At 20% axial strain, the values of a, b and c are 1.48×10^{-3}, 5.25×10^{-4} and 0.26, respectively.

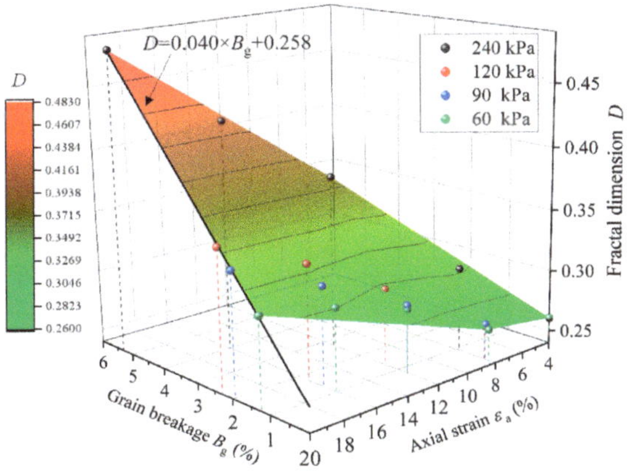

Figure 14. Results of fractal dimension D against grain breakage B_g.

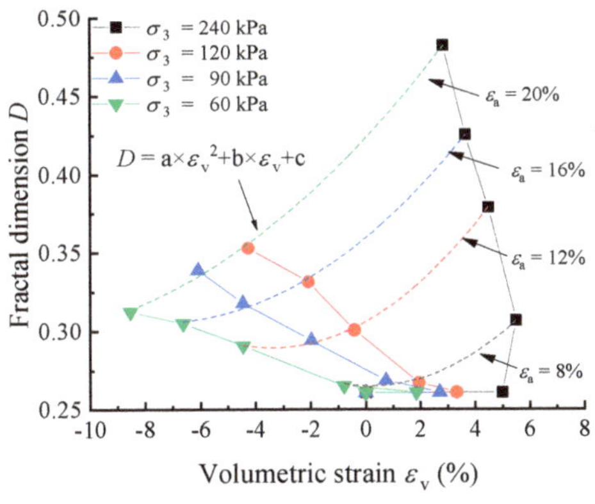

Figure 15. Results of fractal dimension D against volumetric strain ε_v.

5. Conclusions

A DEM model of irregular-shaped ballast particles using the fractal replacement method (FRM) and a FDM flexible membrane model using shell elements are presented in this paper. A series of DEM–FDM coupled simulations of triaxial testing under monotonic loading and with different confining pressures were performed to study the ballast breakage and fractal behavior of PSD. We draw the following main conclusions:

(1) The proposed FRM model can simulate the ballast breakage and capture the essential features of the stress-dilatancy in the monotonic triaxial test. During the loading process, the particles within the top and bottom half moved toward the horizontal plane in the middle. With axial strains above 4%, due to mutual resistance, the intermediate particles started to move outward to produce a shear sliding surface, where the angle between the sliding surface and the horizontal direction gradually increased.

(2) More particle breakage occurred under higher confining pressure or larger axial strain and most such breakage was in the second generation or earlier. The number of fragments could be ignored at the initial compaction stage and the majority of particle breakage occurred with axial strain of 8% or above. Compared with the experimental results, the simulation underestimated the degree of particle breakage. This could be explained by that the simple splitting model of FRM did not take particle abrasion and small corner breakage into consideration.

(3) With increasing axial strain or confining pressure, the percentage of small particle fragments increased in correspondence with the PSD curves which remained concave

upwards, leading to an increasing fractal dimension. However, a different situation applies to granular materials in natural state because the grading ballast used in railway is usually qualified by the ballast specification, leading to much smaller values of D.

(4) The evolution of fractal dimension against grain breakage showed a linear relation, Conversely, a quadratic curve relation was obtained between the fractal dimension and volumetric strain under different axial strain stages. Therefore, *D* has the potential to be a key indicator to evaluate the degree of ballast crushing and PSD degradation, which can contribute to better decision making concerning railway track bed maintenance in practice.

Author Contributions: Conceptualization, C.C. and Y.S.; methodology, C.C. and Z.W.; software, C.C. and X.Z.; validation, C.C., X.Z. and Y.S.; formal analysis, Y.S. and R.R.; investigation, L.Z. and R.R.; resources, L.Z. and Z.W.; data curation, Y.S.; writing—original draft preparation, C.C. and X.Z.; writing—review and editing, Z.W. and Y.S.; visualization, R.R.; supervision, C.C. and Z.W.; project administration, Z.W.; funding acquisition, Z.W. All authors have read and agreed to the published version of the manuscript.

Funding: This research was funded by the National Natural Science Foundation of China (No. 51708438) and the Self-determined and Innovative Research Funds of Wuhan University of Technology (No. 2022CG030).

Institutional Review Board Statement: Not applicable.

Informed Consent Statement: Not applicable.

Data Availability Statement: Not applicable.

Conflicts of Interest: The authors declare no conflict of interest.

References

1. Raymond, G.P.; Diyaljee, V.A. Railroad ballast load ranking classification. *J. Geotech. Eng. Div.* **1979**, *105*, 1133–1153. [CrossRef]
2. Selig, E.T.; Waters, J.M. *Track Geotechnology and Substructure Management*; Thomas Telford: London, UK, 1994; pp. 16–26. [CrossRef]
3. Turcotte, D.L. Fractals and fragmentation. *J. Geophys. Res.* **1986**, *91*, 1921–1926. [CrossRef]
4. Zhao, W.J.; Cui, Z.; Ma, H. Fractal features of soil particle-size distributions and their relationships with soil properties in gravel mulched fields. *Arab. J. Geosci.* **2017**, *10*, 211–218. [CrossRef]
5. Tao, G.; Wang, Q.; Chen, Q.; Nimbalkar, S.; Peng, Y.; Dong, H. Simple graphical prediction of relative permeability of unsaturated soils under deformations. *Fractal Fract.* **2021**, *5*, 153. [CrossRef]
6. Bai, Y.; Qin, Y.; Lu, X.; Zhang, J.; Chen, G.; Li, X. Fractal dimension of particle-size distribution and their relationships with alkalinity properties of soils in the western Songnen Plain, China. *Sci. Rep.-UK* **2020**, *10*, 20603. [CrossRef]
7. He, S.H.; Ding, Z.; Hu, H.B.; Gao, M. Effect of grain size on microscopic pore structure and fractal characteristics of carbonate-based sand and silicate-based sand. *Fractal Fract.* **2021**, *5*, 152. [CrossRef]
8. Sui, L.; Yu, J.; Cang, D.; Miao, W.; Wang, H.; Zhang, J.; Yin, S.; Chang, K. The fractal description model of rock fracture networks characterization. *Chaos Soliton Fract.* **2019**, *129*, 71–76. [CrossRef]
9. Xiao, Y.; Meng, M.; Daouadji, A.; Chen, Q.; Wu, Z.; Jiang, X. Effects of particle size on crushing and deformation behaviors of rockfill materials. *Geosci. Front.* **2020**, *11*, 375–388. [CrossRef]
10. Zhang, K.; Liu, X.; Liu, W.; Zhang, S. Influence of weak inclusions on the fracturing and fractal behavior of a jointed rock mass containing an opening: Experimental and numerical studies. *Comput. Geotech.* **2021**, *132*, 104011. [CrossRef]
11. Tyler, S.W.; Wheatcraft, S.W. Fractal scaling of soil particle-size distributions: Analysis and limitations. *Soil Sci. Soc. Am. J.* **1992**, *56*, 362–369. [CrossRef]
12. Tang, l.; Song, Y.; Jiang, Z.; Jiang, S.; Li, Q. Pore structure and fractal characteristics of distinct thermally mature shales. *Energy Fuels* **2019**, *33*, 5116–5128. [CrossRef]
13. *Railway Ballast: TB/T2140-2018*; China Railway Specifications. China Railway Administration: Beijing, China, 2018.
14. Tavarez, F.A.; Plesha, M.E. Discrete element method for modelling solid and particulate materials. *Int. J. Numer Meth. Eng.* **2007**, *70*, 379–404. [CrossRef]
15. Adhikari, S.; You, Z. 3D discrete element models of the hollow cylindrical asphalt concrete specimens subject to the internal pressure. *Int. J. Pavement Eng.* **2010**, *11*, 429–439. [CrossRef]
16. McDowell, G.R.; De Bono, J.P. On the micro mechanics of one-dimensional normal compression. *Géotechnique* **2013**, *63*, 895–908. [CrossRef]

17. Ciantia, M.O.; Arroyo, M.; Calvetti, F.; Gens, A. An approach to enhance efficiency of DEM modelling of soils with crushable grains. *Géotechnique* **2015**, *65*, 91–110. [CrossRef]
18. Zhou, W.; Wang, D.; Ma, G. Discrete element modeling of particle breakage considering different fragment replacement modes. *Powder Technol.* **2020**, *360*, 312–323. [CrossRef]
19. Vallejo, L.E.; Lobo-Guerrero, S. Fractal fragmentation of granular materials under compression. *AIP Conf. Proc.* **2009**, *1145*, 847–850. [CrossRef]
20. De Bono, J.P.; McDowell, G.R. Particle breakage criteria in discrete-element modelling. *Géotechnique* **2016**, *66*, 1014–1027. [CrossRef]
21. Li, H.Y.; Chai, H.W.; Xiao, X.H.; Huang, J.Y.; Luo, S.N. Fractal breakage of porous carbonate sand particles: Microstructures and mechanisms. *Powder Technol.* **2020**, *363*, 112–121. [CrossRef]
22. Lu, M.; McDowell, G.R. The importance of modelling ballast particle shape in the discrete element method. *Granul. Matter* **2007**, *9*, 69–80. [CrossRef]
23. Sun, Y.; Indraratna, B.; Nimbalkar, S. Three-dimensional characterization of particle size and shape for ballast. *Géotechnique Lett.* **2014**, *4*, 197–202. [CrossRef]
24. Jaeger, J.C. Failure of rocks under tensile conditions. *Int. J. Rock Mech. Min. Sci.* **1967**, *4*, 219–227. [CrossRef]
25. Lim, W.L.; McDowell, G.R. Discrete element modelling of railway ballast. *Granul. Matter* **2003**, *7*, 19–29. [CrossRef]
26. Indraratna, B.; Ionescu, D.; Christie, H.D. Shear behaviour of railway ballast based on large scale triaxial testing. *J. Geotech. Geoenviron.* **1998**, *5*, 439–449. [CrossRef]
27. Roner, C.J. Some Effect of Shape, Gradation and Size on the Performance of Railroad Ballast. Master's Thesis, Report No. AAR85-324P. Department of Civil Engineering, University of Massachusetts, Amherst, MA, USA, 1985.
28. Marsal, R.J. Mechanical properties of rockfill: In embankment-dam engineering. *Int. J. Rock Mech. Min. Sci.* **1975**, *12*, 67. [CrossRef]

fractal and fractional

Article

Effects of Relative Density and Grading on the Particle Breakage and Fractal Dimension of Granular Materials

Gui Yang *, Zhuanzhuan Chen, Yifei Sun and Yang Jiang

Key Laboratory of Ministry of Education for Geomechanics and Embankment Engineering, College of Civil and Transportation Engineering, Hohai University, Nanjing 210024, China; czzjya@163.com (Z.C.); sunsunyifei@126.com (Y.S.); m18251975654@163.com (Y.J.)

* Correspondence: ygheitu@163.com

Abstract: Particle breakage was reported to have great influence on the mechanical property of granular materials. However, limited studies were conducted to quantify the detailed effects of relative density and initial grading on the particle breakage behaviour of granular materials under different confining pressures. In this study, a series of monotonic drained triaxial tests were performed on isotropically consolidated granular materials with four different initial gradings and relative densities. It is observed that particle breakage increases as the confining pressure or relative density increases, whereas it decreases with the increasing coefficient of uniformity. Due to particle breakage, the grading curves of granular materials after triaxial tests can be simulated by a power-law function with fractal dimension. As the confining pressure increases, the fractal dimension approaches the limit of granular materials, i.e., 2.6. A unique normalized relation between the particle breakage extent and confining pressure by considering relative density and grading index was found.

Keywords: granular materials; fractal dimension; particle breakage; relative density; grading curve

Citation: Yang, G.; Chen, Z.; Sun, Y.; Jiang, Y. Effects of Relative Density and Grading on the Particle Breakage and Fractal Dimension of Granular Materials. *Fractal Fract.* **2022**, *6*, 347. https://doi.org/10.3390/fractalfract6070347

Academic Editor: Norbert Herencsar

Received: 28 March 2022
Accepted: 20 June 2022
Published: 22 June 2022

1. Introduction

Owing to the convenience of lower procurement cost, good compaction ability and high shear strength, granular materials have been widely used in hydraulic engineering, civil engineering and transport engineering, among others [1,2].

Due to the development of the rolling equipment technology, various types and grades of granular materials have been used for engineering construction, where the compactness of granular materials has been significantly improved, which effectively reduced the deformation of engineering facilities, e.g., a rockfill dam [3]. In a rockfill dam, granular materials often account for 60–90% of the total filling volume. Therefore, the strength and deformation of rockfill has an important impact on the stability and deformation analysis. However, upon external loading, significant particle breakage of granular materials can occur, which would have an important impact on the stability and deformation of the engineering facilities during both construction and operating periods. In particular, the grading curve changed evidently before and after loading [4]. Such degradation behaviour of granular materials was found to depend not only on stress history [5,6], but also on grain size distribution curve [7,8], parent rock type, particle size [9,10], particle shape [11,12] and relative density [13,14]. For example, particle shape has a significant influence on the particle breakage, which increases with the increasing in shape index sphericity, aspect ratio, convexity and overall regularity [12]. The critical state parameters (M, ϕ_{cs}, e_Γ, and λ_c) decrease with increasing aspect ratio, sphericity and convexity [8].

The relative density and grading curve are important control indicators at engineering site, e.g., the relative density of sand should be at least 0.70 [15]. It was also found to have a great influence on the shear strength and deformation of granular materials. The greater the relative density, the greater the initial elastic modulus and peak friction angle and the smaller the volume strain [16,17]. However, previous studies mainly focused on the effect of strength and deformation of granular materials [18,19]. The fundamental physical properties that trigger particle breakage were not investigated in depth. For example, at large relative density, the stress–strain relationship usually exhibited strain softening. The stress–strain curve exhibited strain hardening with the decrease of relative density. The effect of initial void ratio on the stress–strain relationship decreases with increasing confining pressure [20,21]. With the increase in confining pressure, the difference of initial void decreases and reaches essentially the same value [20]. The residual strength of rockfill at different relative density was observed to be the same [22]. The critical state parameters (M and λ_c) are less affected by relative density and particle grading [7,13]. However, the critical state parameters (e_Γ) decrease with increasing relative density. The relative breakage index decreases with increasing relative density [14]. Extensive research has also been carried out on the impact of granular materials particle gradation [23–25]. For example, model parameters varied linearly with the coefficient of uniformity. The larger the coefficient of uniformity, the smaller the peak intensity and the larger volumetric strain.

In this study, a comprehensive study on the effect of relative density and initial grading on the particle breakage behavior of granular materials will be carried out, by using drained triaxial tests. An attempt is also made to propose a unique normalized relation between the particle breakage extent, applied pressure, relative density and grading index. The research results play an important role in understanding and mastering the gradation change law and strength characteristics of granular materials before and after loading, and provide an important reference for the stable design parameters of the structure.

2. Laboratory Tests

The granular materials, as shown in Figure 1, were collected from a quarry near Nanjing, China. Aggregates were derived from the parent of sandstone rock. The grain size distribution of the material and triaxial apparatus are shown in Figure 2. The dry density, coefficient of uniformity (C_u) and curvature coefficient (C_c) are listed in Table 1.

Figure 1. Test materials.

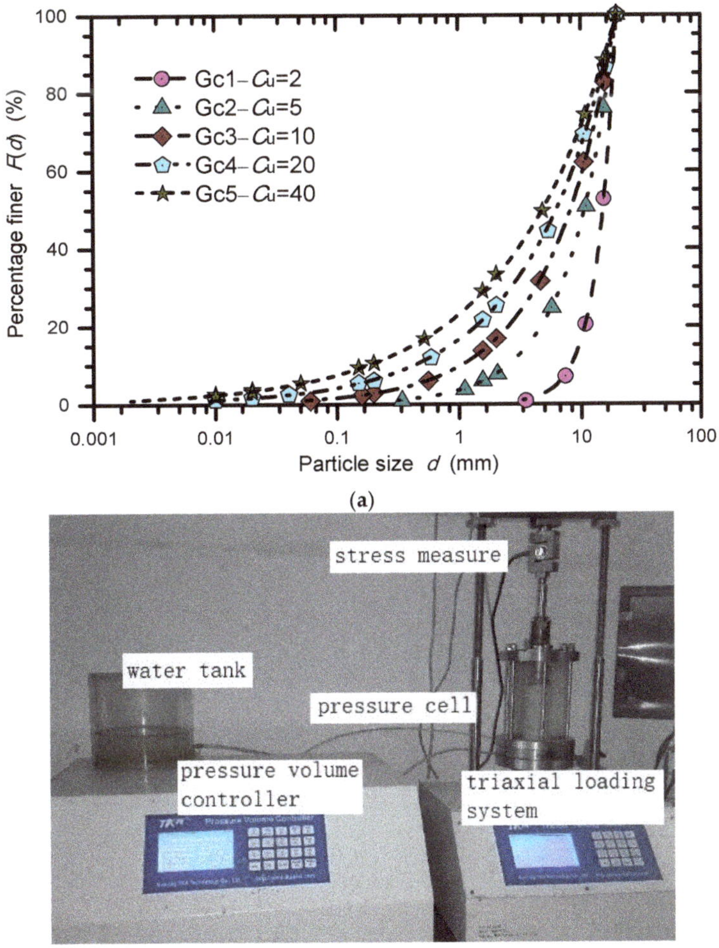

(**b**)

Figure 2. (**a**) Grading curve; (**b**) Triaxial apparatus. Grain size distribution and triaxial apparatus for tested granular materials.

Table 1. The basic characteristics of grading curves.

	Gc1	Gc2	Gc3	Gc4	Gc5
Coefficient of uniformity	2	5	10	20	40
Curvature coefficient	1.17	1.44	1.68	1.97	2.30
Maximum dry density (g/cm^3)	1.71	1.78	1.94	2.12	2.23
Minimum dry density (g/cm^3)	1.47	1.57	1.68	1.78	1.79

The experimental program comprised a consolidated drained triaxial shear test with an initial sample size of a 100 mm diameter and a 200 mm height. The specimen preparation and loading process were carried out step by step with reference to the specification of soil test (SL237-1999) [26]. Aggregates were weighed separately and mixed together before being split into five equal portions. Each portion was then compacted inside a split cylindrical mould. The monotonic shearing rate of displacement was determined to be 0.6 mm/min. Before shearing, the sample was saturated by allowing water to pass through the base of the triaxial cell under a back pressure of 50 kPa until Skempton's B-value exceeded 0.95. In order to analyse the particle breakage, four different relative densities (i.e., Rd = 0.6, 0.7, 0.8 and 0.9 of grading curve Gc3) and five kinds of grading curves (i.e., Gc1, Gc2, Gc3, Gc4 and Gc5 in Rd = 0.9) were selected under four values of confining pressures (i.e., σ3 = 0.2, 0.4, 0.6 and 0.8 MPa). A load cell and pore-pressure sensor were used to measure the deviator load and drainage volume, respectively, through the electronic display system. All the tests were conducted up to a maximum axial strain of 25%.

3. Analysis of Test Results

3.1. Particle Breakage under Different Relative Density

The particle size distribution of natural granular materials in this study can be described by using the Talbot grading curve [27]. Blasting granular materials can meet the Talbot grading curve by adjusting the blasting parameters, and subsequently it will represent fractal characteristics and be easier to achieve the maximum dry density. The proportion of particle mass can be calculated by using the following formula:

$$\frac{W(\delta > d)}{W_0} = 1 - F(d) \tag{1}$$

$$F(d) = \left(\frac{d}{d_M}\right)^{3-D} \tag{2}$$

where d is the diameter of the particle, $W(\delta > d)$ is the mass of particle with diameter larger than d, W_0 is the total mass of granular materials, d_M is the particle maximum diameter, $F(d)$ is the mass ratio of granular materials with diameter less than d and D is the fractal dimension.

The stress–strain behaviour of granular materials tested at different Rd of 0.6, 0.7, 0.8 and 0.9 is shown in Figure 3, respectively. The peak shear stress increases (such as 2.42, 2.43, 2.53 and 2.78 MPa) with the increase in Rd, for a given effective confining pressure (σ_3 = 0.6 MPa). However, the residual shear stresses remain approximately the same for the different Rd for a given confining pressure. The stress–strain behaviour of granular materials transforms from a strain hardening type to a strain softening type for all the tested specimens. As the shear strain further increases, the residual shear stresses corresponding to the rockfill with different Rd become stable. All the granular materials tested in this study exhibit characteristics similar to those observed by Lade [28] for sands. This can be attributed to the observation that the greater the relative density, the greater the interlocking between the particles; and the greater the loading, the greater was the peak strength of the specimen under the same confining pressure. As the load increased, particle breakage occurred, the interlocking force between particles would decrease and the shear strength would decrease.

Furthermore, a general compression followed by the dilatancy can be observed in all granular materials tested at low confining pressures. Granular materials gradually became more compressive with the increasing confining pressure. The larger the relative density was, the dilatancy became much more pronounced. The volumetric strain became stable as the strain increased, whereas it decreased as the relative density increased. This can be attributed to the enhanced sliding rather than rotation between aggregates when the compressive pressure increases.

At the monotonic loading test, the grading curve of granular materials changed due to the occurrence of particle breakage. Thus, the fractal dimensions of each grading curve changed after each test. Figure 4 shows the fractal dimension obtained under different confining pressures after monotonic loading. There is a good linear relationship between ln$F(d)$ and ln(d/d_M) after loading under different confining pressures. The values of R^2 are larger than 0.97, and the Root mean squared Error (Re) is smaller than 0.23. The fractal dimension increases with confining pressure during of particle breakage. For example, the fractal dimension increases from 2.33 (σ_3 = 0.2 MPa) to 2.37 (σ_3 = 0.8 MPa) when relative density Rd = 0.6. However, relative density has little influences on the fractal dimension, especially under higher relative density. For example, the fractal dimensions are almost the same for Rd = 0.8 and 0.9.

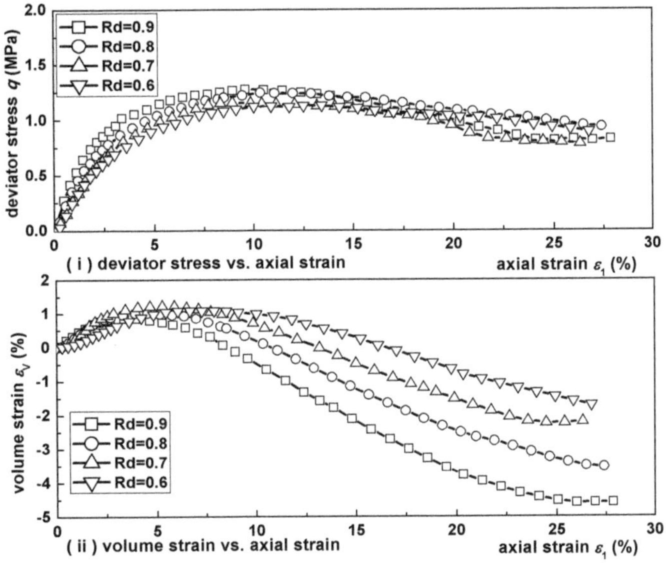

(a)

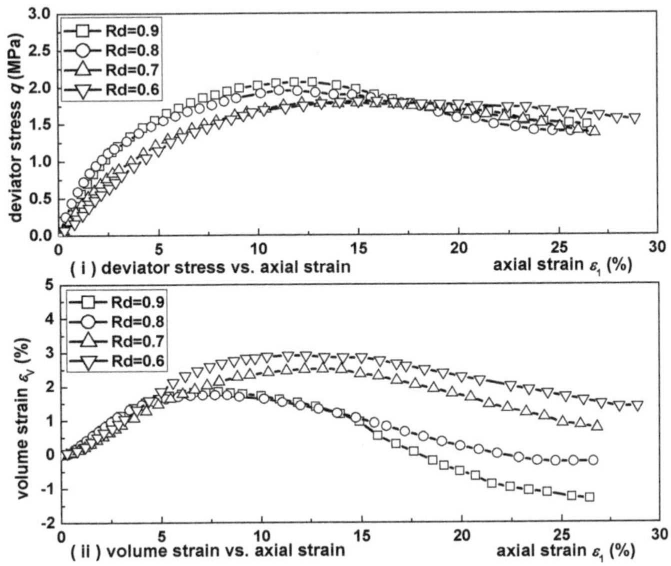

(b)

Figure 3. *Cont.*

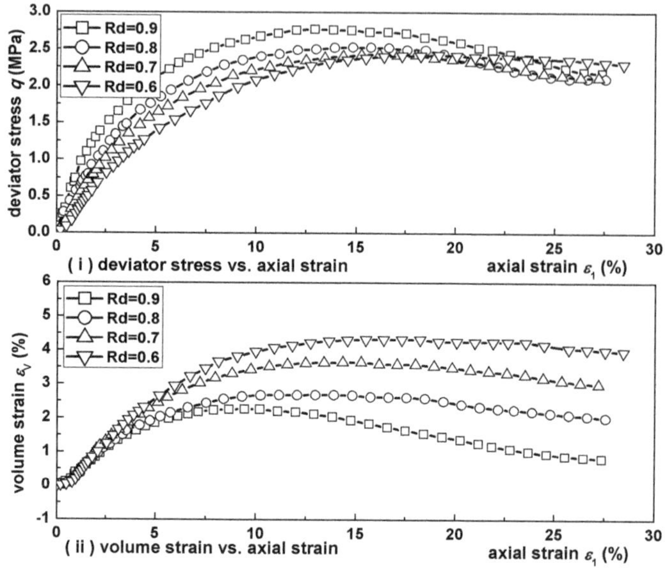

(c)

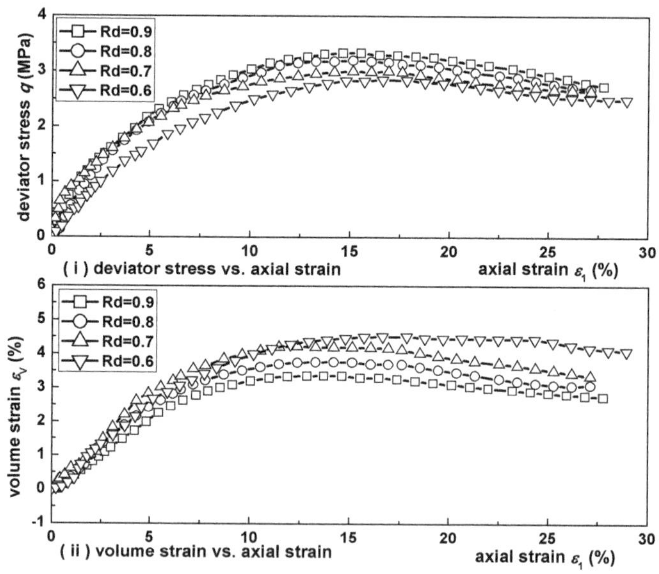

(d)

Figure 3. (**a**) confining pressure of 0.2 MPa; (**b**) confining pressure of 0.4 MPa; (**c**) confining pressure of 0.6 MPa; (**d**) confining pressure of 0.8 MPa. Stress–strain behaviour of granular materials under different confining pressures.

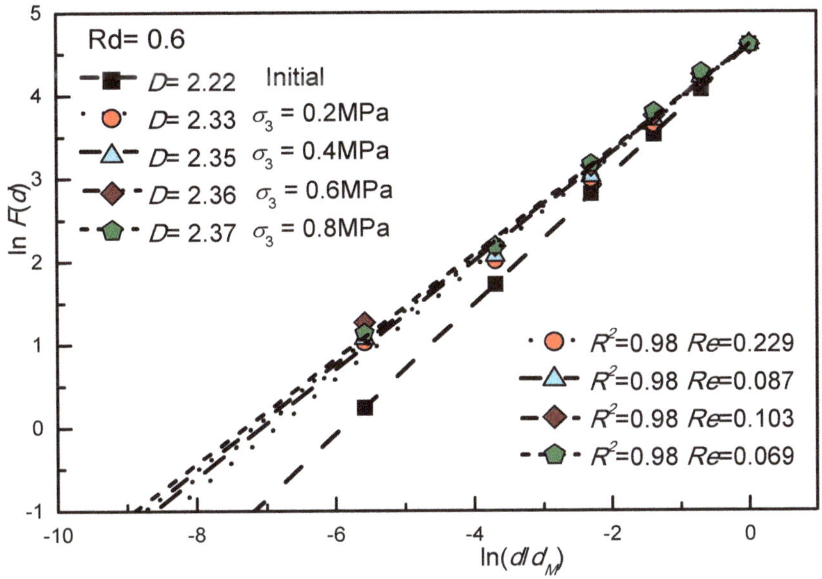

(**a**)

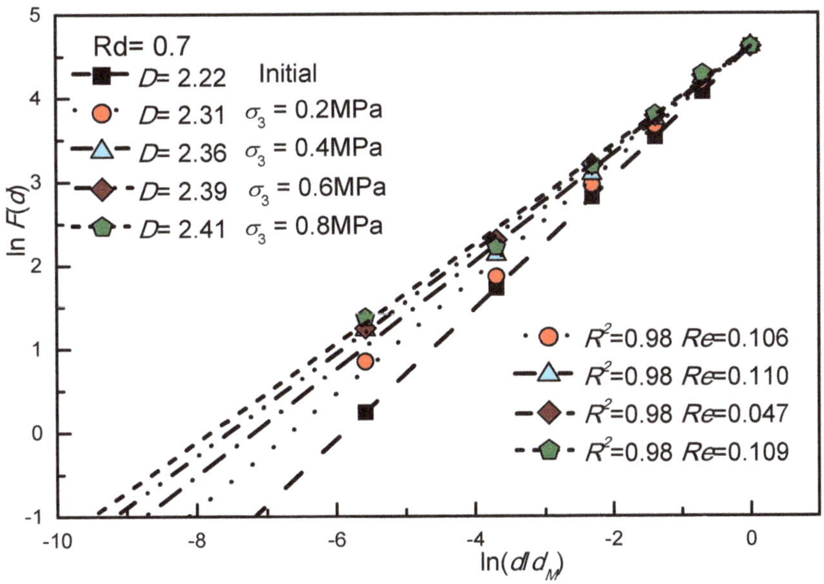

(**b**)

Figure 4. *Cont.*

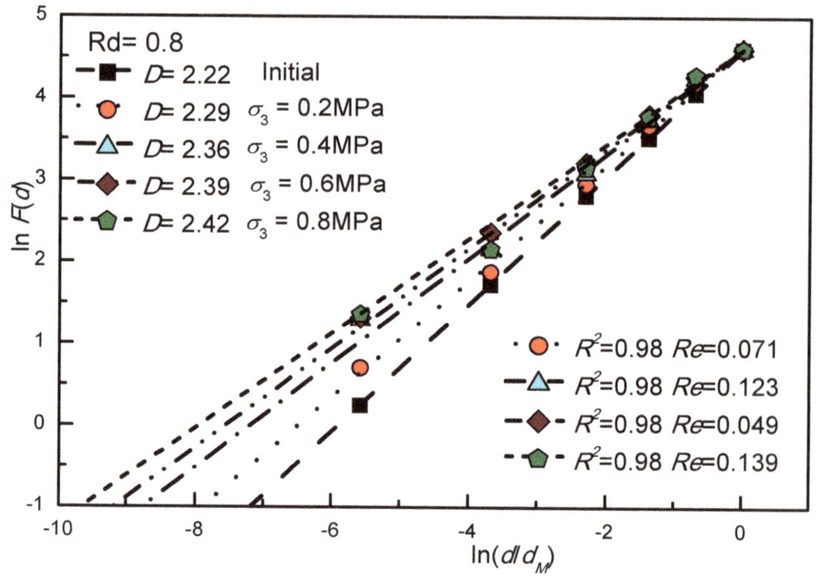

(**c**)

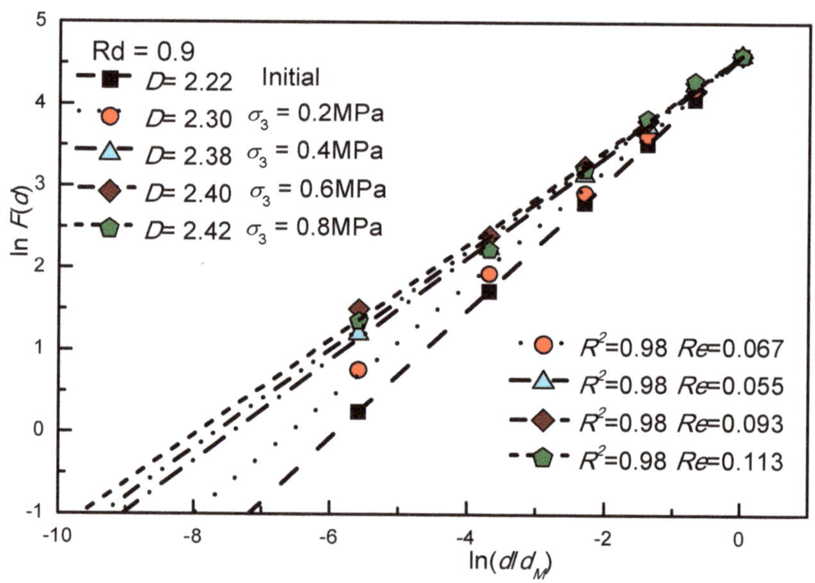

(**d**)

Figure 4. (**a**) Rd = 0.6; (**b**) Rd = 0.7; (**c**) Rd = 0.8; (**d**) Rd = 0.9. Grading curves of different relative densities.

The extent of particle breakage during monotonic drained shearing is further assessed. In this study, the breakage ratio Bg presented by Marsal [29] is used for analysis, which can be expressed as the following:

$$B_g = \sum |P_{test} - P_{ini}| \tag{3}$$

where P_{test} is percentage by mass of particles after the test, and P_{ini} is percentage by mass of particles before the test.

The evolution of breakage ratio at different confining pressures is shown in Figure 5. According to the results, breakage ratio increases with an increase in confining pressure. The relationship between breakage ratio and confining pressure can be simulated by using a power function. The larger the relative density is, the greater the interlocking force between the particles is. Moreover, at the same axial strain level, the larger the shear stress is, the larger the particle breakage ratio is. With the increase of the relative density, breakage ratio increases. A power function is found to be able to fit well the relationship between breakage ratio, relative density and confining pressure, such that:

$$B_g = k_1\left(\frac{\sigma_3}{p_a}Rd\right)^n \tag{4}$$

where k_1 and n are model parameters, determined to be 6.36 and 0.46, respectively; p_a (= 101 kPa) is the atmospheric pressure.

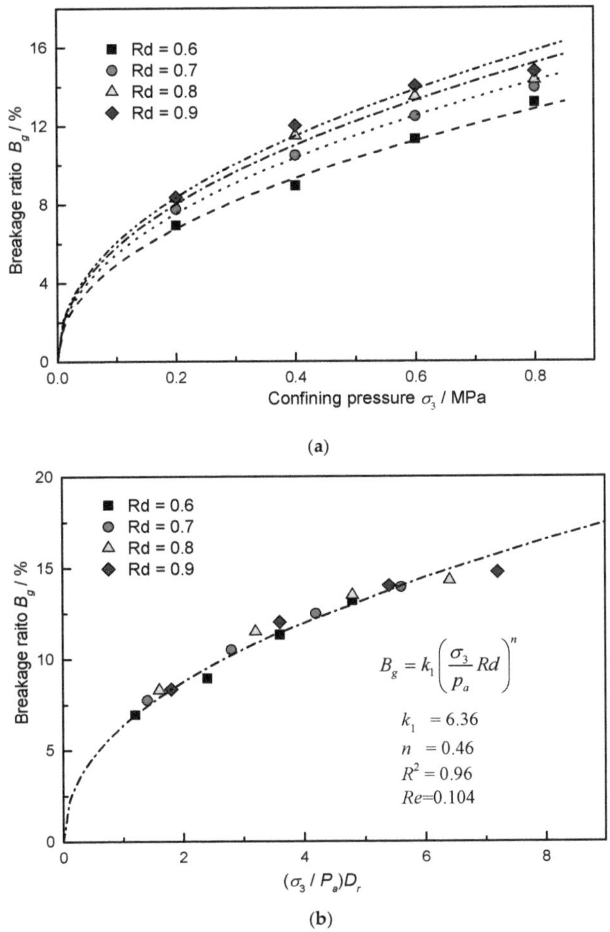

Figure 5. (**a**) Breakage ratio vs. confining pressure; (**b**) Breakage ratio vs. normalized confining pressure. Particle breakage of different relative densities.

3.2. Particle Breakage under Different Grading Curve

Figure 6 shows the fractal dimension under different confining pressures after monotonic loading. There is also a good linear relationship between ln$F(d)$ and ln(d/d_M) after loading at different confining pressures. Compared with those under different relative densities, the initial fractal dimension has great influences on the final fractal dimensions obtained under different confining pressures, as shown in Figure 7. It can also be observed that under low confining pressures, granular materials can undergo significant particle breakage. The smaller the fractal dimension is, the more significant the particle breakage extent will be. The larger the fractal dimension is, the greater the content of fine particles, the smaller the particle crushing rate during shearing and the smaller the effect of confining pressure on particle crushing. As the confining pressure increases, it tends to the final limit fractal dimension 2.6 [30].

The evolution of the breakage ratio at different confining pressures is shown in Figure 8. Unlike the relationship between the relative density and particle breakage ratio in direct proportion, the coefficient of uniformity is inversely proportional to the particle breakage ratio. The greater the coefficient of uniformity, the smaller is the particle breakage ratio. A normalized power function is also found to fit well the relationship between breakage ratio, relative density, uniform coefficient and confining pressure. The expression is shown as follows:

$$B_g = k_2\left(\frac{\sigma_3}{p_a C_u} Rd\right)^n \tag{5}$$

where k_2 and n are model parameters, determined to be 19.1 and 0.46, respectively.

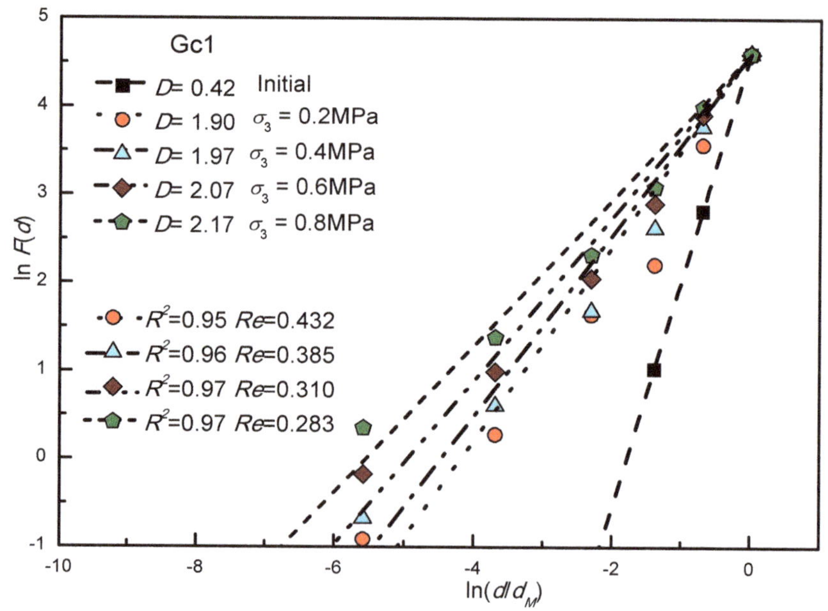

(a)

Figure 6. *Cont.*

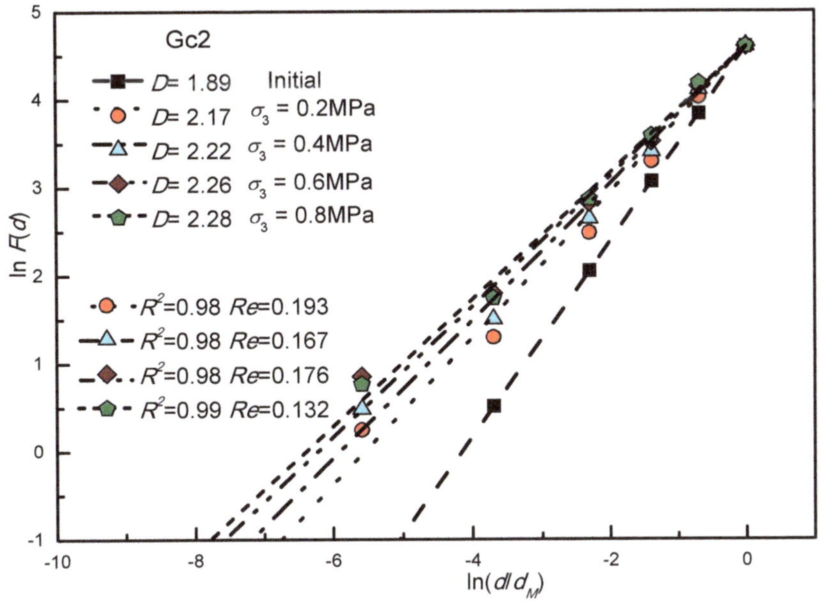

(b)

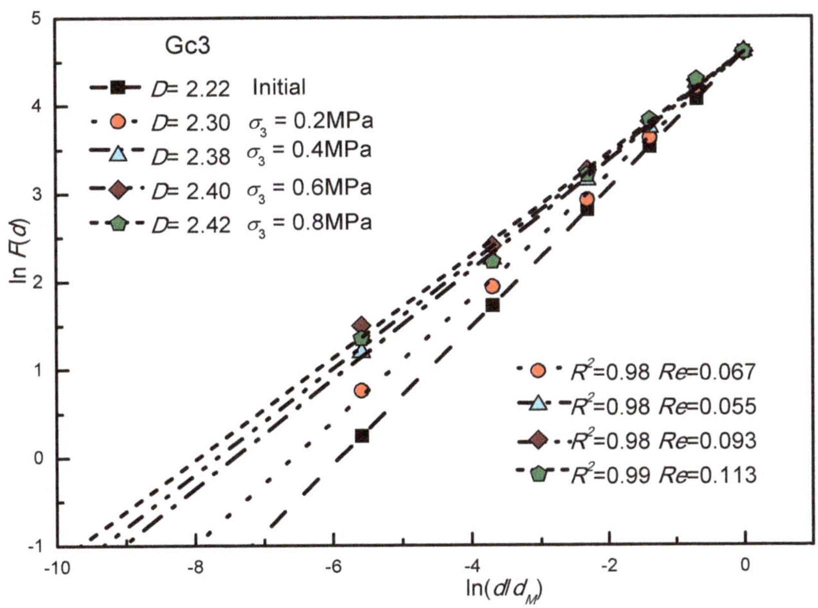

(c)

Figure 6. *Cont.*

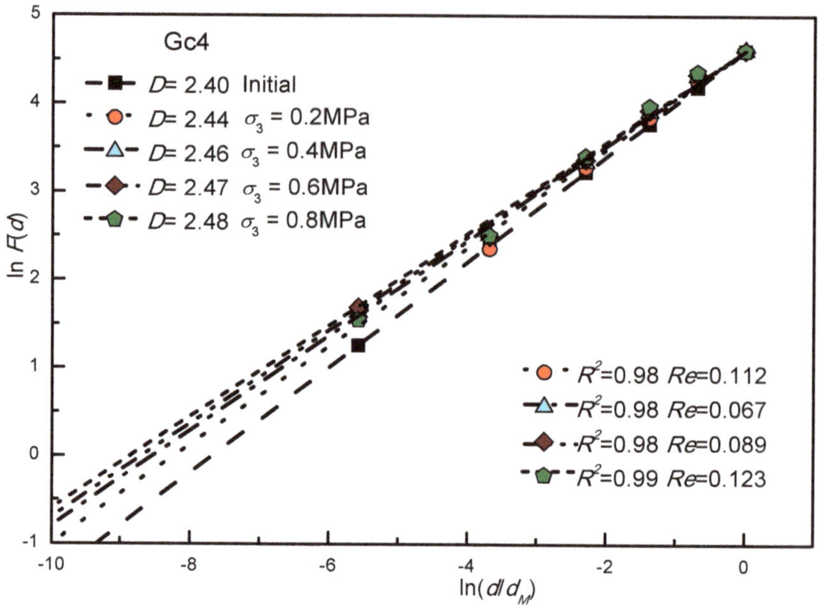

(d)

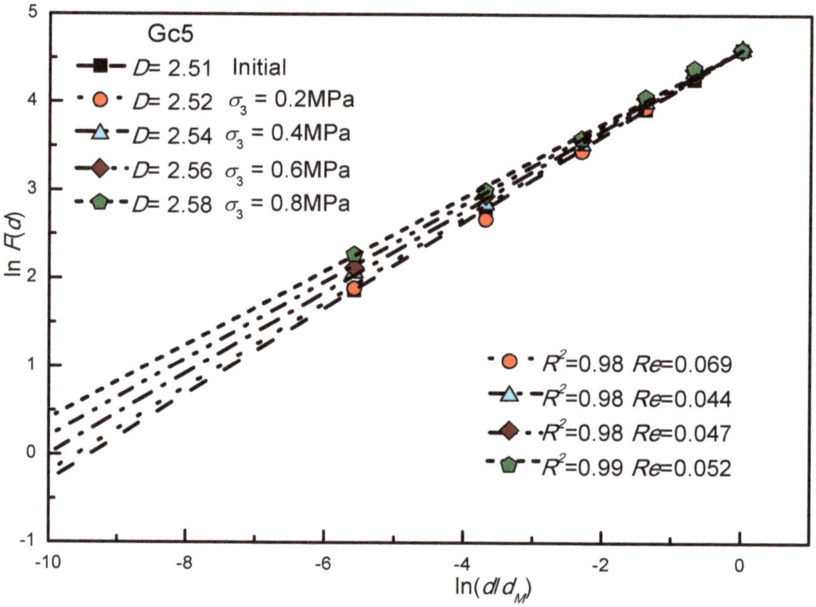

(e)

Figure 6. (**a**) Grading curves of Gc1; (**b**) Grading curves of Gc2; (**c**) Grading curves of Gc3; (**d**) Grading curves of Gc4; (**e**) Grading curves of Gc5. Grading curves of different grading curves.

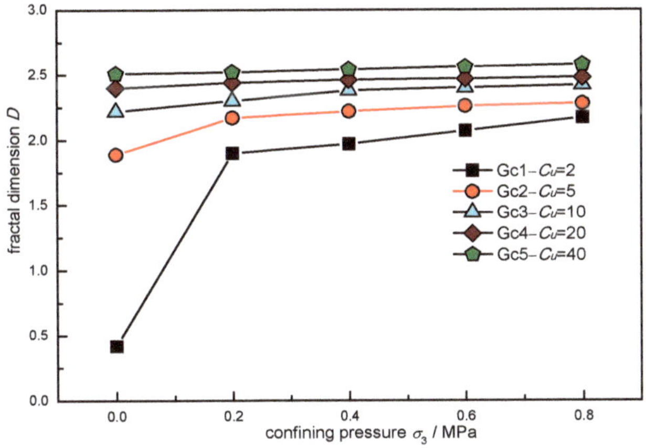

Figure 7. Fractal dimension of different grading curves.

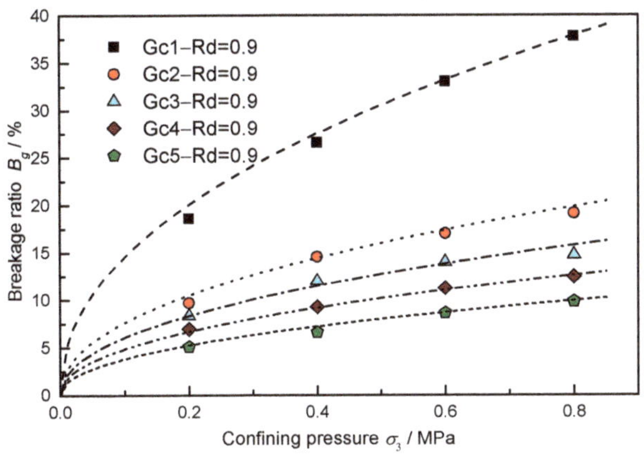

(**a**)

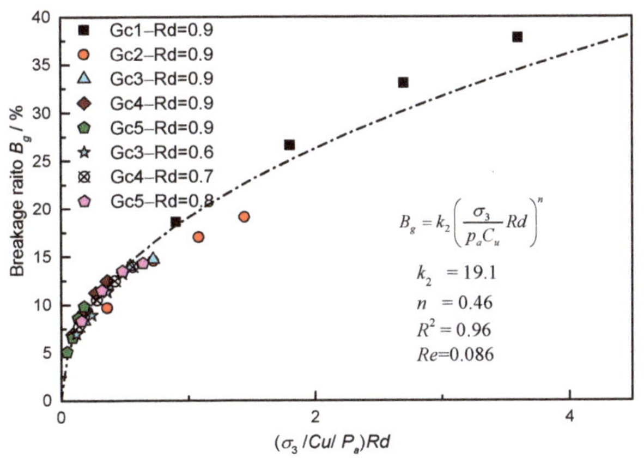

(**b**)

Figure 8. (**a**) Breakage ratio vs. confining pressure; (**b**) Breakage ratio vs. normalized confining pressure. Particle breakage of different grading curves.

Median diameter d_{50} is an important soil grading curve index. The larger the median diameter, the higher the shear strength and the more obvious is the shear expansion effect [31]. Figure 9 shows the relationship between the particle breakage ratio with d_{50} under different confining pressures after monotonic loading. The greater d_{50}, the greater is the particle breakage ratio. According to previous research, the confining pressure effect can be expressed by a power function. A normalized power function is also found to fit well the relationship between breakage ratio, d_5 and confining pressure. The expression is shown as follows:

$$B_g = k_3 \frac{d_{50}}{d_M} \left(\frac{\sigma_3}{p_a} \right)^{n_3} \quad (6)$$

where k_2 and n_3 are model parameters, determined to be 16.5 and 0.46 (which is the same as the previous value of the coefficient of uniformity).

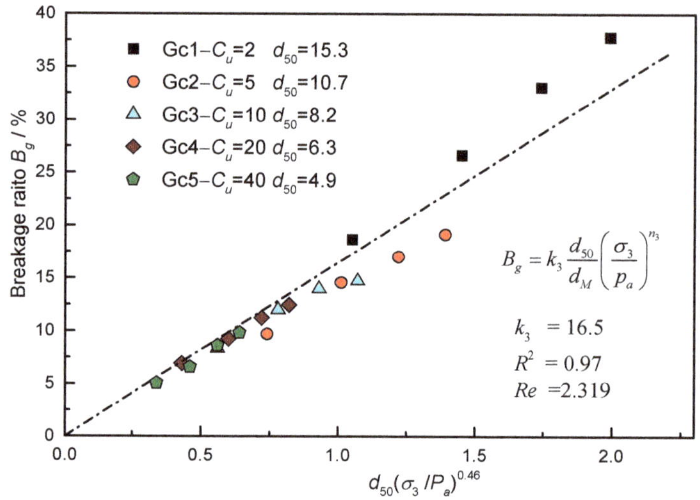

Figure 9. Breakage ratio vs. median diameter under different confining pressures.

The particle breakage is related to the relative density, uniform coefficient and stress level of the granular materials. Figure 10 shows the percentage increment of particle weight, ΔF, under different confining pressures.

$$\Delta F = P_{test} - P_{ini} \quad (7)$$

During the shear process, the percentage increment in the range of 5–10 mm is found to be almost the same under different confining pressures except the grading curve Gc1. However, the percentage increment in the range of 10–20 mm shows considerable increase with an increase in the confining pressure, relative density and decrease in coefficient of uniformity. Compared with particle percentage increment, particle breakage mainly occurs in large particles during shear tests, indicating that the particles in the range of 5–10 mm exhibit limited breakage, because the large particles can break into middle and small particles. However, further mechanisms should be analysed, combined with discrete element model (DEM) simulations.

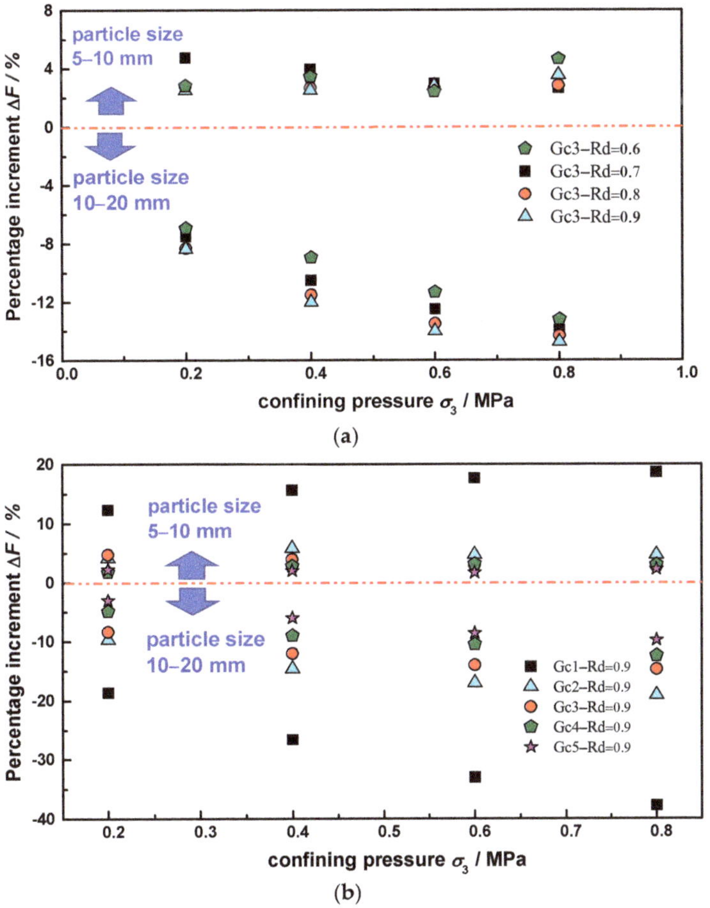

Figure 10. (**a**) Different relative density; (**b**) different grading curves. Particle percentage increment ($\Delta F > 0$, particle percentage increasing).

4. Conclusions

In this study, the influence of relative density and grading curve on the particle breakage of granular material was investigated. A series of consolidated drained triaxial tests were performed on granular materials subjected to monotonic loading. The major findings of this study are summarized below:

(1) The particle size distribution exhibited good fractal characteristics after monotonic loading of rockfill at different confining pressures. The fractal dimension increased with the increase in confining pressure. The coefficient of uniformity exhibited a greater effect on the fractal dimension than relative density.

(2) During the shearing process, the main occurrence of breakage was found to be in large particles. The extent of particle breakage increased with the increase of confining pressure and relative density, whereas it decreased with an increase in the coefficient of uniformity, which can be well described by a normalized power function. The relationship between the breakage ratio and the median diameter can be described by a linear function.

(3) The conclusions are mainly based on the results after the test. In fact, fractal dimension and particle breakage change with axial loading. Future research should focus on the

results during the shear test. The relationship between fractal dimension and particle breakage with shear modulus and volume strain should be investigated.

Author Contributions: Conceptualization, G.Y. and Y.J.; methodology, G.Y.; formal analysis, Y.S. and Y.J.; writing—original draft preparation, G.Y. and Z.C.; writing—review and editing, Y.S. and Z.C.; funding acquisition, G.Y. All authors have read and agreed to the published version of the manuscript.

Funding: This research was funded by the National Natural Science Foundation of China (Grant No. 51479059).

Informed Consent Statement: Informed consent was obtained from all subjects involved in the study. Written informed consent has been obtained from the patient(s) to publish this paper.

Data Availability Statement: All data, models, and code generated or used during the study appear in the submitted article.

Acknowledgments: The authors are indebted to the anonymous reviewers for their valuable comments and remarks that helped to improve the presentation and quality of the manuscript. The study was supported by the National Natural Science Foundation of China (Grant No. 51479059). This support is gratefully acknowledged.

Conflicts of Interest: The authors declare no conflict of interest.

References

1. Liu, J.; Liu, F.; Kong, X.; Yu, L. Large-scale shaking table model tests on seismically induced failure of Concrete-Faced Rockfill Dams. *Soil Dyn. Earthq. Eng.* **2016**, *82*, 11–23. [CrossRef]
2. Yang, G.; Yu, T.; Yang, X.; Han, B. Seismic Resistant Effects of Composite Reinforcement on Rockfill Dams Based on Shaking Table Tests. *J. Earthq. Eng.* **2017**, *21*, 1010–1022. [CrossRef]
3. Wei, K.M.; Zhu, S.; Yu, X.H. Influence of the scale effect on the mechanical parameters of coarse-grained soils. *Iran. J. Sci. Technol.—Trans. Civ. Eng.* **2014**, *38*, 75–84.
4. Huang, J.Y.; Hu, S.S.; Xu, S.L.; Luo, S.N. Fractal crushing of granular materials under confined compression at different strain rates. *Int. J. Impact Eng.* **2017**, *106*, 259–265. [CrossRef]
5. Xu, M.; Song, E.; Chen, J. A large triaxial investigation of the stress-path-dependent behavior of compacted rockfill. *Acta Geotech.* **2012**, *7*, 167–175. [CrossRef]
6. Jia, Y.; Xu, B.; Chi, S.; Xiang, B.; Xiao, D.; Zhou, Y. Particle Breakage of Rockfill Material during Triaxial Tests under Complex Stress Paths. *Int. J. Geomech.* **2019**, *19*, 04019124. [CrossRef]
7. Yan, W.; Dong, J. Effect of particle grading on the response of an idealized granular assemblage. *Int. J. Geomech. ASCE.* **2011**, *11*, 276–285. [CrossRef]
8. Yang, J.; Luo, X.D. Exploring the relationship between critical state and particle shape for granular materials. *J. Mech. Phys. Solids.* **2015**, *84*, 196–213. [CrossRef]
9. Ovalle, C.; Frossard, E.; Dano, C.; Hu, W.; Maiolino, S.; Hicher, P.Y. The effect of size on the strength of coarse rock aggregates and large rockfill samples through experimental data. *Acta Mech.* **2014**, *225*, 2199–2216. [CrossRef]
10. Varadarajan, A.; Sharma, K.G.; Venkatachalam, K.; Gupta, A.K. Testing and Modeling Two Rockfill Materials. *J. Geotech. Geoenviron. Eng.* **2003**, *129*, 206–218. [CrossRef]
11. Yang, G.; Yan, X.; Nimbalkar, S.; Xu, J. Effect of Particle Shape and Confining Pressure on Breakage and Deformation of Ar-tificial Rockfill. *Int. J. Geosynth. Gr. Eng.* **2019**, *5*, 15. [CrossRef]
12. Li, X.; Liu, J.; Li, J. Fractal dimension, particle shape, and particle breakage analysis for calcareous sand. *Bull. Eng. Geol. Environ.* **2022**, *81*, 106. [CrossRef]
13. Xiao, Y.; Liu, H.; Chen, Y.; Jiang, J.; Zhang, W. Testing and modeling of the state-dependent behaviors of rockfill material. *Comput. Geotech.* **2014**, *61*, 153–165. [CrossRef]
14. Wang, C.; Ding, X.; Xiao, Y.; Peng, Y.; Liu, H. Effects of relative densities on particle breaking behaviour of non-uniform grading coral sand. *Powder Technol.* **2021**, *382*, 524–531. [CrossRef]
15. USACE. *General Design and Construction Considerations for Earth and Rock-Fill Dams;* USACE: Washington, DC, USA, 2004; 130 p.
16. Honkanadavar, N.P.; Sharma, K.G. Testing and Modeling the Behavior of Riverbed and Blasted Quarried Rockfill Materials. *Int. J. Geomech.* **2014**, *14*, 04014028. [CrossRef]
17. Ueng, T.; Chen, T. Energy aspects of particle breakage in drained shear of sands. *GeoTechnology* **2000**, *50*, 65–72. [CrossRef]
18. Nakata, Y.; Kato, Y.; Hyodo, M.; Hyde, A.F.L.; Murata, H. one-dimensional compression behaviour of uniformly graded sand related to single particle crushing strength. *Soils Found.* **2001**, *41*, 39–51. [CrossRef]
19. Shi, D.; Zheng, L.; Xue, J.; Sun, J. DEM Modeling of Particle Breakage in Silica Sands under One-Dimensional Compression. *Acta Mech. Solida Sin.* **2016**, *29*, 78–94. [CrossRef]
20. Lade, P.V.; Bopp, P.A. Relative density effects on drained sand behavior at high pressures. *Soils Found.* **2005**, *45*, 1–13.

21. Miura, N.; Yamanouchi, T. Effect of water on the behavior of quartz-rich sand under high stresses. *Soils Found.* **1975**, *15*, 23–34. [CrossRef]
22. Kikumoto, M.; Wood, D.M.; Russell, A. Particle Crushing and Deformation Behaviour. *Soils Found.* **2010**, *50*, 547–563. [CrossRef]
23. Indraratna, B.; Sun, Y.; Nimbalkar, S. Laboratory Assessment of the Role of Particle Size Distribution on the Deformation and Degradation of Ballast under Cyclic Loading. *J. Geotechnol. Geoenviron. Eng.* **2016**, *142*, 04016016. [CrossRef]
24. Soroush, A.; Jannatiaghdam, R. Behavior of rockfill materials in triaxial compression testing. *Int. J. Civ. Eng.* **2012**, *10*, 153–161.
25. Yang, G.; Jiang, Y.; Nimbalkar, S.; Sun, Y.; Li, N. Influence of Particle Size Distribution on the Critical State of Rockfill. *Adv. Civ. Eng.* **2019**, *2019*, 8963971. [CrossRef]
26. Ministry of Water Resources of the PRC. *Specification of Soil Test*; China Water Conservancy Hydropower Publishing House: Beijing, China, 1999.
27. Ochiai, M.; Ozao, R.; Yamazaki, Y.; Holz, A. Self-similarity law of particle size distribution and energy law in size reduction of solids. *Phys. A Stat. Mech. Its Appl.* **1992**, *191*, 295–300. [CrossRef]
28. Lade, P.V.; Yamamuro, J.A.; Bopp, P.A. Relative density effects on drained and undrained strengths of sand at high pressures. In Proceedings of the 16th International Conference on Soil Mechanics and Geotechnical Engineering, Osaka, Japan, 12–16 September 2005; pp. 537–541.
29. Marsal, R.J. Large scale testing of rockfill materials. *Soil Mech Found Div.* **1967**, *93*, 27–43. [CrossRef]
30. Einav, I. Breakage mechanics—Part I: Theory. *J. Mech. Phys. Solids.* **2007**, *55*, 1274–1297. [CrossRef]
31. Zhu, Z.; Zhang, F.; Dupla, J.; Canou, J.; Foerster, E. Investigation on the undrained shear strength of loose sand with added materials at various mean diameter ratios. *Soil Dyn. Earthq. Eng.* **2020**, *137*, 106276. [CrossRef]

fractal and fractional

MDPI

Article

Influence of the Fractal Distribution of Particle Size on the Critical State Characteristics of Calcareous Sand

Xue Shen [1], Yang Shen [1,*], Junhong Xu [2] and Hanlong Liu [1,3]

[1] Key Laboratory of Ministry of Education for Geomechanics and Embankment Engineering, Hohai University, Nanjing 210024, China; 170804010002@hhu.edu.cn (X.S.); hliu@hhu.edu.cn (H.L.)
[2] School of Civil Engineering, Nanjing Forestry University, Nanjing 210037, China; jhxu@njfu.edu.cn
[3] School of Civil Engineering, Chongqing University, Chongqing 400044, China
* Correspondence: 20070060@hhu.edu.cn

Abstract: To study the influence of the fractal distribution of particle size on the critical state characteristics of calcareous sand, a type of calcareous sand from a certain reef of the South China Sea was used in this study. For comparison, standard quartz sand was also used. A series of drained shear tests on the two sands were then conducted to investigate their critical state characteristics. It was demonstrated that the fractal dimension is suitable for characterizing the particle size distribution (PSD) of calcareous sand with different fine sand content. The critical state equation of sand proposed by Li and Wang (1998) is suitable for fitting the critical state line of calcareous sand. In the plane of deviatoric stress versus the effective confining pressure (q–p' plane) and the plane of void ratio versus $(p'/p_a)^\alpha$, the critical state lines of calcareous sand are always above those of quartz sand. The critical state lines of calcareous sand with different fractal dimensions in the q–p' plane are unique. However, in the e–$(p'/p_a)^\alpha$ plane, the critical state lines appear to rotate anticlockwise as the fractal dimension increases. In addition, there is an "intersection" in the e–$(p'/p_a)^\alpha$ plane. Considering the influence of the fractal distribution of particle size, an expression for the critical state line of calcareous sand in the e–$(p'/p_a)^\alpha$ plane was proposed. The related constitutive model was also revised, where a complete set of model parameters suitable for modeling calcareous sand was provided.

Keywords: calcareous sand; fractal dimension; particle size distribution; critical state; shear characteristic

Citation: Shen, X.; Shen, Y.; Xu, J.; Liu, H. Influence of the Fractal Distribution of Particle Size on the Critical State Characteristics of Calcareous Sand. *Fractal Fract.* **2022**, 6, 165. https://doi.org/10.3390/fractalfract6030165

Academic Editor: Zine El Abiddine Fellah

Received: 16 February 2022
Accepted: 14 March 2022
Published: 17 March 2022

1. Introduction

As the main foundation material for construction in the South China Sea, calcareous sand is a type of marine biogenic sand. Its chemical composition is different from that of common terrestrial sand and marine sand; the calcium carbonate content is more than 50%. It is characterized by an extremely irregular particle shape, high friction angle, high porosity (including internal porosity), high compressibility, brittleness of grains, and so on [1–10]. The calcareous sand from the South China Sea is composed of coral debris, and the calcium carbonate content is mostly above 90% [4]. These characteristics make its basic mechanical properties more complicated than those of quartz sand [11–13]. It is of great significance to study its mechanical properties. Recent research on the calcareous sand in the South China Sea found that its particle breakage is far less than that in previous research under the same load condition [14]. This may be related to its generation environment, hydraulic fill method, and so on [5,15]. Compared with the original calcareous sand, the dredger fill calcareous sand is not easily broken because of the breakage during hydraulic fill. In addition, the confining pressure plays an important role in the drained shear properties of calcareous sand [16–18]. Compared with quartz sand, the volumetric deformation, internal friction angle, and shear strength of calcareous sand are larger [19–21].

The critical state refers to the state in which the axial deformation of soil develops continuously without volumetric change under constant stress [22]. It is often used to design and evaluate the long-term strength of buildings (structures) in engineering practice. The design of filled islands also relies on the concept of critical state strength [23,24]. Early studies based on clay assumed that the critical state line is a straight line in the plane of void ratio versus the logarithm of the effective confining pressure (e–lgp' plane) [22]. However, for granular materials, it has been found that the critical state line is not a straight line. Thus, a series of critical state equations have been introduced [25–27]. Li and Wang [28] suggested that the critical state line of sandy soil is approximately a straight line in the e–$(p'/p_a)^{\alpha}$ plane based on the data from previous research. Based on this, the critical state line of quartz sand can be well fit within the stress range of 10~500 kPa. This conclusion is widely used [29]. However, calcareous sand is multangular and irregular, leading to more complicated mechanical properties compared to quartz sand. It remains to be studied whether the existing critical state theory is suitable for calcareous sand.

Recently, fractal theory has been widely used in respectively describing and studying the particle size distribution, the pore size distribution, the particle morphology, and so on [30–35]. The particle size distribution (PSD) has a significant impact on the critical state characteristics of sand [36,37]. Particle breakage can lead to a sharp increase in fine sand content. Moreover, seepage can lead to a significant decrease in fine sand content. Both of these two conditions will significantly change the fine sand content of calcareous sand. Then, the critical state characteristics of calcareous sand are affected. Therefore, it is necessary to study the effect of fine sand content on the triaxial shear and critical state characteristics of calcareous sand. However, the effect of the fractal distribution of particle size on the critical state characteristics of calcareous sand is still poorly understood.

To reveal the particularity of calcareous sand, quartz sand was used as a reference for comparison in this study. Triaxial tests of the two sands were performed to investigate their critical state characteristics. The applicability of the critical state theory to calcareous sand was analyzed. The influence of the fractal distribution of particle size on the critical state characteristics of calcareous sand was studied. The relationship between the fractal dimension and the critical state characteristics of calcareous sand was established. Based on this, the related constitutive model was revised. A complete set of model parameters suitable for calcareous sand was given.

2. Materials and Methods

The calcareous sand used in this study was from a certain reef of the South China Sea. The influence of the content of particles smaller than 0.25 mm ($P_{0.25}$) is researched in this paper. When $P_{0.25}$ is changed, the relative content of the rest of the particle fractions remains the same. According to this method, the designed gradation is similar to the actual gradation. Table 1 shows the test grain size distribution of the calcareous sand and quartz sand. The gradation in bold is the actual gradation. The gradation curves of the calcareous sand with different $P_{0.25}$ are shown in Figure 1.

Table 1. Test grain size distribution of the calcareous sand and quartz sand.

Particle Fraction (mm)	5~2	2~1	1~0.5	0.5~0.25	0.25~0.075	<0.075
	9.9%	7.3%	33.5%	34.3%	14.2%	0.8%
	8.8%	6.5%	29.5%	30.2%	23.7%	1.3%
Calcareous sand	**7.6%**	**5.6%**	**25.6%**	**26.2%**	**33.1%**	**1.9%**
	6.4%	4.7%	21.7%	22.2%	42.6%	2.4%
	5.3%	3.9%	17.7%	18.1%	52.1%	2.9%
Quartz sand	7.6%	5.6%	25.6%	26.2%	33.1%	1.9%

To study the specific calcareous sand, quartz sand was adopted as a reference for comparison. Table 2 shows the basic property parameters of the calcareous sand and quartz sand with the same gradation ($P_{0.25}$ = 35%).

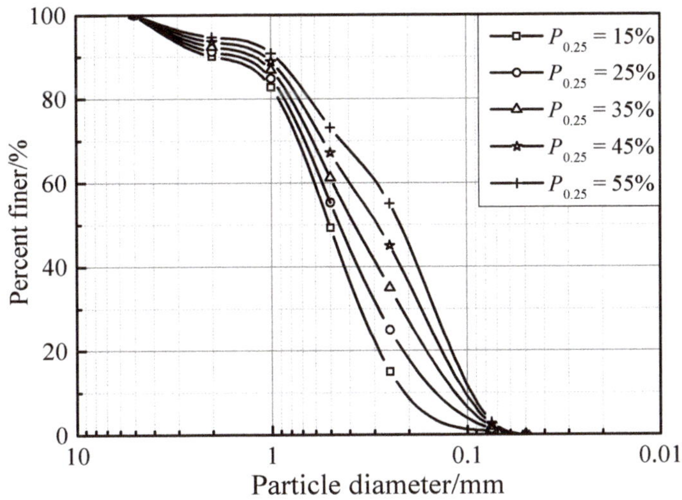

Figure 1. Grain size distribution curve of the tested calcareous sand.

Table 2. Basic property parameters of the test sands ($P_{0.25}$ = 35%).

Sample	Coefficient of Uniformity (C_u)	Coefficient of Curvature (C_c)	Maximum Void Ratio (e_{max})	Minimum Void Ratio (e_{min})	Specific Gravity (Gs)
Calcareous sand	4.8	0.9	1.19	0.80	2.70
Quartz sand	4.8	0.9	0.75	0.39	2.67

Figure 2 presents the 3-D morphology characteristics of the tested calcareous sand and quartz sand particles according to the CT scanner. It can be seen that, compared with quartz sand, the particles of calcareous sand are multiangular, flatter, and more irregular. The calcareous sand particles have more internal pores and their surface is uneven. When the calcareous sand is compressed with a vertical load of 4000 kPa, the relative breakage B_r [38] is only 0.7% (0~1) [39]. This shows that the particle breakage of this type of calcareous sand is insignificant and can be ignored. This conclusion is similar to that of the aforementioned research [14]. Therefore, this paper focuses on exploring the influence of the fractal distribution of particle size on the critical state characteristics of calcareous sand, without considering the impact of particle breakage.

A strain path triaxial apparatus was used to conduct a series of drained shear tests (CD tests). The shear characteristics of the calcareous sand with different fractal distributions of particle size were studied. The diameter and height of the test samples were 61.8 and 135 mm, respectively. The calcareous sand used in the test was saturated by applying head pressure and back pressure. When the B-value (i.e., saturation) [40] was greater than 0.98, the sample was considered saturated. Four effective confining pressures (100, 200, 300, and 400 kPa) were conducted. The tests were conducted using a constant vertical shear rate of 0.016mm/min until the axial strain achieved 25%. It should be noted that the average value of the critical state data in the last 3% strain range was used as the final value.

Four drained shear tests for quartz sand ($P_{0.25}$ = 35%) with different effective confining pressures were conducted as comparisons. The relative densities of the sand samples were all 40%.

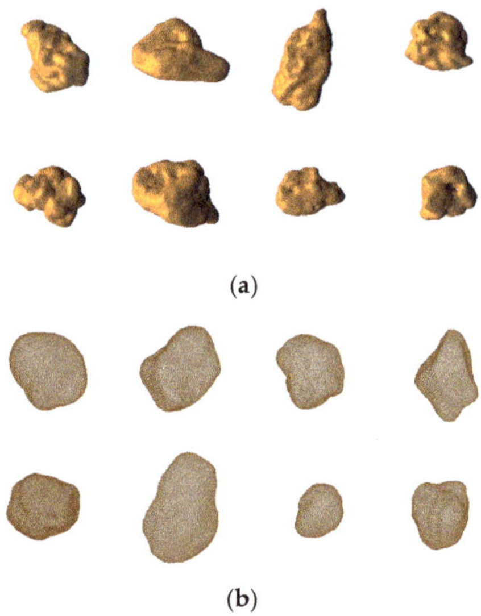

Figure 2. CT scan images of the tested calcareous sand and quartz sand particles: (**a**) calcareous sand particles; (**b**) quartz sand particles.

3. Experiments

3.1. Fractal Behavior of Particle Size Distribution

Recently, fractal theory has been widely used in respectively describing and studying the geometric features of the PSD. Tyler et al. [41] proposed the fractal model of particle size distribution. It was expressed with the relationship between the cumulative mass and particle size of the particles, as follows:

$$\frac{M(\delta < d_i)}{M_t} = \left(\frac{d_i}{d_{\max}}\right)^{3-D_{\mathrm{m}}} \tag{1}$$

The fractal dimension D_{m} can be determined by:

$$\lg\left(\frac{M(\delta < d_i)}{M_t}\right) = (3 - D_{\mathrm{m}})\lg(d_i) + (D_{\mathrm{m}} - 3)\lg(d_{\max}) \tag{2}$$

where δ is the particle size, d_i is the diameter of the ith (i = 1, 2, ..., n) sieve pore, $M(\delta < d_i)$ is the cumulative mass of the sand particles whose grain size is smaller than d_i, M_t is the total mass of sand particles, $d_{\max}$ is the maximum grain size, and D_{m} is the fractal dimension.

However, only one fractal dimension is not enough to describe the distribution for the entire range of particle sizes [42,43]. Posadas et al. [42] applied multifractal techniques to describe contrasting PSDs. The dominant particles of calcareous sand are distributed in the large size range (i.e., the particle size greater than 0.075 mm). As shown in Figure 3, the fractal dimension of the PSD (i.e., the particle size greater than 0.075 mm) was obtained through the regression analysis based on Equation (2). The fitting results of fractal dimensions reveal that the particle size distribution of calcareous sand complies with the fractal law.

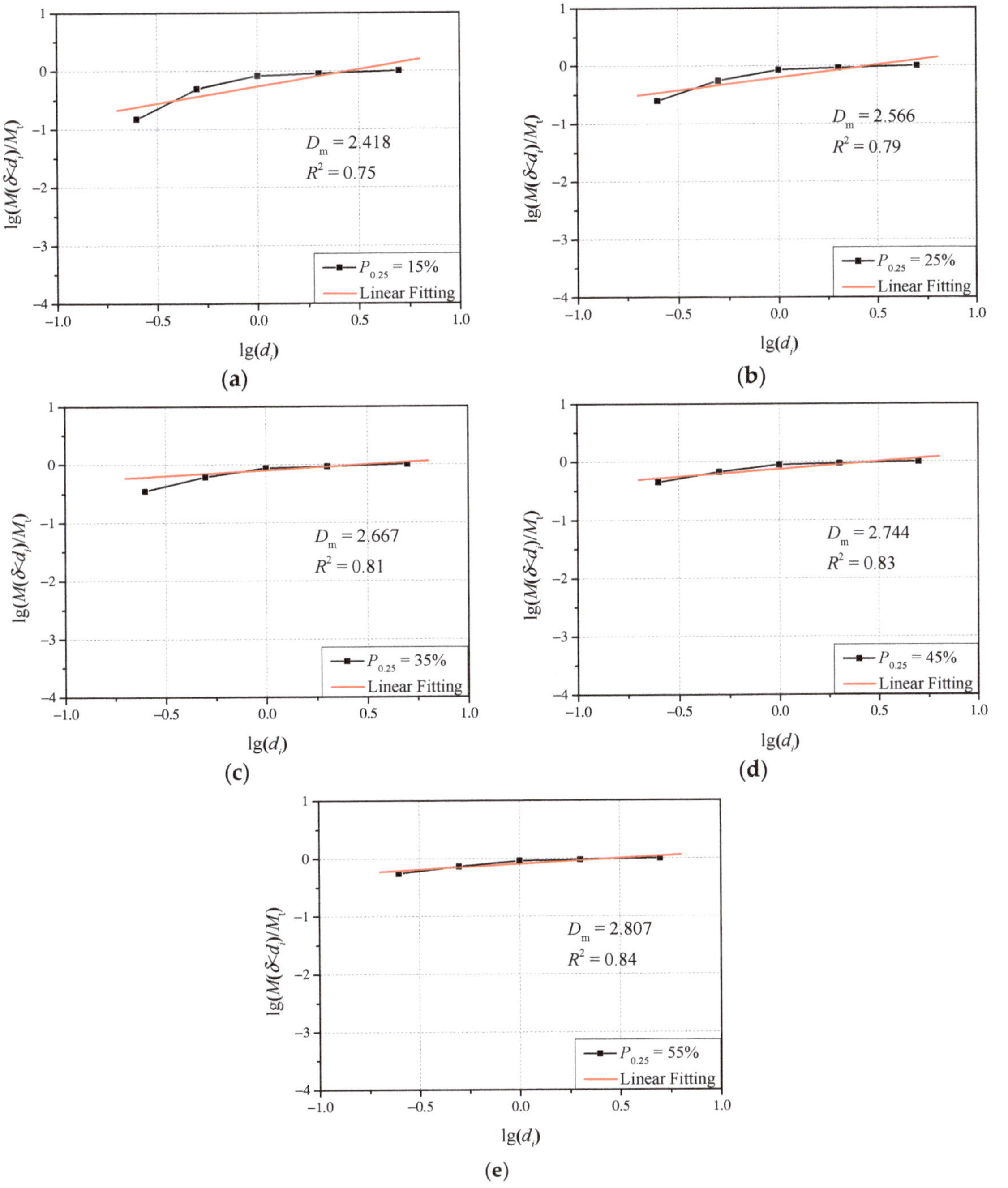

Figure 3. Determination of fractal dimensions of calcareous sand with different gradation: (**a**) content of particles smaller than 0.25 mm is 15%; (**b**) content of particles smaller than 0.25 mm is 25%; (**c**) content of particles smaller than 0.25 mm is 35%; (**d**) content of particles smaller than 0.25 mm is 45%; (**e**) content of particles smaller than 0.25 mm is 55%.

Figure 4 shows the fractal dimensions of the tested calcareous sand with different gradation. As is shown, the higher the content of fine materials in the PSD is, the larger the value of D_m is. The relationship between the $P_{0.25}$ and D_m is almost linear ($R^2 = 0.96$). Therefore, the fractal dimension can be well used to characterize the gradation of the calcareous sand with different fine sand content.

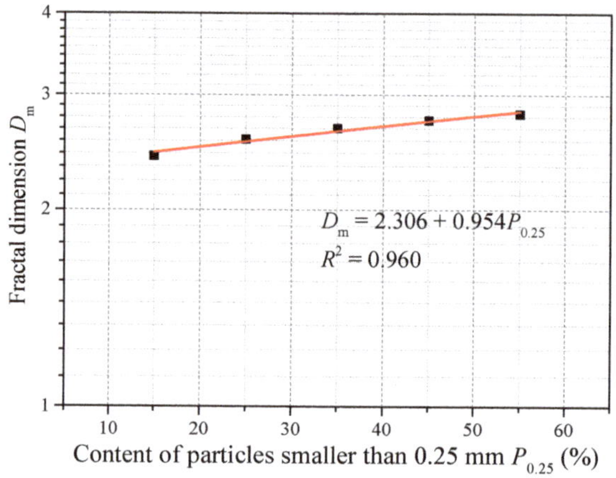

Figure 4. Fractal dimensions of the calcareous sand with different gradation.

3.2. Critical State Characteristics of the Calcareous Sand

Figure 5 shows the critical state lines in the plane of deviatoric stress versus the effective confining pressure (q–p' plane) of the calcareous sand and quartz sand ($D_m = 2.667$).

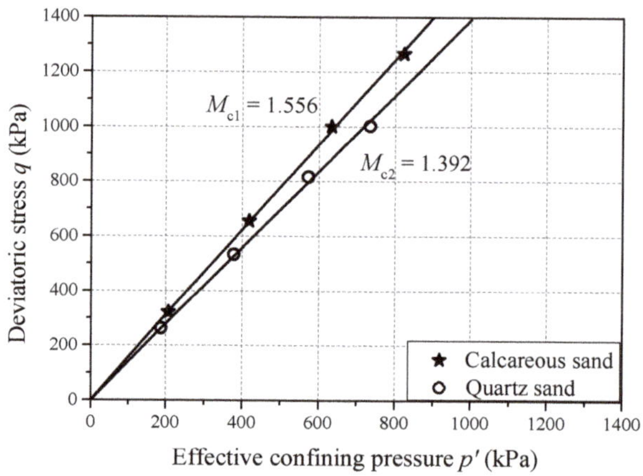

Figure 5. The critical state lines in the q–p' plane of the calcareous sand and quartz sand ($D_m = 2.667$).

As is shown, the critical stress ratio of calcareous sand ($M_{c1} = 1.556$) is approximately 12% higher than that of quartz sand ($M_{c2} = 1.392$). The critical stress ratio reflects the particle interlocking. Its value depends primarily on the particle mineralogy and shape (angularity) [44]. The particles of calcareous sand are more angular than those of quartz sand, leading to stronger particle interlocking. As the damaged calcareous sand reached the critical state, q reached a constant value. The initial sand fabric was destroyed at this condition. The residual strength of calcareous sand is larger due to the stronger particle interlocking.

The sand critical state equation put forward by Li and Wang [28] is used to fit the critical state line of calcareous sand in the e–p' plane, as shown in the following Equation (3):

$$e_c = e_{c0} - \lambda\left(\frac{p'}{p_a}\right)^{\alpha} \tag{3}$$

where e_c is the critical void ratio; e_{c0}, λ, and α are the material constants determining the critical state line in the e–p' plane [28], and p_a is the atmospheric pressure for normalization. α is a material parameter affected by the material type and the modulus type. The effect of the value of α on the straightness of the lines is mild; α can be simply chosen as a default value [28]. For the calcareous sand analyzed in this article, α can be taken as 0.55. For the quartz sand analyzed in this article, α can be taken as 0.71. Based on this, the critical state lines in the e–$(p'/p_a)^{\alpha}$ plane of the calcareous sand and quartz sand (D_m = 2.667) are as shown in Figure 6.

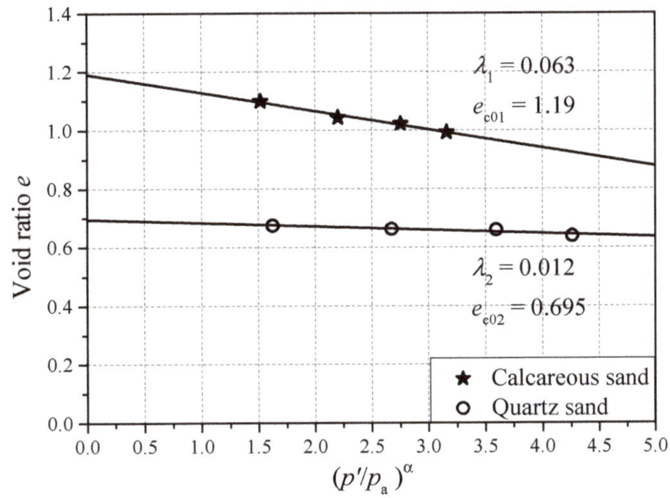

Figure 6. The critical state lines in the e–$(p'/p_a)^{\alpha}$ plane of the calcareous sand and quartz sand (D_m = 2.667).

As is shown in Figure 6, the critical state lines of the calcareous sand and quartz sand can be plotted as nearly straight in the e versus $(p'/p_a)^{\alpha}$ plane proposed by Li and Wang [28]. The critical state line of calcareous sand is always above that of quartz sand with the same gradation. The critical state parameter λ and e_{c0} values of the calcareous sand are all larger than those of the quartz sand. The higher initial void ratio of the calcareous sand results in a larger e_{c0}. Moreover, the larger slope λ indicates the greater influence of confining pressure on the critical state void ratio of calcareous sand [39].

3.3. Influence of the Fractal Distribution of Particle Size on the Critical State Line in the q–p′ Plane of the Calcareous Sand

Figure 7 shows the critical state line in the q–p' plane of the calcareous sand with different fractal dimensions.

As is shown, the critical state points of calcareous sand with different fractal dimensions basically fall along the same critical state line (M_c = 1.575) in the q–p' plane (R^2 = 92.61%). Therefore, the critical state lines of calcareous sand with different fractal dimensions in the q–p' plane are unique. The critical stress ratio M_c of the calcareous sand with different fractal dimensions is a constant. This conclusion is the same as those of other sands in related research [45,46].

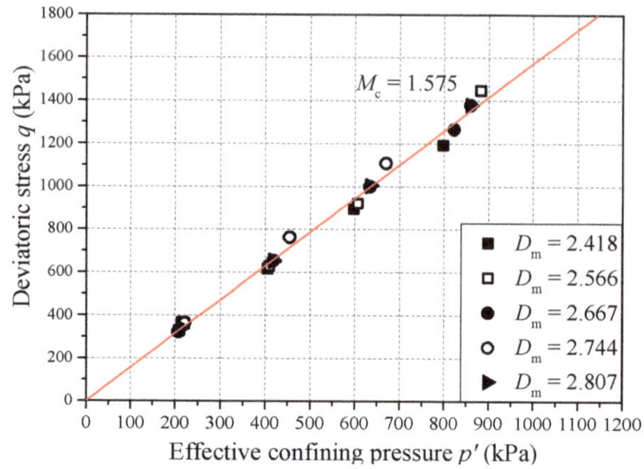

Figure 7. The critical state line in the q–p' plane of the calcareous sand with different fractal dimensions.

3.4. Influence of the Fractal Distribution of Particle Size on the Critical State Line in the e–p′ Plane of the Calcareous Sand

The critical state lines in the e–$(p'/p_a)^\alpha$ plane of the calcareous sand with different fractal dimensions are shown in Figure 8.

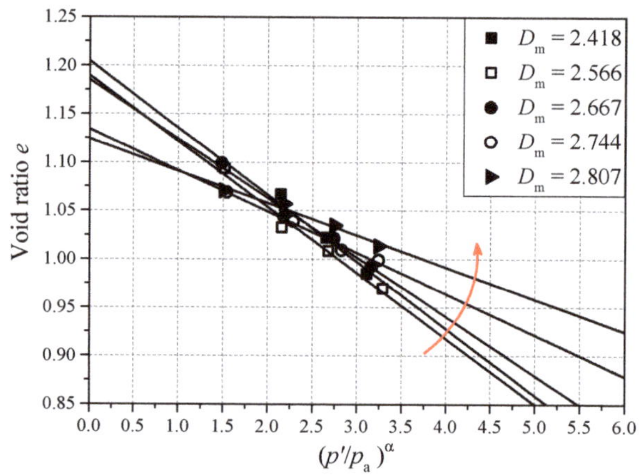

Figure 8. The critical state lines in the e–$(p'/p_a)^\alpha$ plane of the calcareous sand with different fractal dimensions.

It can be found that with the increase in fractal dimension, the critical state parameter λ and e_{c0} of calcareous sand both decrease. The same is true for the Hoston sand. However, the slope λ of the critical state lines of the calcareous sand changes more significantly than that of the Hoston sand [36].

With the increase in fractal dimension, the fine sand content of calcareous sand increases. The pores formed by the coarse particles are filled by the fine particles. The initial void ratio of the calcareous sand decreases and the critical state parameter e_{c0} decreases. The smaller initial void ratio leads to a weaker influence of confining pressure on the critical state void ratio of calcareous sand, and the slope λ of the critical state line of the calcareous sand decreases.

Figure 8 also shows that the critical state lines in the e–$(p'/p_a)^\alpha$ plane appear to rotate anticlockwise approximately as the fractal dimension increases. When $(p'/p_a)^\alpha$ is equal to 2.2, the critical state void ratios of calcareous sand with different fractal dimensions are around 1.05. This means that there is an "intersection" in the e–$(p'/p_a)^\alpha$ plane. When the effective confining pressure p' is equal to 425 kPa, the critical void ratios of the calcareous sand with different fractal dimensions are approximately the same.

Based on this, the critical state of calcareous sand in the e–$(p'/p_a)^\alpha$ plane can be expressed as:

$$e_c - b = -\lambda\left[\left(\frac{p\prime}{p_a}\right)^\alpha - a\right] \tag{4}$$

$$e_c = b + a\lambda - \lambda\left(\frac{p\prime}{p_a}\right)^\alpha \tag{5}$$

where a and b are material constants. They are also the "intersection point" coordinates of the critical state lines in the e–$(p'/p_a)^\alpha$ plane.

By combining Equations (3) and (5), the parameter e_{c0} can be written as:

$$e_{c0} = b + a\lambda \tag{6}$$

Figure 9 shows the relationship between the reference void ratio e_{c0} and slope of critical state lines λ.

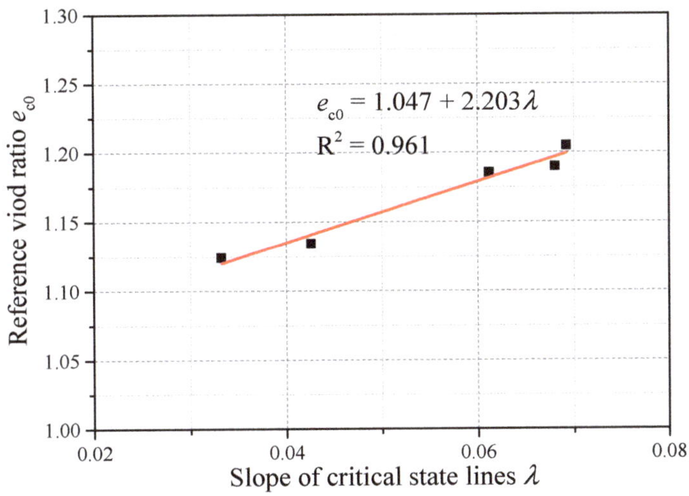

Figure 9. Relationship between reference void ratio e_{c0} and slope of critical state lines λ.

As is shown in Figure 9, there is indeed a linear relationship between the reference void ratio e_{c0} and slope of critical state lines λ ($R^2 = 0.961$). For the calcareous sand analyzed in this article, a is 2.203 and b is 1.047.

Figure 10 presents the relationship between the slope of critical state lines λ and fractal dimension D_m.

Second-order functions can be applied to fit the relationship, as follows:

$$\lambda = cD_m{}^2 + dD_m + f \tag{7}$$

where c, d, and f are material constants. For the calcareous sand analyzed in this paper, c is -0.384, d is 1.908, and f is -2.302.

Figure 11 shows the comparison of material constants determined by the model with experimental data.

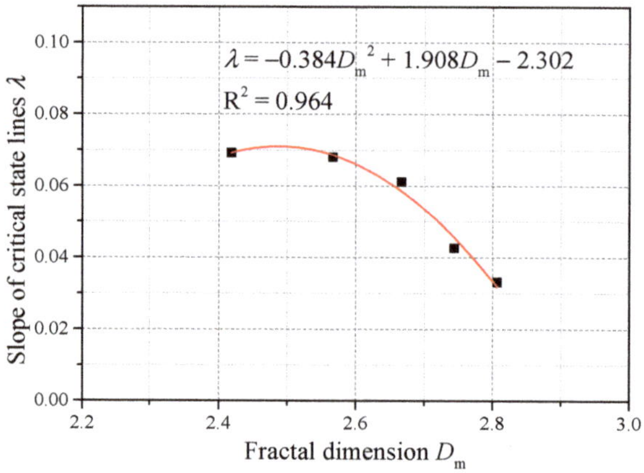

Figure 10. Relationship between slope of critical state lines λ and fractal dimension D_m.

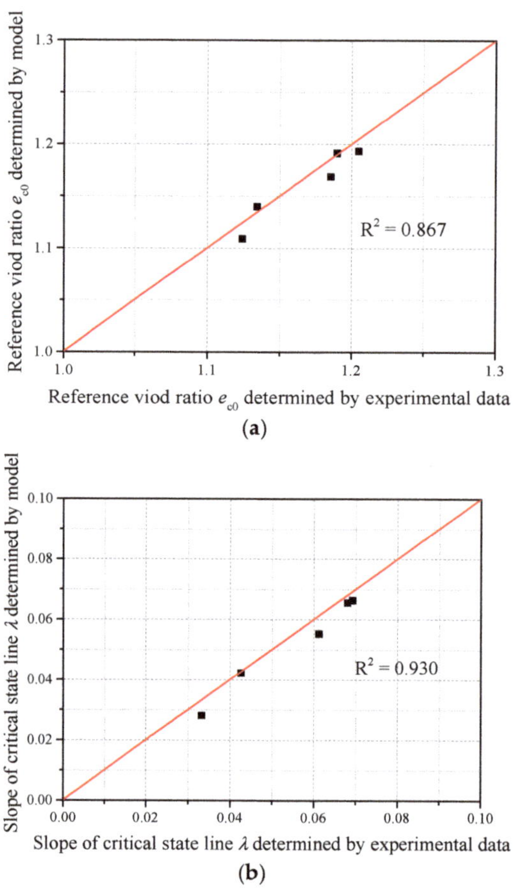

Figure 11. Comparison of material constants determined by the model with experimental data: (**a**) reference void ratio e_{c0}; (**b**) slope of critical state lines λ.

As can be seen, the material constants determined by the model and those determined by the experimental data are highly correlated. The model expressed as Equations (4)–(7) can be used to predict the material parameters of calcareous sand with different fractal dimensions.

By combining Equations (5) and (7), considering the influence of the fractal distribution of particle size, an expression for the critical state of calcareous sand in the e–$(p'/p_a)^\alpha$ plane is proposed:

$$e_c = b + a\left(cD_{\mathrm{m}}^2 + dD_{\mathrm{m}} + f\right) - \left(cD_{\mathrm{m}}^2 + dD_{\mathrm{m}} + f\right)\left(\frac{p\prime}{p_a}\right)^\alpha \tag{8}$$

4. Theoretical Model

The expression of state-dependent dilatancy put forward by Li et al. [47] was used in this paper, as follows:

$$d = d_0\left(e^{m\psi} - \frac{\eta}{M}\right) \tag{9}$$

where d_0 and m are two modeling parameters, η is the stress ratio, and Ψ is a state parameter. Ψ is defined as:

$$\psi = e - e_c \tag{10}$$

Considering the influence of the fractal distribution of particle size, e_c can be expressed by Equation (8). The state parameter Ψ is proposed:

$$\psi = e - b - a\left(cD_{\mathrm{m}}^2 + dD_{\mathrm{m}} + f\right) + \left(cD_{\mathrm{m}}^2 + dD_{\mathrm{m}} + f\right)\left(\frac{p\prime}{p_a}\right)^\alpha \tag{11}$$

The state-dependent constitutive model established by Li et al. [47] was used in this paper, as follows:

$$\left\{\begin{array}{c} dq \\ dp\prime \end{array}\right\} = \left[\left(\begin{array}{cc} 3G & 0 \\ 0 & K \end{array}\right) - \frac{h(L)}{K_p + 3G - K\eta d}\left(\begin{array}{cc} 9G^2 & -3KG\eta \\ 3KGd & -K^2\eta d \end{array}\right)\right]\left(\begin{array}{c} d\varepsilon_q \\ d\varepsilon_v \end{array}\right) \tag{12}$$

where G is the elastic shear modulus, K is the elastic bulk modulus, L is a loading index, K_p is the plastic modulus, and $h(L)$ is a Heaviside function with $h(L) = 1$ for $L > 0$ and $h(L) = 0$ otherwise.

The elastic shear modulus G can be calculated by the following empirical equation [48]:

$$G = G_0\frac{(2.97 - e)^2}{1 + e}\sqrt{p\prime p_a} \tag{13}$$

where G_0 is a material constant, and e_{c0} is the initial void ratio. The elastic bulk modulus K can be calculated based on elasticity theory, as follows:

$$K = G\frac{2(1 + v)}{3(1 - 2v)} \tag{14}$$

where ν is the Poisson's ratio.

The constitutive relation is put forward to express the plastic modulus K_p:

$$K_p = hG\left(\frac{M}{\eta} - e^{n\psi}\right) \tag{15}$$

$$h = h_1 - h_2 e \tag{16}$$

where h_1, h_2, and n are three material constants.

5. Model Validation

Figure 12 presents the relationship between the phase transformation state stress ratio and peak state stress ratio and the state parameter of calcareous sand. As is shown, as the ψ increases, the M_d increases and the M_p decreases. Equations proposed by Li and Dafalias [47] are used to quantify such evolutions, as follows:

$$M_d = A_d M_c \exp(m\psi) \quad (17)$$

$$M_p = A_p M_c \exp(-n\psi) \quad (18)$$

where A_d and A_p are the correction factor. m and n are material constants. For the calcareous sand analyzed in this article, m is 0.7438 and n is 0.5698.

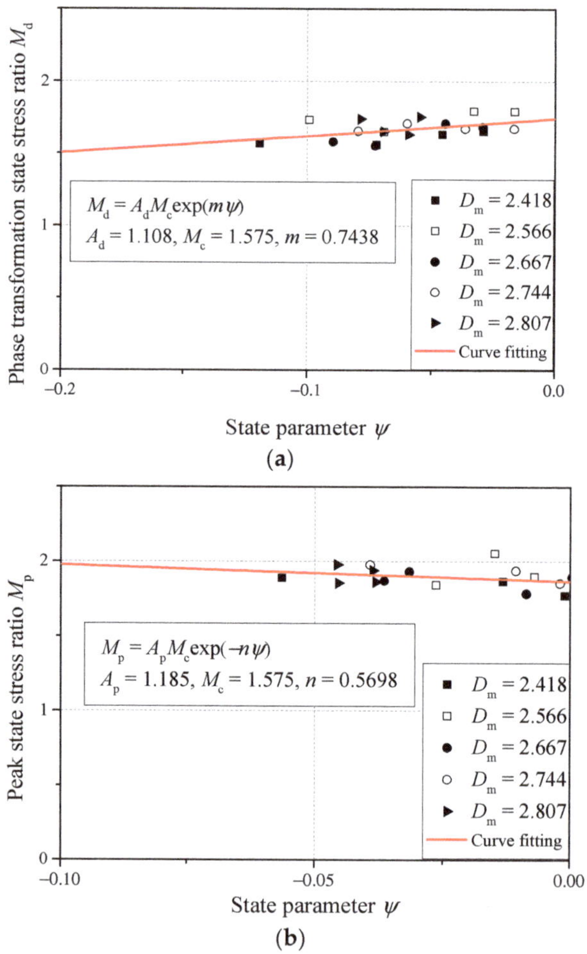

Figure 12. Relationship between state stress ratio and state parameter: (**a**) phase transformation state stress ratio; (**b**) peak state stress ratio.

As shown in Table 3, fourteen material constants are used in the model. All these parameters can be calibrated by a systematic procedure based on triaxial data.

Table 3. Model parameters calibrated for calcareous sand.

Elastic Parameters	Critical State Parameters		Dilatancy Parameters	Hardening Parameters
G_0 =125	M = 1.58	d = 1.91	d_0 = 6.05	h_1 = 3.35
ν = 0.30	a = 2.20	f = −2.30	m = 0.74	h_2 = 3.05
	b = 1.05	α = 0.55		n = 0.57
	c = −0.38			

6. Conclusions

To study the effect of the fractal distribution of particle size on the critical state characteristics of calcareous sand, a type of calcareous sand from the South China Sea was adopted in this study. For comparison, standard quartz sand was also used. A series of drained shear tests on the two sands were performed to research their critical state characteristics. Conclusions were drawn as follows:

(1) The particle size distribution of calcareous sand can obey the fractal law. The higher the fine sand content in the PSD was, the larger the value of D_{m} was. The fractal dimension can be used to characterize the particle size distribution of calcareous sand with different fine sand content.
(2) The critical stress ratio of calcareous sand (M_c = 1.556) was around 12% higher than that of quartz sand (M_c = 1.392), and its critical state line parameters λ and e_{c0} were also larger than those of quartz sand. In the q–p' plane and e–$(p'/p_a)^{\alpha}$ plane, the critical state lines of calcareous sand were always above those of the quartz sand.
(3) The critical state lines of calcareous sand with different fractal dimensions in the q–p' plane were unique.
(4) The critical state equation of sand proposed by Li and Wang [28] was suitable for the fitting of the critical state line of calcareous sand in the e–$(p'/p_a)^{\alpha}$ plane.
(5) In the e–$(p'/p_a)^{\alpha}$ plane, the critical state lines appeared to rotate anticlockwise as the fractal dimension increased. There was an "intersection" in the e–$(p'/p_a)^{\alpha}$ plane. When the effective confining pressure p' was equal to 425 kPa, the critical void ratios of the calcareous sand with different fractal dimensions were approximately the same. Considering the influence of the fractal distribution of particle size, an expression for the critical state of calcareous sand in the e–$(p'/p_a)^{\alpha}$ plane was proposed.
(6) Based on the critical state characteristics of calcareous sand with different fractal dimensions, the dilatancy equation and constitutive model suitable for calcareous sand were proposed. Moreover, a complete set of model parameters suitable for calcareous sand was given.

Author Contributions: Conceptualization, Y.S. and H.L.; methodology, X.S. and Y.S.; formal analysis, X.S. and J.X.; investigation, X.S. and J.X.; resources, Y.S. and H.L.; writing—original draft preparation, X.S. and Y.S.; funding acquisition, Y.S. and X.S. All authors have read and agreed to the published version of the manuscript.

Funding: This work was financially supported by the National Natural Science Foundation of China (No. 51979087), the Postgraduate Research and Practice Innovation Program of Jiangsu Province (KYCX20_0435), the Fundamental Research Funds for the Central Universities (No. B200203080), and the China Scholarship Council (No. 202006710143).

Institutional Review Board Statement: Not applicable.

Informed Consent Statement: Not applicable.

Data Availability Statement: Not applicable.

Acknowledgments: The authors would like to thank Yifei Sun from Ruhr-University Bochum for his kind suggestions and help.

Conflicts of Interest: The authors declare no conflict of interest.

References

1. Coop, M.R. The mechanics of uncemented carbonate sands. *Géotechnique* **1990**, *40*, 607–626. [CrossRef]
2. Datta, M.; Gulhati, S.K.; Rao, G.V. Crushing of calcareous sands during drained shear. *Soc. Pet. Eng. J.* **1980**, *20*, 77–85. [CrossRef]
3. Yasser, D.; Habib, S.; Hossein, S.; Reza, R. Compressibility and undrained behavior of hormuz calcareous sand. *Electron. J. Geotech. Eng.* **2010**, *15*, 1684–1702.
4. Liu, C.Q.; Yang, Z.Q.; Wang, R. Preliminary research on physical and mechanical properties of calcareous sand. *Rock Soil Mech.* **1998**, *19*, 32–37.
5. Shahnazari, H.; Rezvani, R. Effective parameters for the particle breakage of calcareous sands: An experimental study. *Eng. Geol.* **2013**, *159*, 98–105. [CrossRef]
6. Wang, G.; Wang, Z.; Ye, Q.; Zha, J. Particle breakage evolution of coral sand using triaxial compression tests. *J. Rock Mech. Geotech. Eng.* **2021**, *13*, 321–334. [CrossRef]
7. Cao, Z.; Chen, J.; Ye, X.; Gu, C.; Guo, Z.; Cai, Y. Experimental study on particle breakage of carbonate gravels under cyclic loadings through large-scale triaxial tests. *Transp. Geotech.* **2021**, *30*, 100632. [CrossRef]
8. Xiao, Y.; Yuan, Z.; Chu, J.; Liu, H.L.; Huang, J.; Luo, S.N.; Wang, S.; Lin, J. Particle breakage and energy dissipation of carbonate sands under quasi-static and dynamic compression. *Acta Geotech.* **2019**, *14*, 1741–1755. [CrossRef]
9. Altuhafi, F.; O'Sullivan, C.; Cavarretta, I. Analysis of an image-based method to quantify the size and shape of sand particles. *J. Geotech. Geoenviron. Eng.* **2013**, *139*, 1290–1307. [CrossRef]
10. Wang, X.Z.; Jiao, Y.Y.; Wang, R.; Hu, M.J.; Meng, Q.S.; Tan, F.Y. Engineering characteristics of the calcareous sand in Nansha Islands, South China Sea. *Eng. Geol.* **2011**, *120*, 40–47. [CrossRef]
11. Cui, M.J.; Zheng, J.J.; Chu, J.; Wu, C.C.; Lai, H.J. Bio-mediated calcium carbonate precipitation and its effect on the shear behaviour of calcareous sand. *Acta Geotech.* **2021**, *16*, 1377–1389. [CrossRef]
12. Hasanlourad, M.; Salehzadeh, H.; Shahnazari, H. Dilation and particle breakage effects on the shear strength of calcareous sands based on energy aspects. *Int. J. Civ. Eng.* **2008**, *6*, 108–119.
13. Li, Y.; Lin, Z.; Li, B.; He, L.; Gong, J. Effects of gradation and grain crushing on the liquefaction resistance of calcareous sand. *Geomech. Geophys. Geo-Energy Geo-Resour.* **2021**, *7*, 12. [CrossRef]
14. Wang, Y.Q.; Hong, Y.; Guo, Z.; Wang, L.Z. Micro- and macro-mechanical behavior of crushable calcareous sand in South China Sea. *Rock Soil Mech.* **2018**, *39*, 199–206.
15. Qin, Y.; Yao, T.; Wang, R.; Zhu, C.Q.; Meng, Q.S. Particle breakage-asad analysis of deformation law of calcareous sediments under high-pressure consolidation. *Rock Soil Mech.* **2014**, *35*, 3123–3128.
16. Lagioia, R.; Nova, R. An experimental and theoretical study of the behaviour of a calcarenite in triaxial compression. *Géotechnique* **1995**, *45*, 633–648. [CrossRef]
17. Sharma, S.S.; Fahey, M. Degradation of Stiffness of Cemented Calcareous Soil in Cyclic Triaxial Tests. *J. Geotech. Geoenviron.* **2003**, *129*, 619–629. [CrossRef]
18. Sharma, S.S.; Fahey, M. Deformation characteristics of two cemented calcareous soils. *Can. Geotech. J.* **2004**, *41*, 1139–1151. [CrossRef]
19. Zhang, J.M.; Jiang, G.S.; Wang, R. Research on influences of particle breakage and dilatancy on shear strength of calcareous sands. *Rock Soil Mech.* **2009**, *30*, 2043–2048.
20. He, S.H.; Shan, H.F.; Xia, T.D.; Liu, Z.J.; Ding, Z.; Xia, F. The effect of temperature on the drained shear behavior of calcareous sand. *Acta Geotech.* **2021**, *16*, 613–633. [CrossRef]
21. He, S.H.; Zhang, Q.; Ding, Z.; Xia, T.D.; Gan, X. Experimental and Estimation Studies of Resilient Modulus of Marine Coral Sand under Cyclic Loading. *J. Mar. Sci. Eng* **2020**, *8*, 287. [CrossRef]
22. Schofield, A.N.; Wroth, P. *Critical State Soil Mechanics*; McGraw-Hill: London, UK, 1968.
23. Sladen, J.A.; D'Hollander, R.D.; Krahn, J.; Mitchell, D.E. Back analysis of the Nerlerk berm liquefaction slides. *Can. Geotech. J.* **1985**, *22*, 579–588. [CrossRef]
24. Cai, Z.Y.; Hou, H.Y.; Zhang, J.X.; Zhang, L.; Guan, Y.F.; Cao, Y.Y. Critical state and constitutive model for coral sand considering particle breakage. *Chin. J. Geotech. Eng.* **2019**, *41*, 7.
25. Riemer, M.F.; Seed, R.B. Factors affecting apparent position of steady-state line. *J. Geotech. Geoenviron.* **1997**, *123*, 281–288. [CrossRef]
26. Manzari, M.; Dafalias, Y. A critical state two-surface plasticity model for sands. *Géotechnique* **1997**, *47*, 255–272. [CrossRef]
27. Maeda, K.; Muir Wood, D.; Nukudani, E. Modelling mechanical consequences of erosion. *Géotechnique* **2010**, *60*, 447–457.
28. Li, X.S.; Wang, Y. Linear Representation of Steady-State Line for Sand. *J. Geotech. Geoenviron.* **1998**, *124*, 1215–1217. [CrossRef]
29. Sun, Y.F.; Wichtmann, T.; Sumelka, W.; Kan, M.E. Karlsruhe fine sand under monotonic and cyclic loads: Modelling and validation. *Soil Dyn. Earthq. Eng.* **2020**, *133*, 106119. [CrossRef]
30. Turcotte, D.L. Fractals and fragmentation. *J. Geophys. Res.* **1986**, *91*, 1921–1926. [CrossRef]

31. Zhang, X.; Hu, W.; Scaringi, G.; Baudet, B.A.; Han, W. Particle shape factors and fractal dimension after large shear strains in carbonate sand. *Geotech. Lett.* **2018**, *8*, 73–79. [CrossRef]
32. Sun, Y.F.; Liu, H.L.; Yang, G. Yielding function for coarse aggregates considering gradation evolution induced by particle breakage. *Rock Soil Mech.* **2013**, *34*, 3479–3484.
33. Li, Y.; Zhang, H.; Huang, M.H.; Yin, H.B.; Jiang, K.; Xiao, K.T.; Tang, S.W. Influence of Different Alkali Sulfates on the Shrinkage, Hydration, Pore Structure, Fractal Dimension and Microstructure of Low-Heat Portland Cement, Medium-Heat Portland Cement and Ordinary Portland Cement. *Fractal Fract.* **2021**, *5*, 79. [CrossRef]
34. He, S.H.; Ding, Z.; Hu, H.B.; Gao, M. Effect of Grain Size on Microscopic Pore Structure and Fractal Characteristics of Carbonate-Based Sand and Silicate-Based Sand. *Fractal Fract.* **2021**, *5*, 152. [CrossRef]
35. Ioelovich, M. Study of Fractal Dimensions of Microcrystalline Cellulose Obtained by the Spray-Drying Method. *Fractal Fract.* **2019**, *3*, 3. [CrossRef]
36. Li, G.; Liu, Y.J.; Yin, Z.Y.; Dano, C.; Hicher, P.Y. Grading effect on critical state behavior of granular materials. *Chin. J. Geotech. Eng.* **2014**, *36*, 452–457.
37. Liu, Y.J.; Wang, J.H.; Yin, Z.Y.; Li, G.; Xia, X.H. Constitutive modeling for granular materials considering grading effect. *Chin. J. Geotech. Eng.* **2015**, *37*, 299–305.
38. Hardin, B.O. Crushing of Soil Particles. *Int. J. Geotech. Eng.* **1985**, *111*, 1177–1192. [CrossRef]
39. Shen, Y.; Shen, X.; Liu, H.L.; Ge, H.Y.; Rui, X.X. Gradation affects basic mechanical characteristics of Chinese calcareous sand as airport subgrade of reefs. *Mar. Georesour. Geotechnol.* **2020**, *38*, 706–715.
40. Skempton, A.W. The pore-pressure coefficients A and B. *Géotechnique* **1954**, *4*, 143–147. [CrossRef]
41. Tyler, S.W.; Wheatacraft, S.W. Fractal scaling of soil particle-size distributions: Analysis and limitations. *Soil Sci. Soc. Am. J.* **1992**, *56*, 362–369. [CrossRef]
42. Posadas, A.N.D.; Giménez, D.; Bittelli, M.; Vaz, C.M.P.; Flury, M. Multifractal characterization of soil particle-size distributions. *Soil Sci. Soc. Am. J.* **2001**, *65*, 1361–1367. [CrossRef]
43. Wang, J.; Cao, L.; Wang, B.; Gong, H.T.; Tang, W. Overview of One-Dimensional Continuous Functions with Fractional Integral and Applications in Reinforcement Learning. *Fractal Fract.* **2022**, *6*, 69.
44. Sadrekarimi, A.; Olson, S. Critical state friction angle of sands. *Géotechnique* **2011**, *61*, 771–783. [CrossRef]
45. Bandini, V.; Coop, M. The influence of particle breakage on the location of the critical state line of sands. *Soil Found.* **2011**, *51*, 591–600. [CrossRef]
46. Luzzani, L.; Coop, M.R. On the relationship between particle breakage and the critical state of sands. *Soil Found.* **2002**, *42*, 71–82. [CrossRef]
47. Li, X.S.; Dafalias, Y.F. Dilatancy for cohesionless soils. *Géotechnique* **2000**, *50*, 449–460. [CrossRef]
48. Richart, F.E.; Hall, J.R.; Woods, R.D. Vibrations of soils and foundations. *Princ. Neurodyn. Spartan* **1970**, *209*, 137.

fractal and fractional

MDPI

Article

Macro- and Micromechanical Assessment of the Influence of Non-Plastic Fines and Stress Anisotropy on the Dynamic Shear Modulus of Binary Mixtures

Meisam Goudarzy * and Debdeep Sarkar

Chair of Soil Mechanics, Foundation Engineering and Environmental Geotechnics, Ruhr-Universität Bochum, 44801 Bochum, Germany; debdeep.sarkar@rub.de

* Correspondence: meisam.goudarzyakhore@rub.de; Tel.: +4-923-4322-6078

Abstract: Resonant column tests were carried out on Hostun sand mixed with 5%, 10% and 20% non-plastic fines (defined as grains smaller than 0.075 mm) in order to quantify the combined influence of the void ratio (e), anisotropic stress state (defined as σ_v'/σ_h') and fines content (f_c) on the maximum small-strain shear modulus G_{max}. A significant reduction in the G_{max} with increasing f_c was observed. Using the empirical model forwarded by Roesler, the influence of e and σ_v'/σ_h' on G_{max} was captured, although the model was unable to capture the influence of varying fines content using a single equation. From the micro-CT images, a qualitative observation of the initial skeletal structure of the 'fines-in-sand' grains was performed and the equivalent granular void ratio e^* was determined. The e was henceforth replaced by e^* in Roesler's equation in order to capture the variation in f_c. The new modification was quantified in terms of the mean square error R^2. Furthermore, the G_{max} of Hostun sand–fine mixtures was predicted with good accuracy by replacing e with e^*. Additionally, a micromechanical interpretation based on the experimental observation was developed.

Keywords: anisotropic stress; Hostun sand; resonant column; maximum shear modulus; fines content; void ratio; equivalent granular void ratio

Citation: Goudarzy, M.; Sarkar, D. Macro- and Micromechanical Assessment of the Influence of Non-Plastic Fines and Stress Anisotropy on the Dynamic Shear Modulus of Binary Mixtures. *Fractal Fract.* **2022**, *6*, 205. https://doi.org/10.3390/fractalfract6040205

Academic Editor: Wojciech Sumelka

Received: 28 February 2022
Accepted: 4 April 2022
Published: 6 April 2022

1. Introduction

The determination of soil stiffness is of primary importance, considering its application in practical fields of geotechnical engineering, including foundation settlements, deformations caused by excavations or wave propagation in the ground due to vibration. In addition, all geotechnical sub/superstructure designs require the soil stiffness parameters in order to estimate the resistance of a structure against dynamic motions, such as earthquakes. In recent years, the development of elasto-plastic and advanced constitutive models (e.g., hardening soil model with small-strain stiffness, SANISAND) exclusively requires the small-strain soil stiffness parameter G_{max} as an input, underlining its importance in terms of research and practice in the field of soil dynamics.

Soil stiffness typically has a maximum value at low strains (strain < 10^{-6}), denoted either G_{max} or G_0, and decreases by increasing the strain amplitude. Since the 1960s, several researchers ([1–3]) identified the influence of the void ratio and the effective stress on the small-strain shear modulus G_{max}. The authors of [1] were the earliest to formulate an empirical equation to capture the influence of the void ratio and effective stress under isotropic stress conditions (Equation (1)):

$$G_{max} = A\, p_a\, f(e) \left(\frac{p\prime}{p_a} \right)^n \tag{1}$$

where A is a material constant depending on the soil type, p_a is the atmospheric pressure (≈100 kPa), p′ is the effective stress, n is a stress exponent and $f(e)$ is the void ratio function, which popularly takes the form of Equation (2) ([1]) or Equation (3) ([4]):

$$f(e) = \frac{(c-e)^2}{1+e} \tag{2}$$

$$f(e) = e^{-d} \tag{3}$$

c and d are fitting parameters. The authors of [5,6] studied the influence of the grain size distribution on G_{max}, where they reported a significant influence of the uniformity coefficient C_u and the mean grain size D_{50}, which they characterized via empirical relations. Other than grain size, the fines content (i.e., grains passing through a standard US 200 sieve and smaller than 0.075 mm) can also influence the mechanical behavior of soils, as was evidenced in previous studies, e.g., refs. [7–11]. Systematic studies on the effect of fines content and isotropic stress on G_{max} have been published by [8–18]. These studies showed a strong influence of the non-cohesive fines fraction on G_{max}—a decrease in G_{max} was observed with increasing fines content. Therefore, ref. [12] proposed a reduction factor for Equation (1), as shown in Figure 1.

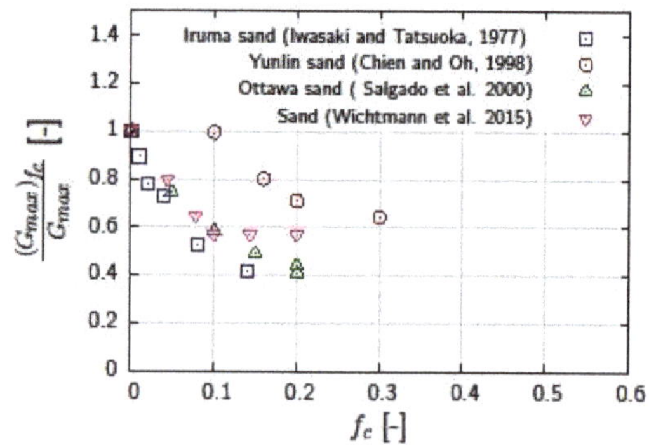

Figure 1. Reduction factor accompanying Equation (1) recommended by [12] to capture the influence of non-plastic fines on the maximum small-strain shear stiffness G_{max}.

It must be noted that Equation (1), in conjunction with the reduction factor, is not very suitable since it assumes different values for each soil type, as demonstrated by [17–19]. The authors of [9,11] reported that the fitting parameters of Equations (1)–(3) are influenced by fines content, which means for every binary mixture, the fitting parameters must be separately determined.

It is possible to capture the influence of the fines content on the mechanical behavior of granular soils by utilizing the concept of equivalent granular void ratio, as defined by [7]. Furthermore, ref [19] defined that a certain portion of fines actively participate in the granular structure contributing to force chains, based on which Equation (4) was suggested, where a boundary was formulated, below which the soil behavior would be dominated by sand (defined as 'fines-in-sand') and above which by fines ('sand-in-fines) ([19,20]).

$$e^* = \frac{e+(1-b)f_c}{1-(1-b)f_c} \qquad f_c < f_{thr} \tag{4}$$

The b-value holds a value between 0 and 1 and is responsible for denoting the active proportion of fines in the soil structure; therefore, it is a function of the fine content—the higher the value the higher the percentage of fines contributing to force chains in the mixture. In [20,21], the authors developed a semi-empirical equation (Equation 5) to calculate b-parameter.

$$b = \left[1 - exp\left\{-0.3\frac{(f_c/f_{thr})}{k}\right\}\right] \times \left(r\frac{f_c}{f_{thr}}\right)^r \tag{5}$$

where $r = (D_{10}/d_{50})^{-1}$, D_{10} = sand grain size corresponding to 10% finer materials by weight passing through, d_{50} = fine grain size corresponding to 50% finer materials by weight passing through, $k = (1 - r^{0.25})$. The authors of [17,19] had previously used this formulation in their work on binary mixtures.

Ref. [10] carried out resonant column (RC) tests on Hostun sand mixed with a fine fraction. They reported the significant influence of the fines content on the maximum shear modulus (Figure 2a). The concept of equivalent granular void ratio e^* [22,23] was successfully used to predict the maximum shear modulus in Equation (1) under isotropic stress conditions. However, no investigation was done for anisotropic stress states.

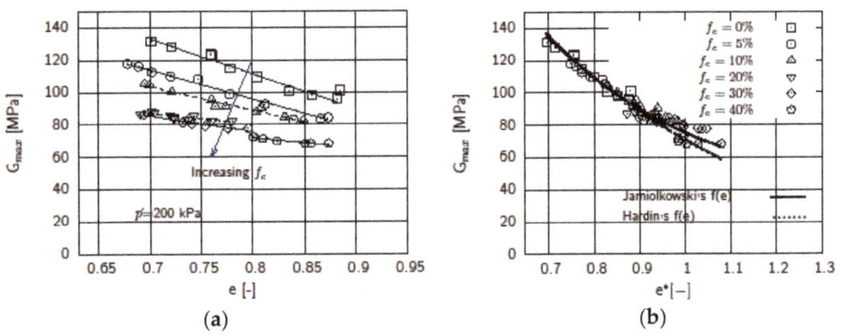

Figure 2. (**a**) Influence of fines content on the G_{max} via RC tests; (**b**) using the concept of equivalent granular void ratio e^* to capture varying percentages of fines content ([10]).

In order to evaluate the effect of anisotropy, numerous laboratory tests using bender elements ([11,24–30]), RC and torsional shear tests ([1,11,31–35]) have already been carried out. Furthermore, [1] and [24] extended the empirical relations to account for anisotropy. [1] believed that shear stress plays an insignificant role on shear modulus and suggested replacing p' in Equation (1) with the average of vertical and horizontal stresses Equation (6).

$$G_{max} = Af(e){p_a}^{(1-n)}\left(\frac{\sigma_v + \sigma_h}{2}\right)^n \tag{6}$$

σ_v and σ_h denotes the vertical and horizontal stresses.

Ref. [24] conducted experimental studies on cubic soil samples to investigate the effects of stress components on shear wave velocity under anisotropic loading, where a significant influence of vertical stress on shear wave velocity and, consequently, on G_{max} was reported. Therefore, Hardin's relationship was modified accordingly by using σ_v' and σ_h' instead of p', as shown in Equation (7):

$$G_{max} = Af(e)p_a\left(\frac{\sigma_v'}{p_a}\right)^{n_v}\left(\frac{\sigma_h'}{p_a}\right)^{n_h}\left(\frac{\sigma_c}{p_a}\right)^{n_c} \tag{7}$$

n_v, n_h and n_c are the stress exponents, while σ_c represents the out-of-plane stress.

Furthermore, it was also found out that the placing technique of the sand does not influence the shear wave velocities. Reference [11] performed tests on glass beads, concluding that reference shear strain was affected by anisotropy, and an empirical relationship was developed, considering confining pressure and anisotropic stress components. Performing tests on Ticino sand under both horizontal and vertical directions using geophones equipped with bender and compression elements, ref. [25] noted stiffnesses in the horizontal plane were larger than in the vertical plane. Using a similar procedure, ref. [26] investigated the interdependence of the active earth pressure coefficient K_0 on stress-induced anisotropy. The work of [24] was further extended by [27], in which the author attempted to establish a difference between the effects of fabric anisotropy and effective stress on soil stiffness. A similar approach using a multiaxial triaxial cell equipped with bender/extender elements was adopted by [28] to assess the evolution of elastic anisotropy under axially symmetric stress conditions. Using results backed by experiments, ref. [29] carried out discrete element method simulations and stated that more contact normal tensors preferred to distribute along the horizontal direction. Reference [30] used three different sample preparation methods to investigate the fabric anisotropy of Hostun sand, where air-pluviated and tamped samples exhibited higher stiffness in the horizontal and vertical directions, respectively.

Most of the previous studies have focused on either of the two aspects: either anisotropy of clean sands or sands with varying fines content. In nature, sands are often mixed with fines, and subjected to anisotropic stress states, possibly due to an existing substructure or repetitive loading over time. Therefore, studies combining both fines content under anisotropic stress conditions are necessary considering their practical importance. Therefore, in the present work, the influence of fines content on G_{max} under anisotropic stress conditions was investigated in detail. In addition, the application of e^* in Equation (7) in specimens under anisotropic stress was inspected.

2. Experimental Procedure and Sample Preparation

The following sub-sections describe the experimental device and procedure followed by the tested materials and the method of sample preparation.

2.1. The RC Device

The RC and Bender element tests are common methods to determine the small-strain dynamic properties of soils (e.g., shear wave velocity, shear stiffness and damping). To determine the shear modulus and damping, the RC device available at the Ruhr-Universität Bochum was used. The device is capable of investigating the influence of small (shear strain $\gamma < 10^{-5}$) and medium shear strains ($10^{-3} < \gamma < 10^{-5}$). This free-free mode of vibration, i.e., the bottom and top plates are freely movable ([36]), was employed. This specimen is enclosed by a latex membrane of thickness 0.4 mm. The inserted cylindrical soil sample is subjected to harmonic torsional vibrations about the longitudinal axis by two electromagnetic excitation heads. The generated acceleration is measured with the help of two accelerometers, which then results in the linearly distributed torsion γ over the specimen height. This recorded acceleration is displayed on an oscilloscope as a sinusoidal excitation signal, the frequency of which is varied using a function generator until the resonant frequency f_R is determined. The dynamic shear modulus for the specific load amplitude can then be calculated from f_R ([36]):

$$G = \left(\frac{2\pi f_R}{a}\right)^2 \tag{8}$$

$$a\,tan(a) - \frac{J^2}{J_0\,J_L}\frac{tan(a)}{a} = \frac{J}{J_0} + \frac{J}{J_L} \tag{9}$$

The polar mass moment of inertia for the bottom and top parts of the device, as well as the sample, are denoted by J_0, J_L and J respectively. ρ denotes the sample density, while $a = \frac{\omega L}{v_s}$, ω is the rotational frequency, L is the sample height and v_s the shear wave velocity. The schematic figure of the RC device is shown in Figure 3.

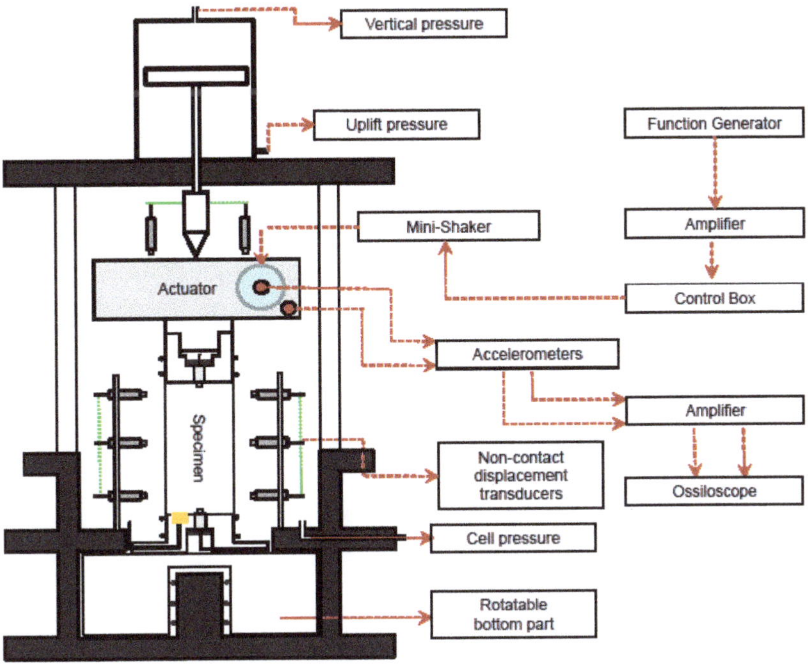

Figure 3. Schematic representation of the RC device used in the present study.

The device was modified for performing tests under anisotropic stress states ([37]). A pneumatic pressure cylinder was mounted on the top of the cell, as shown in Figure 3, for applying vertical pressure on the specimen. The calibration and validation of the device were performed following different methods, further details of which are in [11,37]. At the beginning, the specimen was isotropically loaded with a cell pressure $\sigma_h' = 50$ kPa. From a cell pressure of 200 kPa onwards, the vertical pressure ensures anisotropy in the specimen. The results on isotropically loaded specimens were shown previously in [17]. This paper aimed to describe the results on the same soil mixtures, but under anisotropic loading. The stress path is currently divided into two areas: isotropic condition at $\sigma_h' = 200$ kPa ([11,17]) and anisotropic condition where σ_h' is constant 200 kPa and σ_v' is increased up to 600 kPa (i.e., effective vertical to horizontal stress ratio $\sigma_v'/\sigma_h' = 3$).

2.2. Tested Materials

In the present work, four series of tests were carried out with a mixture of Hostun sand (a quartz sand originally mined in France) and non-plastic quartz (obtained locally) as fines with different proportions by weight—0%, 5%, 10% and 20%. The material mixtures of Hostun sand and the different proportions of fines are shown in Figure 4. The white opaque parts show the quartz powder and the light-colored elements denote sand particles. Hostun sand has been used previously in numerous tests ([37]). Silica is the predominant chemical that constitutes the sand grains ($SiO_2 > 98\%$). The sand is angular with a grain density of 2.65 g/cm^3. The mean grain size of Hostun sand and quartz powder (also consisting of SiO_2 as the main component) is 0.375 mm and 0.004 mm, respectively. Figure 5 shows the grain size distribution of the tested materials obtained following [38].

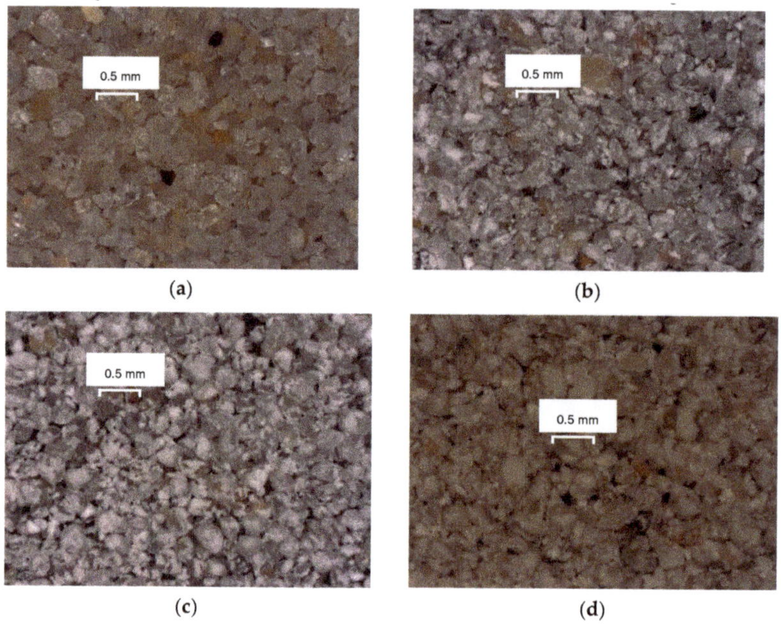

Figure 4. (**a**) Clean Hostun sand and sand mixed with (**b**) 5%, (**c**) 10% and (**d**) 20% quartz fines used in the present study.

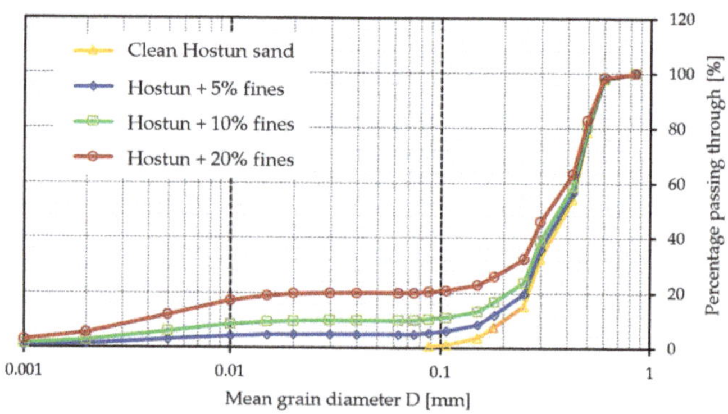

Figure 5. Grain size distribution of the tested materials.

3. Experimental Results

From Figure 6a–d, the influence of the anisotropic stress on the variation in void ratio e over different fines content is shown. The results show that with an increase in the effective anisotropic deviatoric stress σ'_v/σ'_h, the e decreases with increasing fines, whereas the decrease is minimal for the clean sand (Figure 6a). This implies a higher compressibility of the mixtures with higher fines.

Furthermore, Figure 7 shows the relationship between G_{max}, e and σ'_v/σ'_h. It is clear that the G_{max} increases with decreasing e and increasing $\frac{\sigma'_v}{\sigma'_h}$. It is interesting to note the relative magnitudes of the G_{max} for differing fines content—it is higher for the clean Hostun sand (Figure 7a) but reduces with increasing fines content. In addition, the reduction in G_{max} is much larger at higher fines content, although the specimens are prepared at an initially dense state (see Figure 7d). Practically, the influence of e is diminished at f_c = 20%, while it is largest at f_c = 0%.

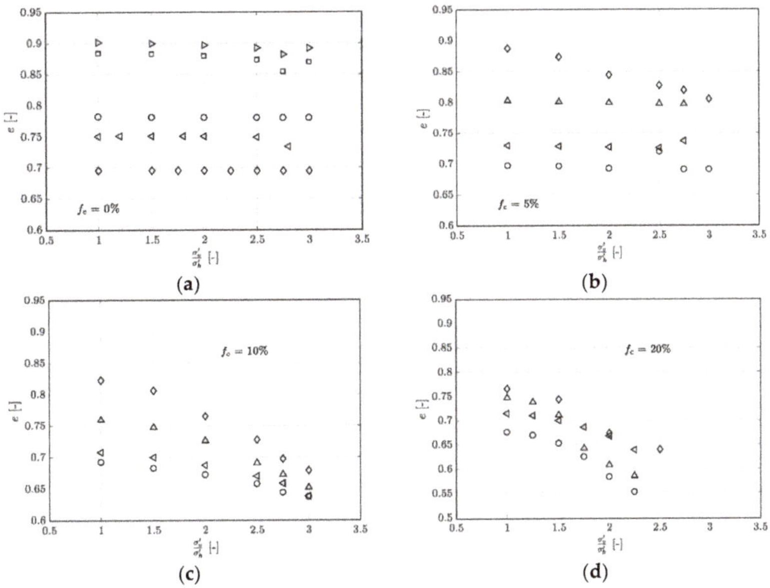

Figure 6. Void ratio e vs. anisotropic stress for (**a**) clean Hostun sand and sand mixed with (**b**) 5%, (**c**) 10% and (**d**) 20% fines, respectively.

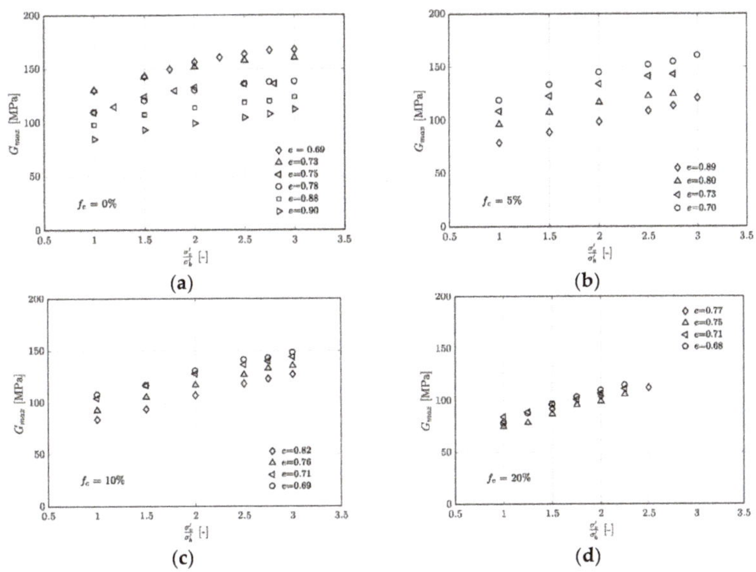

Figure 7. The influence of the anisotropic stress on the G_{max} for (**a**) clean Hostun sand and sand mixed with (**b**) 5%, (**c**) 10% and (**d**) 20% fines, respectively.

In Figure 8, the variation of the G_{max} with different e and f_c values is shown along with isotropic (Figure 8a) and anisotropic (Figure 8b) states. For all specimens with 0%, 5%,

10% and 20% fines content, G_{max} decreases with the increase in e and increase of vertical stress, σ'_v. In comparison to the isotropic specimen with initial e = 0.80 and containing 5% fines, the 20% fines specimen under the same boundary conditions shows a lower G_{max}- for 5% f_c, which is close to 100 MPa, while for the other, it measures around 75 MPa. The same trend can be noted for the anisotropic specimen in Figure 8b, which implies an increase in anisotropic stress load causes an increase in the maximum shear modulus, whereas the increase in f_c causes a reduction in the maximum shear modulus.

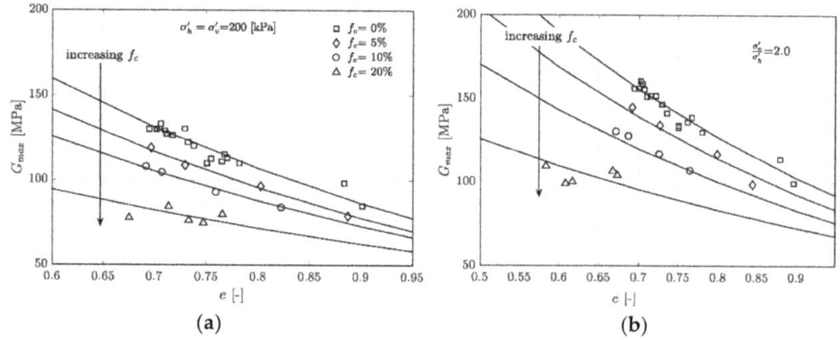

Figure 8. Variation of the G_{max} with different void ratios and fines content under (**a**) isotropic and (**b**) anisotropic stress states.

For $\sigma'_v/\sigma'_h = 2$, the G_{max} for clean sand measures approximately 160 MPa at e = 0.70 (Figure 8b), while under isotropic conditions (Figure 8a), it measures around 135 MPa. This reinforces the observation that at anisotropic stress conditions, an increase in the e as well as fines content results in a significant decrease in the G_{max}.

4. Analysis of Results

Using the popular models of [1,23], the variation in the G_{max} with e, p′ and $\sigma'_v/\sigma'_h = 1$ (isotropic) and 2 (anisotropic) was quantified as shown in Figure 9. In general, a good fit with the experimental data can be seen. Later, the G_{max} was normalized with the void ratio function employing an average value of the fitting parameter c (= 2.12, 2.15, 2.37 and 3.25 for 0%, 5%, 10% and 20% fines, respectively). To capture the variation in the anisotropic stress, the G_{max} was initially normalized with the Hardin's void ratio function, considering the average value of the fitting parameter c (= 2.12, 2.15, 2.37 and 3.25 for 0%, 5%, 10% and 20% fines, respectively), following which the normalized G_{max} was further plotted against the σ'_v/σ'_h (shown only for the clean sand case in Figure 10a) to obtain the best fit magnitudes of the fitting parameters n_v and n_h. Figure 10b presents a 3D overview of the variation of G_{max} with regards to e, σ'_v and σ'_h (Equation (10) in the case of the clean sand only, shown in Figure 10b below, where the proposed model shows a good fit with the experimental data with R^2 = 0.95 (K_0 denotes the ratio σ'_h/σ'_v).

$$\frac{G_{max}}{f(e)} = 0.836 p_a \left(\frac{\sigma'_v}{p_a}\right)^{0.21} \left(\frac{\sigma'_h}{p_a}\right)^{0.23} \tag{10}$$

However, it is not possible to capture in a single curve the influence of varying fines content. Therefore, as suggested by various studies, the global void ratio can also be replaced by the equivalent granular void ratio e^*, which would have the same magnitude as e for clean sand but would depend on the parameter b for fine content below the threshold fines. Based on the suggestion of [20] from their experiments with 10 sands, a value of 30% was deemed suitable for the f_{thr} since it was able to capture many characteristic responses of undrained soil behavior regardless of host sands. The b value was calculated accordingly in order to convert the e to e^* up to a fines content of 20%. The resulting $G_{max} - e^*$ curves for both isotropic and anisotropic stress states are shown in Figure 11 below. The data at higher e^* and lower G_{max} show relatively greater scatter than at lower e^* and higher G_{max}. Hardin's relationship also shows a slight deviation to higher e^* and lower G_{max}, which may be attributed to the inherent variability of G_{max}. Figure 12 shows a comparison of the measured and predicted G_{max} with the proposed Equation (10), where a good fit is evident (R^2 = 0.80) for different f_c values ranging between 0–20% as well as different degrees of anisotropy achieved by loading in the vertical direction ($1 \leq \sigma'_v/\sigma'_h \leq 3$).

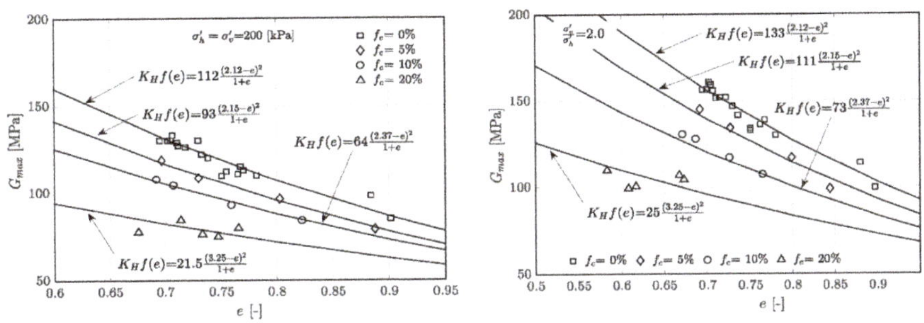

(a) (b)

Figure 9. Prediction of the G_{max} with different void ratios and fines content under (**a**) isotropic and (**b**) anisotropic stress states. The lines denote the best fits obtained using the Hardin void ratio equation, $K_H f(e)$.

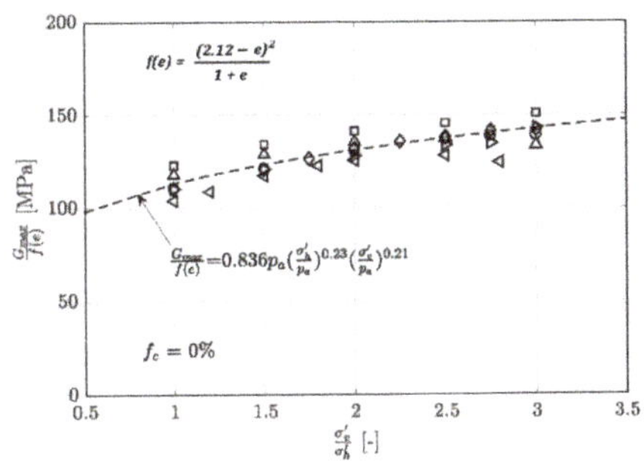

(a)

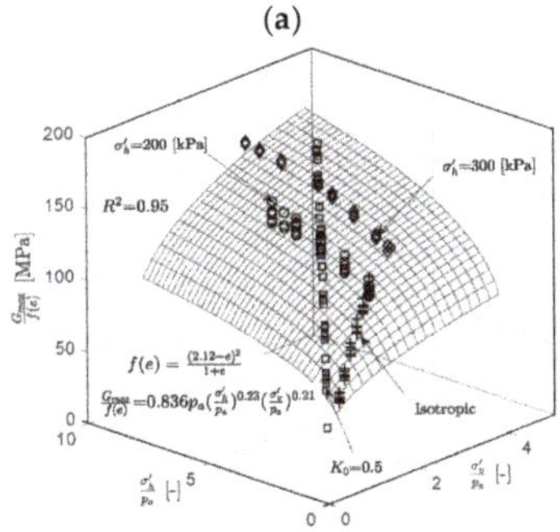

(b)

Figure 10. (**a**) Variation of the normalized G_{max} (using the Hardin void ratio function) with the horizontal and vertical stresses accounting for both isotropic and anisotropic states; (**b**) 3D representation of the same (also shown in [39]).

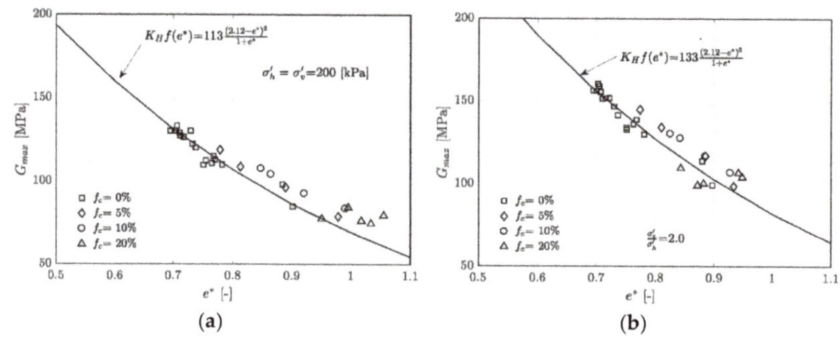

Figure 11. Using e^* instead of e to predict G_{max} under (a) isotropic and (b) anisotropic stress states.

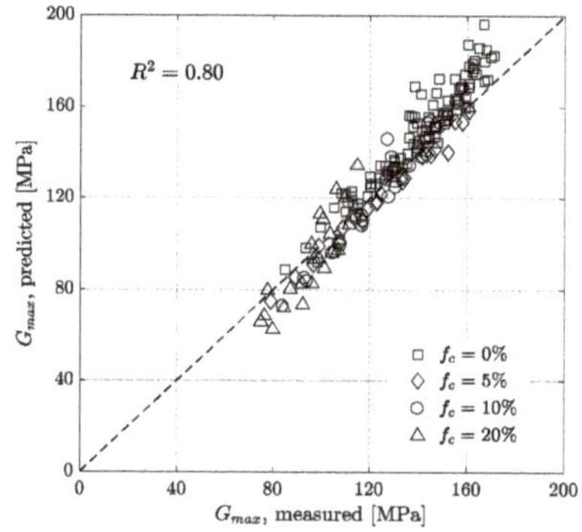

Figure 12. Comparison of the measured and predicted G_{max} values considering different fines content.

5. Discussion

As shown in the previous section, by using e^* instead of e in the Hardin's model, it was possible to capture the influence of varying fines lower than the threshold value. The biggest advantage afforded by using e^* is that one can avoid the numerous back calculations to determine the fitting parameters for each fines content. In addition, a solid micromechanical interpretation based on certain existing studies employing similar concepts [40–44] asserts a more logical approach, which, presently, is achieved via the b parameter—a micromechanical parameter specially designed to account for the positive or negative effects of f_c or the micromechanical influence of f_c in fine sand mixtures, and can be roughly estimated if the grain size distribution is available.

Micromechanical Interpretation

The stiffness of soil samples depends on microstructural properties, mainly grain-to-grain contacts that dictate the stiffness of the material. According to the theory of Hertz–Mindlin, the tangential or shear stiffness K_T between two grains is a function of the normal stiffness K_N, contact forces f_T and f_N and the elastic properties of the grains ([45,46]), and is mathematically represented as:

$$K_T = C_2\, K_N \left[1 - \frac{f_T}{f_N\, tan\varphi}\right]^{\eta} \quad (11)$$

$$K_N = C_1 f_N^n \quad (12)$$

$$\eta = n = \frac{1}{3} \quad (13)$$

n and η are fitting parameters. With the increase in the vertical load f_N, the normal stiffness increases, which results in increments in the friction and, consequently, the tangential stiffness and shear modulus G_{max} between particles. An additional load applied in the vertical direction results in an increase in f_T. However, according to Equation (11), an increase in f_T should result in a decrease in K_T, which explains the slower increase in G_{max} for anisotropic stresses compared to G_{max} for isotropic cases.

Presently, to reinforce our observations, a micro-CT scan was performed on the sample having 10% fines content in Kumamoto University in Japan (Figure 13a). The sample height and diameter for the CT scan was 40 mm and 7 mm, respectively (Figure 13b). Three densities were selected: loose ($\rho = 1.207$ g/cm^3, Figure 14a), medium-dense ($\rho = 1.312$ g/cm^3, Figure 14b) and dense ($\rho = 1.437$ g/cm^3, Figure 14c), details of which are in Table 1. The white particles denote sand grains, while those in gray denote fines, i.e., the quartz powder. From Figure 14a, the fines cover some of the coarse particles merely like a coating. Some of the fine grains arrange themselves between the coarse particles, i.e., these fine grains actively interact with the sand grains. The remaining fine grains distribute themselves in the voids between the coarse grains, i.e., these parts of the fine grains act inactively ([17]). An external load causes friction between the individual grains. For large grains, the friction surface is correspondingly large, making the grain-to-grain contact system relatively stable. Adding more fine grains (Figure 14b,c) reduces the friction area between the individual grains, causing the system to lose stability. Mathematically, this results in a lower contact force K_T and thus results in a lower maximum shear stiffness ([10]).

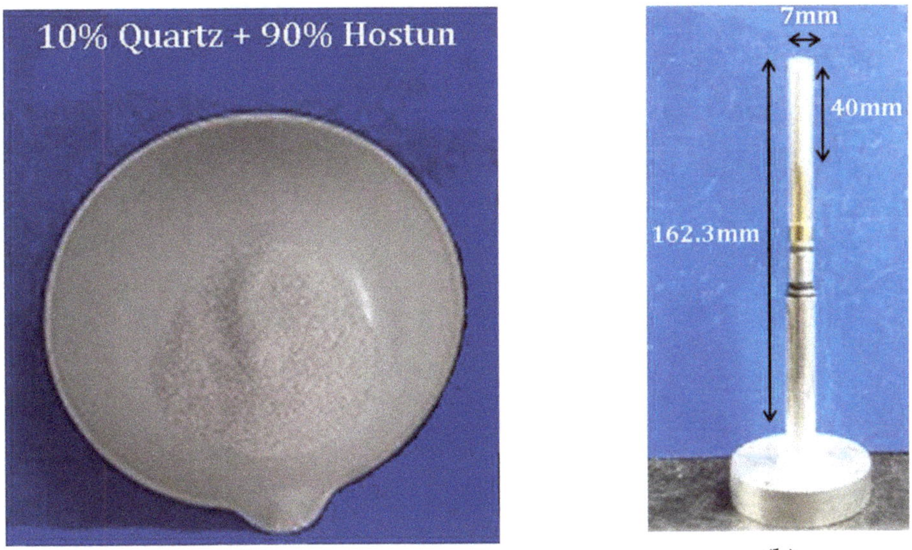

(a) (b)

Figure 13. (**a**) Mixture of 10% quartz and 90% Hostun sand; (**b**) typical sample for micro-CT analysis.

Table 1. Schedules of the various samples for micro-CT analysis for different densities.

Case	Material	Weight Ratio (%)		Sample Height (mm)	Weight (g)	Dry Density (g/cm^3)
		Quartz	Hostun			
1				37.5		1.207
2	Quartz/Hostun	10	90	34.5	1.742	1.312
3				31.5		1.437

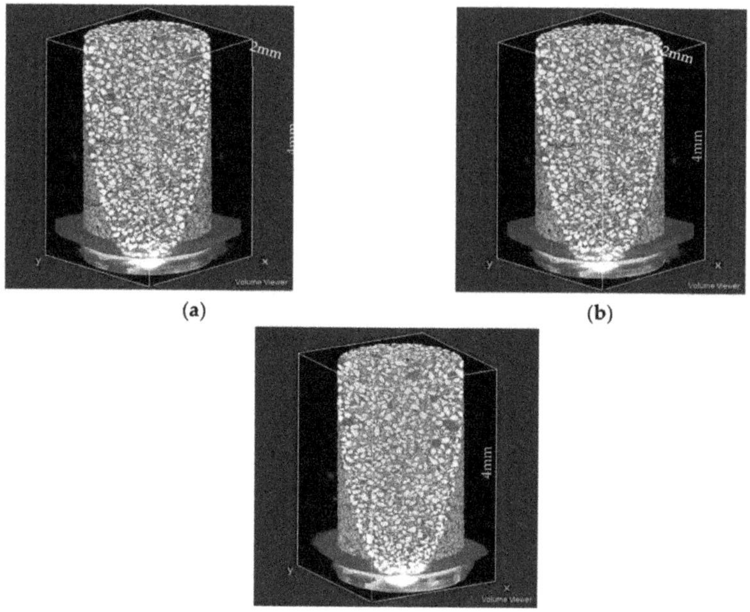

Figure 14. Micro-CT scan images for (**a**) loose, (**b**) medium-dense and (**c**) dense specimens for 10% fines content.

6. Conclusions

The aim of the study was to demonstrate the variation of the maximum shear modulus in binary mixtures of sand–non-plastic fines considering stress anisotropy due to additional vertical loading. As already mentioned, individual studies considering stress anisotropy or mixtures were conducted in previous research, however this work presents one of the first considering both these influences. This paper briefly summarizes the results and interpretation of the tests with respect to the influence of stress anisotropy and fine grain content on the G_{max}. The major findings are summed up below:

- Based on a series of resonant column tests with different initial global void ratios and confining pressures, it was found that G_{max} is lower for increasing fines content as well as higher void ratios. Using a popular empirical model, the variation of G_{max} with void ratio, confining pressure and stress state was captured with a good accuracy.
- Furthermore, to include fines content, the concept of equivalent intergranular void ratio was used where the global void ratio e was replaced by e^* in the various formulations with a higher degree of accuracy, particularly for lower magnitudes of e^*.
- Micro-CT scans of the binary mixtures were additionally made to enable a closer look into the microstructure, e.g., grain-to-grain contacts. From the skeletal structure, the contact forces between the individual grains were evident. In the case of anisotropic stress, a shear force acts in addition to the normal force for isotropically loaded specimens. An increase in fines results in a reduction in the friction between sand grains which leads to lower contact stiffness.

The results of the present work will provide practical knowledge for design engineers to predict the shear modulus in binary mixtures considering the designed overburden vertical stress leading to a stress anisotropy, and therefore, allow them to reliably calculate ground settlements/factor of safety of structures under dynamic loads. Furthermore, advanced constitutive models can also incorporate both these influencing factors in complex multi-variable numerical simulations.

For future work, the influence of stress anisotropy and grain distribution on elastic modulus E, grain shape or size and grain minerals can be investigated. In addition, the influence of other stress paths can be checked. Furthermore, a higher percentage of fines greater than 20% can be investigated. For a more accurate interpretation of the results, various methods, such as the discrete element method (DEM), to determine the number of grain contacts or the force distribution on the grains can be applied.

Author Contributions: Conceptualization, M.G.; methodology, M.G.; validation, M.G.; formal analysis, M.G.; writing—original draft preparation, D.S.; writing—review and editing, M.G. and D.S. All authors have read and agreed to the published version of the manuscript.

Funding: This research received no external funding.

Institutional Review Board Statement: Not applicable.

Data Availability Statement: Not applicable.

Acknowledgments: The authors would like to acknowledge the student Huma Randjoor for assisting in the experiments. Furthermore, the authors would like to dedicate this work in the memory of the late Tom Schanz, who was instrumental with his ideas, and without whose support, the research would not have been successfully accomplished. The authors are further grateful to Kumamoto University, Japan for providing us the micro-CT scans of the materials in 2014.

Conflicts of Interest: The authors declare no conflict of interest.

References

1. Hardin, B.O.; Black, W.L. Sand stiffness under various triaxial stresses. *J. Soil Mech. Found. Div.* **1966**, *92*, 27–42. [CrossRef]
2. Drnevich, V.P. *Resonant-Column Testing: Problems and Solutions*; Technical Report; Dynamic Geotechnical Testing: ASTM International, USA, 1978; pp. 384–398.
3. Seed, B.; Wong, R.; Idriss, I.; Tokimatsu, K. *Moduli and Damping Factors for Dynamic Analyses of Cohesionless Soils*; Technical Report; National Science Foundation: Alexandria, VA, USA, 1984.
4. Jamiolkowski, M.; Lancellotta, R.; Lo Presti, D.C.F. Remarks on the stiffness at small strains of six Italian clays. In Proceedings of the International Symposium, Sapporo, Japan, 12–14 September 1994; pp. 817–836.
5. Darendeli, M.B. Development of a New Family of Normalized Modulus Reduction and Material Damping Curves. Ph.D. Thesis, University of Texas at Austin, Austin, TX, USA, 2001.
6. Wichtmann, T.; Triantafyllidis, T. On the influence of the grain size distribution curve of quartz sand on the small strain shear modulus. *J. Geotech. Geoenviron. Eng.* **2009**, *135*, 1404–1418. [CrossRef]
7. Thevanayagam, S. Effect of fines and confining stress on undrained shear strength of silty sands. *J. Geotech. Geoenviron. Eng.* **1998**, *124*, 479–491. [CrossRef]
8. Salgado, R.; Bandini, P.; Karim, A. Shear strength and stiffness of silty sand. *J. Geotech. Geoenviron. Eng.* **2000**, *126*, 451–462. [CrossRef]
9. Wichtmann, T.; Navarrete Hernández, M.A.; Triantafyllidis, T. On the influence of a non-cohesive fines content on small strain stiffness, modulus degradation and damping of quartz sand. *Soil Dyn. Earthq. Eng.* **2015**, *69*, 103–114. [CrossRef]
10. Goudarzy, M.; König, D.; Schanz, T. Small strain stiffness of granular materials containing fines. *Soils Found.* **2016**, *56*, 756–764. [CrossRef]
11. Goudarzy, M.; König, D.; Santamarina, J.C.; Schanz, T. Influence of anisotropic stress state on the intermediate strain behaviour of granular materials. *Geotechnique* **2018**, *68*, 221–232. [CrossRef]
12. Iwasaki, T.; Tatsuoka, F. Effects of grain size and grading on dynamic shear moduli of sands. *Soils Found.* **1977**, *17*, 19–35. [CrossRef]
13. Tao, M.; Figueroa, J.; Saada, A. Influence of non-plastic fines content on the liquefaction resistance of soils in terms of the unit energy. In *Cyclic Behavior of Soils and Liquefaction Phenomena*; CRC Press: Boca Raton, FL, USA, 2004.
14. Chien, L.K.; Oh, Y.N. Influence of fines content and initial shear stress on dynamic properties of hydraulic reclaimed soil. *Can. Geotech. J.* **2002**, *39*, 242–253. [CrossRef]
15. Carraro, J.A.H.; Prezzi, M.; Salgado, R. Shear strength and stiffness of sands containing plastic of non-plastic fines. *J. Geotech. Geoenviron. Eng.* **2009**, *135*, 1167–1178. [CrossRef]
16. Yang, J.; Liu, X. Shear wave velocity and stiffness of sand: The role of non-plastic fines. *Geotechnique* **2016**, *66*, 500–514. [CrossRef]
17. Goudarzy, M.; Rahman, M.M.; König, D.; Schanz, T. Influence of non-plastic fine particles on maximum shear modulus of granular materials. *Soils Found.* **2016**, *56*, 973–983. [CrossRef]
18. Goudarzy, M.; König, D.; Schanz, T. Small and intermediate strain properties of sands containing fines. *Soil Dyn. Earthq. Eng.* **2018**, *110*, 110–120. [CrossRef]

19. Rahman, M.M.; Cubrinovski, M.; Lo, S.R. Initial shear modulus of sandy soils and equivalent granular void ratio. *Geomech. Geoengin.* **2012**, *7*, 219–226. [CrossRef]
20. Rahman, M.M.; Lo, S.C.R.; Gnanendran, C.T. On equivalent granular void ratio and steady state behavior of loose sand with fines. *Can. Geotech. J.* **2008**, *45*, 1439–1455. [CrossRef]
21. Rahman, M.M.; Lo, S.C.R.; Gnanendran, C.T. Reply to the discussion by Wanatowski and Chu on "On equivalent granular void ratio and steady state behaviour of loose sand with fines". *Can. Geotech. J.* **2008**, *46*, 483–486. [CrossRef]
22. Thevanayagam, S.; Shenthan, T.; Mohan, S.; Liang, J. Undrained fragility of clean sands, silty sands, and sandy silts. *J. Geotech. Geoenviron. Eng.* **2002**, *128*, 849–859. [CrossRef]
23. Zuo, L.; Baudet, B.A. Determination of the transitional fines content of sand-non plastic fines mixtures. *Soils Found.* **2015**, *55*, 213–219. [CrossRef]
24. Roesler, S. Anisotropic shear modulus due to stress anisotropy. *J. Geotech. Eng. Div.* **1979**, *105*, 871–880. [CrossRef]
25. Bellotti, R.; Jamiolkowski, M.; Lo Presti, D.C.; O'Neill, D.A. Anisotropy of small strain stiffness in Ticino sand. *Geotechnique* **1996**, *46*, 115–131. [CrossRef]
26. Zeng, X.; Ni, B. Stress-induced anisotropic G_{max} of sands and its measurement. *J. Geotech. Geoenviron. Eng.* **1999**, *125*, 741–749. [CrossRef]
27. Fioravante, V. Anisotropy of small strain stiffness of Ticino and Kenya sands from seismic wave propagation measured in triaxial testing. *Soils Found.* **2000**, *40*, 129–142. [CrossRef]
28. Sadek, T.; Lings, M.; Dihoru, L.; Wood, D.M. Wave transmission in Hostun sand: Multi-axial experiments. *Riv. Ital. Geotech.* **2007**, *2*, 69–84.
29. Wang, Y.; Mok, C. Mechanisms of small strain shear modulus anisotropy in soils. *J. Geotech. Geoenviron. Eng.* **2008**, *134*, 1516–1530. [CrossRef]
30. Ezaoui, A.; Di Benedetto, H. Experimental measurements of the global anisotropic elastic behaviour of dry Hostun sand during triaxial tests, and effect of sample preparation. *Geotechnique* **2009**, *59*, 621–635. [CrossRef]
31. Yanagisawa, E. Influence of void ratio and stress condition on the dynamic shear modulus of granular media. *Adv. Mech. Flow Granul. Mater.* **1983**, *2*, 947–960.
32. Yu, P.; Richart, F. Stress ratio effects on shear modulus of dry sands. *J. Geotech. Eng.* **1984**, *110*, 331–345. [CrossRef]
33. Ishibashi, I.; Chen, Y.; Chen, M.T. Anisotropic behavior of Ottawa sand in comparison with glass spheres. *Soils Found.* **1991**, *31*, 145–155. [CrossRef]
34. Santamarina, C.; Cascante, G. Stress anisotropy and wave propagation: A micromechanical view. *Can. Geotech. J.* **1996**, *33*, 770–782. [CrossRef]
35. Payan, M.; Khoshghalb, A.; Senetakis, K.; Khalili, N. Small-strain stiffness of sand subjected to stress anisotropy. *Soil Dyn. Earthq. Eng.* **2016**, *88*, 143–151. [CrossRef]
36. Wichtmann, T.; Sonntag, T.; Triantafyllidis, T. Über das Erinnerungsvermögen von Sand unter zyklischer Belastung. *Bautechnik* **2001**, *78*, 852–865. [CrossRef]
37. Goudarzy, M. Micro and Macro Mechanical Assessment of Small and Intermediate Strain Properties of Granular Material. Ph.D. Thesis, Ruhr-Universität Bochum, Bochum, Germany, 2015.
38. *DIN 18126:1996-11*; Baugrund, Untersuchung von Bodenproben—Bestimmung der Dichte Nichtbindiger Böden bei Lockerster und Dichtester Lagerung. Beuth: Berlin, Germany, 1996. (In German)
39. Goudarzy, M.; König, D.; Schanz, T. Interpretation of small and intermediate strain characteristics of Hostun sand for various stress states. *Soils Found.* **2018**, *58*, 1526–1537. [CrossRef]
40. Bohaienko, V.; Bulavatsky, V. Fractional-Fractal Modeling of Filtration-Consolidation Processes in Saline Saturated Soils. *Fractal Fract.* **2020**, *4*, 59. [CrossRef]
41. He, S.; Ding, Z.; Hu, H.; Gao, M. Effect of Grain Size on Microscopic Pore Structure and Fractal Characteristics of Carbonate-Based Sand and Silicate-Based Sand. *Fractal Fract.* **2021**, *5*, 152. [CrossRef]
42. Fu, X.; Ding, H.; Sheng, Q.; Zhang, Z.; Yin, D.; Chen, F. Fractal Analysis of Particle Distribution and Scale Effect in a Soil–Rock Mixture. *Fractal Fract.* **2022**, *6*, 120. [CrossRef]
43. Xu, J.; Shen, Y.; Sun, Y. Cyclic Mobilisation of Soil–Structure Interface in the Framework of Fractional Plasticity. *Fractal Fract.* **2022**, *6*, 76. [CrossRef]
44. Adeli, E.; Rosic, B.V.; Matthies, H.G.; Reinstädler, S.; Dinkler, D. Bayesian Parameter determination of a CT-Test described by a Viscoplastic-Damage model considering the Model Error. *Metals* **2020**, *10*, 1141. [CrossRef]
45. Yimsiri, S.; Soga, K. Application of micromechanics model to study anisotropy of soils at small strains. *Soils Found.* **2002**, *42*, 15–26. [CrossRef]
46. Johnson, K.L. *Contact Mechanics*; Cambridge University: Cambridge, UK, 1985.

fractal and fractional

MDPI

Article

Cyclic Mobilisation of Soil–Structure Interface in the Framework of Fractional Plasticity

Junhong Xu [1], Yang Shen [2] and Yifei Sun [2,*]

[1] School of Civil Engineering, Nanjing Forestry University, Nanjing 210037, China; jhxu@njfu.edu.cn

[2] Key Laboratory of Ministry of Education for Geomechanics and Embankment Engineering, Hohai University, Nanjing 210098, China; 20070060@hhu.edu.cn

[*] Correspondence: yifei.sun@hhu.edu.cn

Abstract: The strength of the soil–structure interface can be mobilised when subjected to cyclic loading. To capture the cyclic mobilisation of the soil–structure interface, an advanced elastoplastic constitutive model is developed within the framework of fractional plasticity, where no additional use of an additional plastic potential is required. Considering the influence of material state and soil fabric on the plastic response of the soil–structure interface, the state-dependent fractional order and hardening modulus are proposed. Further numerical simulation of the developed model shows that it can reasonably capture the mobilised strength and deformation of the soil–structure interface under cyclic loads.

Keywords: soil–structure interaction; fractional plasticity; cyclic loads

Citation: Xu, J.; Shen, Y.; Sun, Y. Cyclic Mobilisation of Soil–Structure Interface in the Framework of Fractional Plasticity. *Fractal Fract.* **2022**, *6*, 76. https://doi.org/10.3390/fractalfract6020076

Academic Editor: Zine El Abiddine Fellah

Received: 29 December 2021
Accepted: 29 January 2022
Published: 31 January 2022

1. Introduction

Geotechnical engineering usually encounters the interaction between soil and structure [1]. Due to the abrupt change in the stiffness of soil and structure, an interface with reduced stiffness and strength usually exists [2,3]. The load–displacement behaviour, e.g., stiffness degradation by in-situ cyclic loading [4], of the soil–structure interface can significantly influence the ultimate bearing capacity of the geotechnical infrastructures, e.g., tunnels, onshore foundations, and other underground space structures [5,6], especially under earthquake loading [7,8]. To understand the strength and deformation behaviour of the soil–structure interface, many experimental and numerical studies have been carried out [9–14]. For example, Zhang and Zhang [11] carried out a series of direct interface shear tests on a gravelly soil–steel interface under various loading conditions, e.g., constant normal load (CNL) and constant normal stiffness (CNS) shearing. Lashkari and Jamali [15] conducted a series of interface shear tests on different sand–geosynthetic interfaces, with the aid of digital particle image velocimetry to track the local particle movement. From these studies, it was found that the interface thickness increased with the mean particle size of soil; the interface strength mobilised with the cyclic loading path and was influenced by the initial relative density, particle shape, particle size, normal stiffness and effective normal stress. Moreover, through a comprehensive three-dimensional discrete element simulation using spheres with contact moments and walls with varying topography, Grabowski et al. [14] observed significant effects of the contact roughness and sand fabric on the mobilised strength and displacement of the sand–structure interface. Such effect of fabric has also been reported by Wang et al. [16] when evaluating the liquefaction and instability behaviour of sands. In addition, Zhang et al. [17] reported a remarkable 3D effect of the turning corner on the stability of geosynthetic-reinforced soil structures.

It can be found that the strength and deformation of the soil–structure interface are complex and influenced by many physical and mechanical factors. In order to properly instruct the engineering design and construction of geotechnical structures, a sophisticated constitutive model that can consider different influencing factors is of critical importance [18]. In fact, a variety of constitutive models have been developed for different soil–structure interfaces. For example, Saberi et al. [19] developed a unified constitutive model to capture the stress path dependence of a sandy and gravelly soil interface by using the framework of the bounding surface plasticity. Despite the elegant model performance, an additional plastic potential function was required, which resulted in the complexity in their model. Stutz and Mašín [20] developed a hypoplastic interface model for fine-grained soils via reduced incremental stress–strain vectors and redefined tensorial operations. However, their model did not consider the effect of particle breakage and soil fabric on the stress–displacement response of the soil–structure interface. By extending the original fractional plasticity for granular and soft soils [21,22], Sun et al. [18,23] proposed a family of fractional plastic models to capture the load–displacement behaviour of the granular soil–structure interface. Nonetheless, these models did not consider the effect of soil fabric on the cyclic mobilisation of the interface, as well. In engineering practice, a natural deposition or construction process can refabricate the soil by imprinting it with a certain structure. Upon shearing, such structure or fabric can evolve with strain level, which in turn influences the strength and deformation of the interface. To better interpret the load–displacement response of the granular soil–structure interface, a refined constitutive model by adopting the fabric-dependent fractional plastic flow rule will be developed within the framework of fractional plasticity. It is noted that this study only addresses the cyclic mobilisation problem during the soil–structure interaction, where the interface between soil and structure exists. When the soil and structure start to separate from each other, the developed interface model is no longer applicable.

2. Constitutive Model in the Framework of Fractional Plasticity

2.1. Constitutive Relation

This study only deals with soil–structure interfaces under a direct two-dimensional interface or simple shearing state. The constitutive relation for the soil–structure interface can be formulated using the elastoplastic equation as:

$$\dot{\sigma}_i = D_{ij}\dot{\mathrm{U}}_j, \tag{1}$$

where $i, j = 1, 2$; $\dot{\sigma}_i$ and $\dot{\mathrm{U}}_j$ are the incremental stress vector and incremental displacement vector, respectively; the elastoplastic stiffness D_{ij} can be expressed as [23]:

$$D_{ij} = \frac{1}{t_n}\left(D^e_{ij} - D^p_{ij}\right), \tag{2}$$

where t_n is the thickness of the soil–structure interface; D^e_{ij} and D^p_{ij} are the elastic modulus and plastic modulus, respectively. D^e_{ij} can be decomposed and expressed as:

$$D^e_{11} = D^e_{n0}\sqrt{\sigma_n p_a}, \tag{3}$$

$$D^e_{22} = D^e_{t0}\sqrt{\sigma_n p_a}, \tag{4}$$

$$D^e_{12} = D^e_{21} = 0, \tag{5}$$

where D^e_{n0} and D^e_{t0} are material constants; σ_n is the normal stress applied on the interface; $p_a = 101$ kPa is the atmospheric pressure. D^p_{ij} can be obtained through algebraic calculation as:

$$D^p_{ij} = \frac{D^e_{ik} n_k m_l D^e_{lj}}{K_p + m_r D^e_{rs} n_s}, \tag{6}$$

where $i, k, l, j, r, s = 1, 2$; n_k, m_l, and K_p are the plastic flow vector, plastic loading vector, and hardening modulus, respectively, which will be provided in the following section after developing the constitutive model.

2.2. Model Development

To characterise the cyclic mobilisation of the soil–structure interface, the framework of the fractional plasticity [24–26] is adopted. Then, the model will be developed by using three main components, i.e., the plastic loading vector, the fractional plastic loading vector, and the hardening modulus. To obtain the plastic loading vector, one can use the following Cam-clay loading/unloading surface (f):

$$f = \tau + tM\sigma_n \ln \sigma_n - tM\sigma_n \ln \sigma_0 = 0, \tag{7}$$

where σ_0 is the size of f; t = +1 or −1, for loading or unloading, respectively; M is a material constant, defining the residual stress ratio of the interface in the $\sigma_n \sim \tau$ plane. Upon loading/unloading, each current stress step should be on the loading/unloading surface.

Then, the plastic loading vector, m_i, can be obtained by solving Equation (7), such that:

$$\begin{bmatrix} m_1 \\ m_2 \end{bmatrix} = \begin{bmatrix} d_f \\ t \end{bmatrix}, \tag{8}$$

$$d_f = \frac{\partial^1 f / \partial \sigma_n^1}{\partial^1 f / \partial \tau^1} = M - t\eta, \tag{9}$$

where the superscript, 1, indicates first-order differentiation; stress ratio $\eta = \tau / \sigma_n$.

Apart from the loading vector, the plastic flow vector, n_i, should also be defined. In this study, n_i is obtained by conducting fractional order differentiation of the Cam-clay surface, f, such that:

$$\begin{bmatrix} n_1 \\ n_2 \end{bmatrix} = \begin{bmatrix} d_g \\ t \end{bmatrix}, \tag{10}$$

$$d_g = \frac{\partial^\alpha f / \partial \sigma_n^\alpha}{\partial^\alpha f / \partial \tau^\alpha} = (\mu M - t\eta)|\eta|^{1-\alpha}, \tag{11}$$

where the superscript, α, denotes the fractional order; stress ratio $\eta = \tau / \sigma_n$; $\mu = \psi(2) - \psi(2-\alpha)$, in which is the digamma function. Details for deriving Equations (10) and (11) can be found in Sun et al. [23] and, thus, are not repeated here for simplicity. As the fractional order defines the contraction and dilation responses of the soil–structure interface during loading/unloading, Sun et al. [23] suggested using the following plastic work-dependent fractional order to properly represent the cyclic normal displacement after stress reversal:

$$\alpha = \exp\left(\frac{\Delta \Psi}{1 + W_{ir}^\chi} \right), \tag{12}$$

where Δ and χ are material constants; Ψ and W_{ir} are the state parameter and plastic work, respectively, which can be expressed as:

$$\Psi = e - e_\Gamma + \lambda \ln\left(\frac{\sigma_n}{p_a} + \frac{p_s}{p_a} \right), \tag{13}$$

$$W_{ir} = \frac{1}{t_n} \left[\int \sigma_n \langle du^{ir} \rangle + \tau dv^{ir} \right], \tag{14}$$

where e_Γ and λ are material constants that depict the residual state line of the interface; p_s denotes the shift stress that describes the effect of particle breakage on the downward bending of the residual state line [27]; e is the current void ratio that obeys the following evolution law: $de = du(1 + e_0)/t_n$, in which e_0 denotes the initial void ratio calculated from the initial density of the interface; u^{ir} and v^{ir} are the irreversible tangential displacement and normal displacement, respectively, as shown in Figure 1. $\langle\ \rangle$ denotes the McCauley bracket. In order to account for the fabric-dependence of the plastic flow of the soil–structure interface, a fabric variable, z, by analogous to Dafalias and Manzari [28], is introduced into Equation (11), such that:

$$d_g = (1 + \langle tz \rangle)(\mu M - t\eta)|\eta|^{1-\alpha}, \tag{15}$$

where z obeys the following evolution law, i.e., $dz = \langle -dv^{ir} \rangle (-t - z)$.

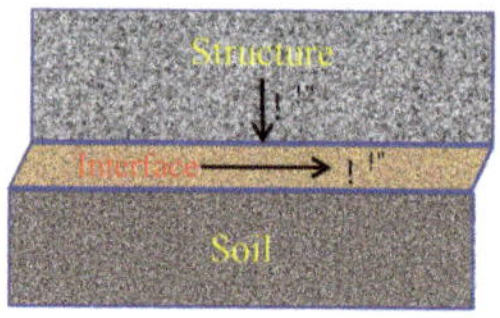

Figure 1. Irreversible displacements of the soil–structure interface.

In addition, the hardening modulus, K_p, can be defined as:

$$K_p = c_0 D^e_{11} \frac{(1 - c_1)M - t(\eta + c_1 M)}{|\eta|}, \tag{16}$$

where c_0 and c_1 are material constants.

2.3. Model Constants

The developed model requires eleven model constants to completely describe the cyclic mobilisation of the soil–structure interface under shearing. t_n is the thickness of the interface, which can be assumed to be $5d_{50}$ of the interface. D^e_{n0} and D^e_{t0} are the elastic constants, describing the degree of difficulty in elastic deformation of the interface; D^e_{n0} and D^e_{t0} can be obtained by fitting the initial stress–displacement curve, as clearly shown in [18]. M indicates the strength of the interface at the residual state; e_Γ is the intercept of the residual state line at $(\sigma_n + p_s) = 1$ kPa; λ is the gradient of the residual state line in the $e - \sigma_n$ plane; p_s indicates the effect of particle breakage on the residual state of the interface; M, e_Γ, λ and p_s can be obtained by fitting the residual state data of the interface. Δ defines the phase transition state of the interface at first loading, which can be obtained from the phase transition state data at first loading. χ defines the phase transition state of the interface at subsequent cyclic loading, which can be obtained by fitting the first unloading stress–displacement curve. c_1 is an empirical parameter, which can be assumed to be 0.01 for general application. c_0 defines the hardening ability of the interface, which is determined by trial and error. For a schematic representation of the parameter identification, one can refer to [18,23]. Table 1 lists the model constant used in this study.

Table 1. Model constants.

t_n	D^e_{n0}	D^e_{t0}	M	e_Γ	λ	p_s	Δ	χ	c_1	c_0
$5d_{50}$	41	35	0.777	0.548	0.0437	1	6	0.46	1.3	0.01

3. Numerical Simulations and Discussions

This section provides the numerical simulations of the cyclic load–displacement behaviour of the Zipingpu rockfill–steel interface [11] tested under CNS and CNL loading conditions. Discussions are to be made on the overall performance in regards to the cyclic mobilisation of the soil–structure interface in the framework of fractional plasticity. Note that the Zipingpu rockfill was reported to have a dry density of 17.5 kg/m^3. All the samples were prepared with an initial size of 500 mm in length, 360 mm in width and 300 mm in height. Then, the TsingHua-20 tonne Cyclic Shear Apparatus for the Soil–Structure Interface (TH-20t CSASSI) was adopted to carry out the interface shear test. During shearing, the length and width of the samples were kept constant whilst the height was variable.

Two boundary conditions, i.e., CNS and CNL, were used for the test. The CNS test was carried out under an initial normal load of 400 kPa, while the CNL test was conducted at a constant normal load of 700 kPa. Displacement-controlled shearing at a loading rate of 1 mm/min was applied, where the corresponding stress and displacement of the interface in tangential and normal directions were automatically measured using preinstalled force or displacement sensors. According to the measured data, the interface exhibited strength degradation or stress–normal dilation/contraction during cyclic loading

According to Zhang and Zhang [9], the Zipingpu rockfill–steel interface had a thickness of 50 mm and an initial void ratio of 0.51. Figures 2 and 3 show the model simulations of the load–displacement behaviour subjected to different initial normal stresses, i.e., 400 and 700 kPa. It can be observed that the fractional plastic interface model can capture the load–displacement loops (Figures 2a and 3a) and normal displacement (Figures 2c and 3c) of the Zipingpu rockfill–steel interface with mobilised tangential displacement in a reasonable manner. At both CNS and CNL loading conditions, the model can characterise the unloading and reloading-induced contraction and dilation response of the interface. After several cycles of mobilisation, a general normal contraction of the interface under CNS loading conditions can be observed, which induces a reduction in the normal stress. Specifically speaking, at the CNS loading condition shown in Figure 2b, the tangential stress as well as the normal stress mobilises towards a smaller value, indicating a reduced resistance of the interface to the cyclic loads. The engineering design in terms of the seismic loads should consider such strength reduction. However, at the CNL loading condition shown in Figure 3b, the tangential stress fluctuates between the residual state line (RSL), while the normal stress remains constant. There is no obvious strength reduction in the interface, since the normal stress was not changed through the entire loading process. The interface resists the cyclic loads through densified normal displacement.

In addition to the load–displacement responses, the void ratio–displacement responses of the interface are also different under CNS and CNL loading conditions, as shown in Figures 2d and 3d. The void ratio of the interface mobilises and decreases with the decreasing normal stress at the CNS loading condition. However, at the CNL loading condition, the normal stress remains constant; even so, the void ratio of the interface mobilises towards a smaller value due to the variation of the normal displacement.

Compared to other existing models using generalised plasticity [19] or a nonlinear incremental approach [29], this model does not require additional plastic potential to capture the nonassociated plastic flow of the soil–structure interface, where the dilatant volumetric response can be well simulated. Even though similar models using fractional plasticity was also developed in [18], their model cannot accurately simulate the strength mobilisation along with the contraction and dilation of the interface under cyclic loading, as shown in Figures 2 and 3.

Figure 4 further evaluates the effect of shear displacement amplitude on the stress–displacement behaviour of the Zipingpu rockfill–steel interface. It can be observed from Figure 4a that the strength of the interface under CNS mobilises negatively under cyclic loads. With the increase in the displacement amplitude, the extent of strength reduction decreases due to the higher extent of normal dilation of the interface, as shown in Figure 4c. This is in accord with the critical state soil mechanics that dilation would contribute to the material strength. In addition, it can be observed from Figure 4b that displacement amplitude has an insignificant influence on strength mobilisation of the interface under CNL. However, a decrease the displacement amplitude would also increase the normal contraction of the interface, as shown in Figure 4d.

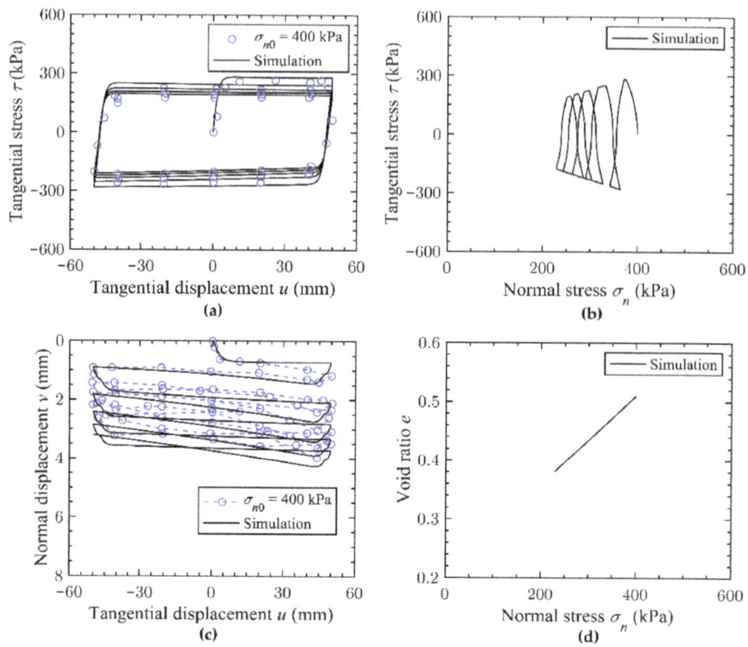

Figure 2. Zipingpu rockfill–steel interface [11] under CNS: (**a**) tangential displacement vs. tangential stress; (**b**) normal stress vs. tangential stress; (**c**) tangential displacement vs. normal displacement; (**d**) normal stress vs. void ratio.

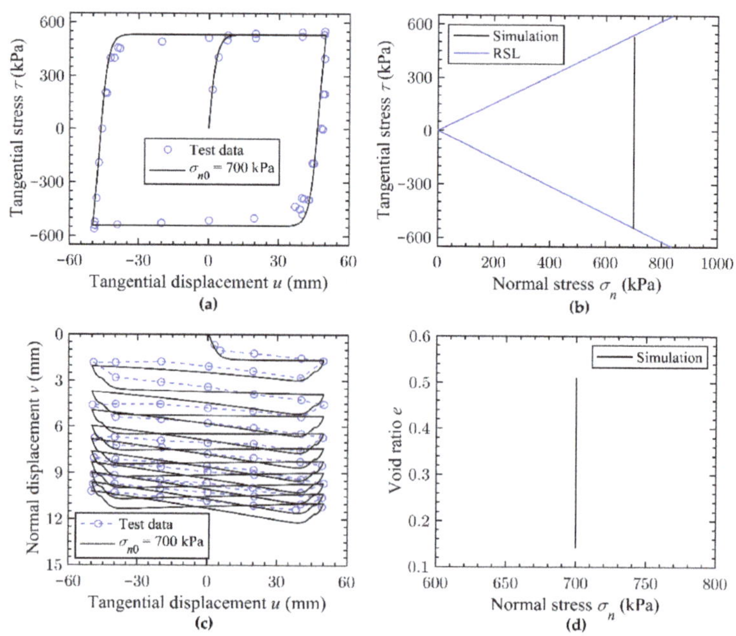

Figure 3. Zipingpu rockfill-steel interface [11] under CNL: (**a**) Tangential displacement v.s. tangential stress; (**b**) Normal stress v.s. tangential stress; (**c**) Tangential displacement v.s. normal displacement; (**d**) Normal stress v.s. void ratio.

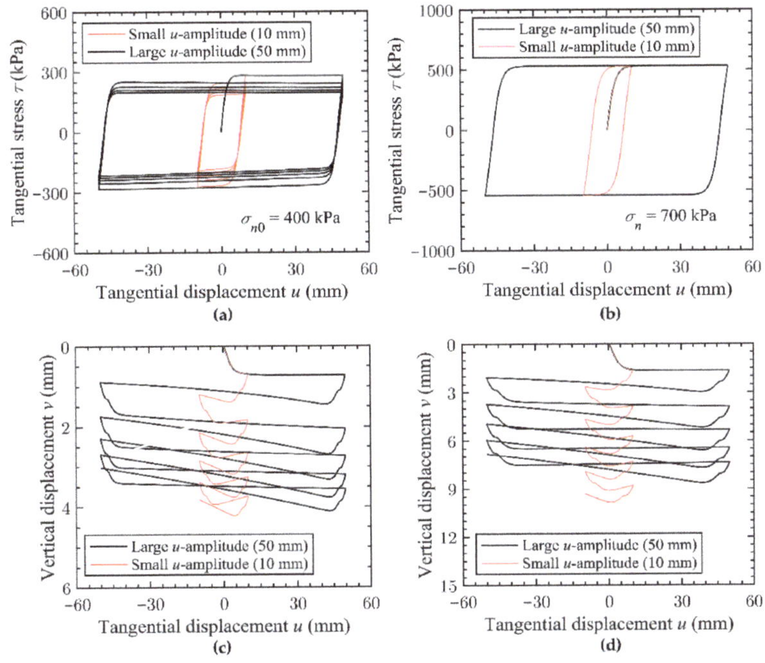

Figure 4. Effect of displacement amplitude on the stress–displacement behaviour of the Zipingpu rockfill–steel interface [11]: (**a**) tangential displacement vs. tangential stress under CNS; (**b**) tangential displacement vs. tangential stress under CNL; (**c**) tangential displacement vs. normal displacement under CNS; (**d**) tangential displacement vs. normal displacement under CNL.

4. Conclusions

This study made an attempt to develop an advanced elastoplastic model for characterising cyclic mobilisation of the granular soil–structure interface by adopting the framework of fractional plasticity. The main conclusions can be drawn as follows:

(a) Without using an additional plastic potential function, the method developed can consider the nonassociated plastic flow response of the soil–structure interface by using fractional order derivatives of the plastic yielding function.

(b) By incorporating the effect of material state and soil fabric on the plastic response of the soil–structure interface, a fabric-enriched state-dependent fractional stress-dilatancy equation and a state-dependent hardening modulus were suggested.

(c) The developed model contained eleven parameters, which can be all determined from laboratory test results. Numerical simulations of cyclic mobilisation of the Zipingpu rockfill–steel interface revealed that the model can characterise the mobilised strength and displacement of the soil–structure interface under different boundary conditions in a reasonable manner.

(d) The variations of tangential stress and void ratio with the normal stress were different under CNS and CNL conditions. The strength of the interface under CNS conditions reduced with the number of load cycles. As the displacement amplitude increased, the extent of the strength reduction decreased. However, displacement amplitude had a limited impact on strength mobilisation of the interface under CNL.

(e) The current model can only simulate the two-dimensional stress-displacement of the soil–structure interface. Further work on developing a multiaxial fractional plastic model for capturing the true three-dimensional response of the interface needs to be carried out.

Author Contributions: Conceptualisation and methodology, Y.S. (Yifei Sun) and Y.S. (Yang Shen); software, validation and data curation, J.X.; writing—original draft preparation, J.X. and Y.S. (Yang Shen); writing—review and editing, Y.S. (Yifei Sun). All authors have read and agreed to the published version of the manuscript.

Funding: This research was funded by the Natural Science Foundation of Jiangsu Province (No. BK20180776).

Institutional Review Board Statement: Not applicable.

Informed Consent Statement: Not applicable.

Data Availability Statement: The data presented in this study are available on request from the corresponding author. The data are not publicly available due to intellectual property.

Acknowledgments: The third author appreciates Wen Chen for his invaluable inspiration.

Conflicts of Interest: The authors declare no conflict of interest.

References

1. Raheem, S.A.; Hayashikawa, T.; Hashimoto, I. Effects of soil–foundation–superstructure interaction on seismic response of cable-stayed bridges tower with spread footing foundation. *J. Struct. Eng.* **2003**, *49*, 475–486.
2. Zaman, M.M.; Desai, C.S.; Drumm, E.C. Interface model for dynamic soil-structure interaction. *J. Geotech. Eng.* **1984**, *110*, 1257–1273. [CrossRef]
3. Desai, C.S. *Mechanics of Materials and Interfaces: The Disturbed State Concept*; CRC Press: Boca Raton, FL, USA, 2000.
4. Kamura, A.; Kazama, M. Assessment of stiffness degradation of soil by in-situ cyclic loading using pressuremeter. In Proceedings of the 6th International Conference on Geotechnical and Geophysical Site Characterization, Budapest, Hungary, 26–29 September 2021.
5. Gao, Y.; Zhang, N.; Li, D.; Liu, H.; Cai, Y.; Wu, Y. Effects of topographic amplification induced by a U-Shaped canyon on seismic waves. *Bull. Seismol. Soc. Am.* **2012**, *102*, 1748–1763. [CrossRef]
6. Zhang, N.; Gao, Y.; Cai, Y.; Li, D.; Wu, Y. Scattering of SH waves induced by a non-symmetrical V-shaped canyon. *Geophys. J. Int.* **2012**, *191*, 243–256. [CrossRef]
7. Gao, Y.; Wu, Y.; Li, D.; Liu, H.; Zhang, N. An improved approximation for the spectral representation method in the simulation of spatially varying ground motions. *Probabilistic Eng. Mech.* **2012**, *29*, 7–15. [CrossRef]
8. Gao, Y.; Wu, Y.; Cai, Y.; Liu, H.; Li, D.; Zhang, N. Error assessment for spectral representation method in random field simulation. *J. Eng. Mech.* **2012**, *138*, 711–715. [CrossRef]
9. Maghsoodi, S.; Cuisinier, O.; Masrouri, F. Non-isothermal soil-structure interface model based on critical state theory. *Acta Geotech.* **2021**, *16*, 2049–2069. [CrossRef]
10. Kishida, H.; Uesugi, M. Tests of the interface between sand and steel in the simple shear apparatus. *Géotechnique* **1987**, *37*, 45–52. [CrossRef]
11. Zhang, G.; Zhang, J. Monotonic and cyclic tests of interface between structure and gravelly soil. *Soils Found.* **2006**, *46*, 505–518. [CrossRef]
12. Martinez, A.; Frost, J.D. Undrained behavior of sand–structure interfaces subjected to cyclic torsional shearing. *J. Geotech. Geoenviron. Eng.* **2018**, *144*, 04018063. [CrossRef]
13. Wang, H.; Zhou, W.; Yin, Z.; Jie, X. Effect of grain size distribution of sandy soil on shearing behaviors at soil–structure interface. *J. Mater. Civ. Eng.* **2019**, *31*, 04019238. [CrossRef]
14. Grabowski, A.; Nitka, M.; Tejchman, J. 3D DEM simulations of monotonic interface behaviour between cohesionless sand and rigid wall of different roughness. *Acta Geotech.* **2021**, *16*, 1001–1026. [CrossRef]
15. Lashkari, A.; Jamali, V. Global and local sand–geosynthetic interface behaviour. *Géotechnique* **2021**, *71*, 346–367. [CrossRef]
16. Wang, R.; Dafalias, Y.F.; Fu, P.; Zhang, J.M. Fabric evolution and dilatancy within anisotropic critical state theory guided and validated by DEM. *Int. J. Solids Struct.* **2019**, *188–189*, 210–222. [CrossRef]
17. Zhang, F.; Gao, Y.; Leshchinsky, D.; Yang, S.; Dai, G. 3D effects of turning corner on stability of geosynthetic-reinforced soil structures. *Geotext. Geomembr.* **2018**, *46*, 367–376. [CrossRef]
18. Sun, Y.; Sumelka, W.; Gao, Y. Reformulated fractional plasticity for soil-structure interface. *Mech. Res. Commun.* **2020**, *108*, 103580. [CrossRef]
19. Saberi, M.; Annan, C.D.; Konrad, J.M. A unified constitutive model for simulating stress-path dependency of sandy and gravelly soil–structure interfaces. *Int. J. Non-Linear Mech.* **2018**, *102*, 1–13. [CrossRef]
20. Stutz, H.; Mašín, D. Hypoplastic interface models for fine-grained soils. *Int. J. Numer. Anal. Meth. Geomech.* **2017**, *41*, 284–303. [CrossRef]
21. Sun, Y.; Gao, Y.; Zhu, Q. Fractional order plasticity modelling of state-dependent behaviour of granular soils without using plastic potential. *Int. J. Plast.* **2018**, *102*, 53–69. [CrossRef]

22. Sun, Y.; Sumelka, W. Multiaxial stress-fractional plasticity model for anisotropically overconsolidated clay. *Int. J. Mech. Sci.* **2021**, *205*, 106598. [CrossRef]
23. Sun, Y.; Sumelka, W.; Gao, Y.; Nimbalkar, S. Phenomenological fractional-order stress-dilatancy model for granular soil and soil-structure interface under monotonic and cyclic loads. *Acta Geotech.* **2021**, *16*, 3115–3132. [CrossRef]
24. Sun, Y.; Gao, Y.; Song, S.; Chen, C. Three-dimensional state-dependent fractional plasticity model for soils. *Int. J. Geomech.* **2020**, *20*, 04019161. [CrossRef]
25. Lu, D.; Liang, J.; Du, X.; Ma, C.; Gao, Z. Fractional elastoplastic constitutive model for soils based on a novel 3D fractional plastic flow rule. *Comput. Geotech.* **2019**, *105*, 277–290. [CrossRef]
26. Lu, D.; Zhou, X.; Du, X.; Wang, G. A 3D fractional elastoplastic constitutive model for concrete material. *Int. J. Solids Struct.* **2019**, *165*, 160–175. [CrossRef]
27. Sun, Y.; Sumelka, W.; Gao, Y. Bounding surface plasticity for sand using fractional flow rule and modified critical state line. *Arch. Appl. Mech.* **2020**, *90*, 2561–2577. [CrossRef]
28. Dafalias, Y.F.; Manzari, M.T. Simple plasticity sand model accounting for fabric change effects. *J. Eng. Mech.* **2004**, *130*, 622–634. [CrossRef]
29. Yang, J.; Yin, Z.Y. Soil-structure interface modeling with the nonlinear incremental approach. *Int. J. Numer. Anal. Meth. Geomech.* **2021**, *45*, 1381–1404. [CrossRef]

fractal and fractional

MDPI

Article

Influence of Groundwater Depth on Pile–Soil Mechanical Properties and Fractal Characteristics under Cyclic Loading

Bingxiang Yuan [1], Zhijie Li [1], Weijie Chen [1], Jin Zhao [1], Jianbing Lv [1,*], Jie Song [2] and Xudong Cao [2]

[1] School of Civil and Transportation Engineering, Guangdong University of Technology, Guangzhou 510006, China; yuanbx@gdut.edu.cn (B.Y.); 2112009177@mail2.gdut.edu.cn (Z.L.); 2112009168@mail2.gdut.edu.cn (W.C.); 2112009070@mail2.gdut.edu.cn (J.Z.)

[2] MCC Capital Engineering & Research Incorporation Limited, Beijing 100176, China; songjie@ceri.com.cn (J.S.); caoxudong@ceri.com.cn (X.C.)

* Correspondence: ljb@gdut.edu.cn

Abstract: The analysis of the behavior of soil and foundations when the piles in offshore areas are subjected to long-term lateral loading (wind) is one of the major problems associated with the smooth operation of superstructure. The strength of the pile-soil system is influenced by variations in the water content of the soil. At present, there are no studies carried out analyzing the mechanical and deformational behavior of both the material of the laterally loaded piles and soil with groundwater level as a variable. In this paper, a series of 1 g model tests were conducted to explore the lateral behavior of both soil and monopile under unidirectional cyclic loading, based on the foundation of an offshore wind turbine near the island. The influence of underground water level and cyclic load magnitude on the performance of the pile–soil system was analyzed. To visualize the movements of soil particles during the experimental process, particle image velocimetry (PIV) was used to record the soil displacement field under various cyclic loading conditions. The relationship curves between pile top displacement and cyclic steps, as well as the relationship curves between cyclic stiffness and cyclic steps, were displayed. Combined with fractal theory, the fractal dimension of each curve was calculated to evaluate the sensitivity of the pile–soil interaction system. The results showed that cyclic loading conditions and groundwater depth are the main factors affecting the pile–soil interaction. The cyclic stiffness of the soil increased in all test groups as loading progressed; however, an increase in the cyclic load magnitude decreased the initial and cyclic stiffness. The initial and cyclic stiffness of dry soil was higher than that of saturated soil, but less than that of unsaturated soil. The ability of the unsaturated soil to limit the lateral displacement of the pile decreased as the depth of the groundwater level dropped. The greater the fluctuation of the pile top displacement, the larger the fractal dimension of each relationship curve, with a variation interval of roughly 1.24–1.38. The average increment of the cumulative pile top displacement between each cycle step following the cyclic loading was positively correlated with fractal dimension. Based on the PIV results, the changes in the pile–soil system were predominantly focused in the early stages of the experiment, and the short-term effects of lateral cyclic loading are greater than the long-term effects. In addition, this research was limited to a single soil layer. The pile–soil interaction under layered soil is investigated, and the results will be used in more complex ground conditions in the future.

Keywords: laterally loaded piles; cyclic loading; unsaturated soil; pile–soil interaction; fractal dimension; model tests; particle image velocimetry

Citation: Yuan, B.; Li, Z.; Chen, W.; Zhao, J.; Lv, J.; Song, J.; Cao, X. Influence of Groundwater Depth on Pile–Soil Mechanical Properties and Fractal Characteristics under Cyclic Loading. *Fractal Fract.* **2022**, *6*, 198. https://doi.org/10.3390/fractalfract6040198

Academic Editors: Zine El Abiddine Fellah and Wojciech Sumelka

Received: 7 February 2022
Accepted: 28 March 2022
Published: 1 April 2022

1. Introduction

The target of this article was to analyze the mechanical and deformational behavior of the soil and the pile under lateral cyclic loading, as well as the results from model tests conducted to estimate the influence of changes in groundwater level and loading conditions on pile–soil interaction. This research can give some reference value to practical engineering.

The monopile is the most popular foundation in offshore wind power projects [1–4]. Currently, 81% of offshore wind turbines have already used monopiles as support structures [5]. One of the competitive advantages of this type is that it has relatively lower installation costs than other foundations and has enough load-bearing capacity in sea areas with water depths of less than 30 m [6–8]. The behavior of monopile foundations has been extensively studied under lateral loading. Due to structural movement, particle migration would occur in the soil around the pile [9–11]. Offshore monopiles are subject to the influence of water for long periods of time, which is different from traditional pile foundations [12–14]. Some tests carried out onshore did not take into account the geotechnical characteristics of the soil in the sea [15–18], but the role of water could not be ignored. The presence of water in sand can easily induce a series of problems that seriously endanger people's lives and property, such as gushing water and surface collapse [19–22]. Compared to clay, water-rich sand layers are more fluid and less cemented, and the separation between water and sand makes the interaction between them more clearly influenced [23–26]. Therefore, it is worth determining the behavior of water-rich soil (sand). The performance of pile foundations in water-rich sand has become a very important issue in engineering applications and academic circles.

Some research and valuable results for monopiles in water-rich strata have been made. The presence of water has multiple effects on soil and structure [27–32]. The dynamic properties of pile foundations in water-rich sand could be profoundly influenced by ground motion parameters [33]. Several analytical solutions for the resistance of saturated soil to a laterally loaded pile have been proposed by many academics to assess the effects of liquid density and interstitial fluid bulk modulus on the stiffness and damping of soil [34,35]. Numerous experiments and engineering examples have shown that the bearing capacity of the pile is mostly determined by the properties of the soil [36]. In contrast, the shaft resistance in saturated sand is less than in dry sand [37,38]. The water flow can also reduce the lateral bearing capacity of the pile [8,12,39,40]. Cui et al. examined the effects of soil porosity, shear modulus, and the thickness of the saturated soil layer at the base of the pile on the dynamic performance of the floating pile [41]. Much of the existing previous work has focused solely on the influence of the soil on the structure under various conditions, neglecting the counteraction of the structure on the soil. Moreover, the soil employed in these studies was saturated. However, for offshore wind turbines adjacent to islands and land, rising and falling tides frequently produce fluctuations in the groundwater levels, and the soil may stay unsaturated for an extended length of time in real settings. Changes in groundwater levels can also have a substantial influence on the mechanical characteristics of the soil.

The soil particles around the pile are constantly rearranged under the lateral cyclic loading. In addition, the plastic deformation of the soil around the pile will accumulate, which leads to the fluctuation of the cumulative deformation, especially at the pile head [42–45]. In previous studies, the analysis of this kind of fluctuation was relatively complicated, and the accumulated deformation was often normalized before discussion [46,47]. It was difficult to directly compare the fluctuation degree about the response of the pile affected by the cycle loading under different test conditions. This paper introduced fractal theory to quantify this fluctuation. The application of fractal theory to analyze the degree of the fluctuation of the curve has become common in a number of research areas. Some researchers have introduced fractal theory to investigate the correlation between the alterations of EMG signals and stride-to-stride variability at various walking speeds [48,49]. In the field of energy, the degree of fluctuation of the daily electrical load curve was successfully calculated under the box-counting method in order to predict future electrical energy demand [50]. Yan et al. and Shu et al. used fractal dimension analysis to provide a comprehensive assessment of wind speed fluctuation over time in various terrain settings [51,52]. Fractals have been widely used in engineering [53–55]. Xiao et al. investigated the various characteristics of the pile–soil interface under sulphury acid corrosion using fractal dimension. The fractal dimension and the contact friction angle have a non-linear connection [56]. Li et al. and

Wang et al. pioneered the use of fractal theory in the study of the microstructure of various cement materials. The findings showed that cement pore structure and crystal shape are linked to hydration [57–59]. Duan et al. have summarized the fractal characteristics about surface cracks of failed coal specimens under different experimental conditions [60]. At present, there is a lack of quantitative analysis using fractal dimension on the fluctuation of the response index of pile–soil systems to cyclic loading.

In this paper, a series of model tests were conducted to simulate a driven monopile at different water levels subjected to lateral cyclic loading. Combined with PIV, the evolution process of the displacement field of soil around a laterally loaded pile under cyclic loading was visualized. The fluctuation of displacement at the pile top and the stiffness of soil around the pile during cyclic loading were successfully quantitatively evaluated by fractals. Then, the potential reasons for the difference in fractal dimension among the experimental groups were discussed. Finally, the relationship between fractal dimension and the variance of cumulative deformation was analyzed.

This study represents the first experimental investigation of the pile–soil interaction of the monopile subjected to lateral cyclic loading under groundwater level variations and introduces fractal theory into the field of model tests of the monopile. The experimental data presented in this paper are particularly useful for modeling calibration and validation in future numerical studies of monopile evaluation in water-rich areas. This research is an important step in developing a technique for assessing the response of the pile–soil system in water-rich sand under cyclic loading.

2. Materials and Methods

2.1. Model Box

This research employed a scaled model test of a single pile under lateral loading. The size of the model box used was 230 mm in length, 180 mm in width, and 320 mm in height. Fixed pulleys were installed on the outer wall of the model box to ensure that the lateral loading application direction is horizontal. A working platform with a laser displacement meter was installed at the elevation of the pile top and on the other side of a fixed pulley device. At the corner of the model box, a water level observation pipe was set up with a 5 mm thick plexiglass plate. The pipe had a permeable hole at the bottom of the model box. The pipe simultaneously served as a channel for groundwater packing. The model box design is shown in Figure 1.

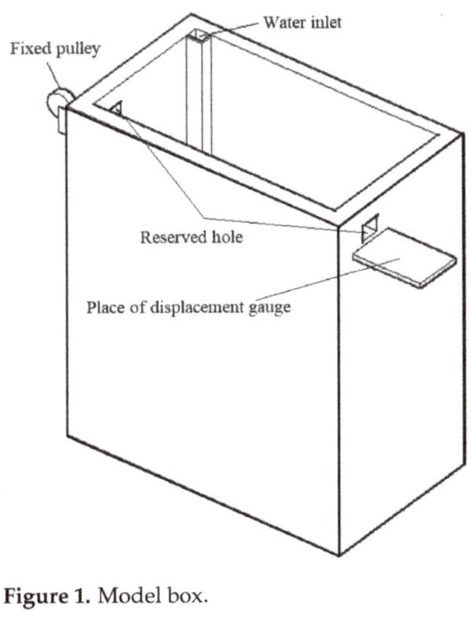

Figure 1. Model box.

2.2. Test Model Pile

The prototype of this test was a C80 large-diameter anti-slide concrete square pile with a width of 3 m, a length of 6 m, and a height of 60 m. For the scaled model test, geometric similarity, motion similarity, dynamic similarity and constitutive similarity are necessary conditions in producing consistent experimental results, and the similarity ratio used for similar experimental results is shown in Table 1 [61]. The test selected a size similarity ratio of 1:300 for the model test, so the section size of the half-mold scaled test model pile was 10 mm in length, 10 mm in width, and 200 mm in height. According to the material and size parameters of the prototype, the bending stiffness of the prototype pile was calculated to be 256.5 GN·m^2, and the bending stiffness of the converted model pile was 31.6 N·m^2. According to this calculated value, a square hollow aluminum tube with an elastic modulus of 60 GPa was selected as the material of the model pile. Its actual bending stiffness was calculated to be 29.5 N·m^2. The specific parameters of the model pile are shown in Table 2. The assembled strain gauge on the side of the model pile is shown in Figure 2.

Table 1. Proportional relation between the model and prototype [61].

Parameter	Scale (Model/Prototype)
Cross-sectional area	$1/n^2$
Pile length	$1/n$
Bending stiffness	$1/n^4$

Table 2. The physical properties of the model pile material.

Pile Length	Width of Pile Section	Pile Thickness	Moment of Inertia I	Bending Stiffness
L (mm)	D (mm)	d (mm)	(mm^4)	EI (N·m^2)
200	10	1	492	29.5

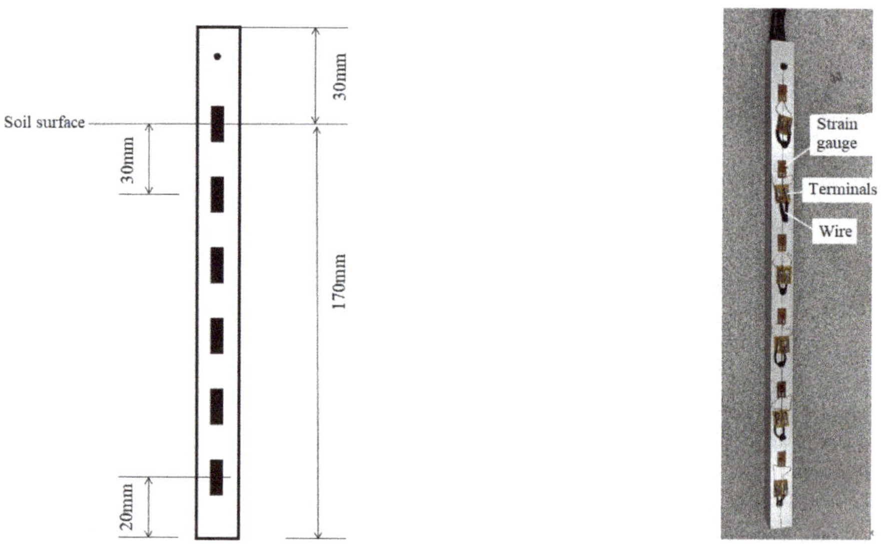

(a) **(b)**

Figure 2. Model pile. (**a**) Layout distance of pile body strain gauge. (**b**) Model pile.

2.3. Test Soil

In order to better control the consistency of the test soil structure when installing the mold and reduce the interference caused by the uneven assembly of soil particles on the test results, this test selected poorly graded Fujian standard sand. The physical properties of the soil particles are shown in Table 3, the relative density is 2.64, the maximum pore ratio is 0.855, and the minimum pore ratio is 0.604. The mass of sand required to achieve a relative compactness of $D_r = 0.3$ can be obtained through Equations (1) and (2). The filling steps are as follows: (1) filling in layers, with each layer filled to a thickness of 50 mm, and the speed of filling was slow enough to obtain a loose soil particle structure. (2) When the filling thickness of the soil in the model box reached 150 mm, the model pile should be set to the inner wall of the model box, the verticality of the model pile should be corrected, and then continue to fill. (3) After filling the soil, the tube was placed at the bottom of the water inlet of the model box, and then the peristaltic pump was used to fill the sand with water. In order to ensure that the compactness of the sand was not affected by the dynamic hydraulic force during the water filling process, the peristaltic pump was set to work at a slow speed of 20 revolutions per minute. The sample is shown in Figure 3.

Table 3. Physical parameters of Fujian standard sand.

Specific Gravity d_s (g/cm^3)	Minimum Dry Density ρ_{dmin} (g/cm^3)	Maximum Dry Density ρ_{dmax} (g/cm^3)	Minimum Void Ratio e_{min}	Maximum Void Ratio e_{max}	Internal Friction Angle (°)
2.64	1.332	1.628	0.604	0.855	36

$$D_r = \frac{e_{max} - e}{e_{max} - e_{min}} \tag{1}$$

$$\rho_d = \frac{d_s \rho_w}{1 + e} \tag{2}$$

D_r—Relative density of soil;
e—The current void ratio of the soil;
ρ_d—Pore volume;
d_s—Relative density of soil;
ρ_w—The density of water.

(a) **(b)**

Figure 3. The sand packed in the box. (**a**) Packed sand. (**b**) Packed sand with groundwater.

2.4. Acquisition System

Currently, in the field of displacement measurement, PIV is the most commonly used soil deformation measurement technology. Stanier et al. [62] and White et al. [63], from the University of Cambridge, first applied this speed measurement technology to geotechnical engineering tests. Gudehus and Nubel [64] used PIV technology in physical models to study the properties of soil particles and visualize the soil shear zones. Because of its technical advantages, such as relatively simple equipment and convenient operation, this technology has gained considerable popularity and additional development in the field of geotechnical engineering over the years [65]. Yuan et al. found that the PIV technology could be used as an application in measuring the displacement field of soil, especially in the study of the soil settlement around the pile and the foundation [66].

The collection of data in this experiment was divided into the collection of movement images of the piles and soil, the collection of pile top displacement, and the collection of pile body strain. In the first phase of data collection, a high-resolution camera was used to make images of the front of the model box. A Panasonic HG-C1030 laser displacement meter was used to collect the change of the pile top displacement during the test loading process. Finally, a computer was connected to the Donghua Test DH3815N static strain collector to collect information relevant to pile strain. The schematic diagram of the overall model test device is shown in Figure 4. The test system mainly consisted of a model box, a model pile, a static strain collector, a high resolution camera, and a computer.

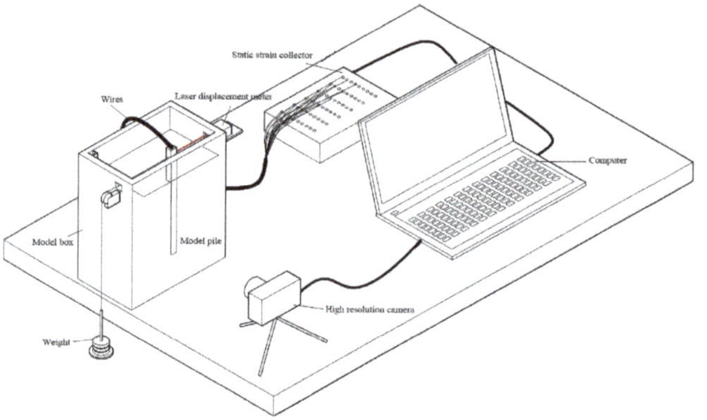

Figure 4. Overall model test device.

2.5. Cyclic Loading Test Plan

The control variables for this test were loading and groundwater depth conditions for ten sets of tests. The number of cyclic loading steps for each set of tests is 20. Five different groundwater depth conditions were set: dry sand, saturated sand, a groundwater depth of 40 mm, a groundwater depth of 90 mm, and a groundwater depth of 140 mm. Two loads with different load peaks were set for cyclic loading, 3 N and 4 N respectively. The test was divided into ten groups. The test plan and test group numbers are shown in Table 4. The specific test process was as follows.

The model box was filled with sand in layers. After the filling, the sand was left to stand for 1 h, allowing the soil particles in the model box to stabilize. Then the camera was turned on to record the front of the box, the laser displacement meter was turned on, the strain collector was turned on, and the relevant parameters were set on the computer. In this study, the weights were used to produce the cyclic loading, and one hundred gram weights were loaded and/or unloaded sequentially, with an interval of about 30 s between each step of loading and unloading. The data were recorded after the displacement of the pile top and the strain of the pile body remained unchanged for 10 s.

Table 4. Cyclic loading test scheme.

Loading Conditions	Group	*H*max = 3 N *H*min = 0 N	Group	*H*max = 4 N *H*min = 0 N
Groundwater depth conditions	1	Saturated sand	6	Saturated sand
	2	Dry sand	7	Dry sand
	3	Groundwater depth 140 mm	8	Groundwater depth 140 mm
	4	Groundwater depth 90 mm	9	Groundwater depth 90 mm
	5	Groundwater depth 40 mm	10	Groundwater depth 40 mm

After the experiment, the model pile was taken out, the water in the soil was pumped out using a peristaltic pump, and the sand was first placed in an oven for 24 h to dry at 105 °C and then placed in a drying oven to air at room temperature. After that procedure, the next experiment could be started.

In order to prevent the occurrence of accidental errors, repeated tests were carried out for each experimental group. If the difference between the two tests (such as the displacement of the pile top and the strain data) was less than 10%, the group with the better photo quality would be selected for analysis. If the error between the two tests was greater than 10%, the test materials and equipment would be necessary to check, and the result would be annulled.

2.6. Theoretical Method

2.6.1. Cyclic Secant Stiffness

For the general research on the cyclic bearing characteristics of laterally loaded piles, the cyclic secant stiffness is a commonly used evaluation parameter in various pile–soil interaction studies. Zhang et al. believed that the cyclic secant stiffness calculated by the cyclic loading amplitude and the full amount of pile deformation was not suitable for evaluating the cyclic bearing characteristics of laterally loaded piles in loose sand foundations [67]. That is because, in the sand, when the pile is subjected to lateral loading, the compression deformation is usually the main factor, and its bearing characteristics increase to a certain extent with the increasing number of cycles. The research of Zhang referred to the previous calculation experience on the cyclic stiffness of the laterally loaded pile and used the cyclic loading amplitude and the lateral deformation increment of the pile to construct the laterally loaded pile cyclic stiffness parameter. The calculation method is shown in Figure 5.

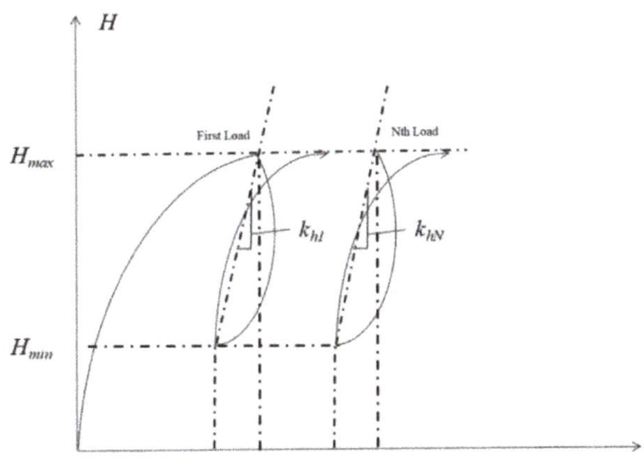

Figure 5. Calculation method for lateral cyclic stiffness of pile.

The lateral cyclic stiffness of the laterally loaded pile under the Nth cyclic loading is:

$$k_{hN} = \frac{H_{\max} - H_{\min}}{y_{N\max} - y_{N\min}} \quad (3)$$

$H_{\max}$—Peak cyclic loading;
$H_{\min}$—Minimum value of cyclic loading;
$y_{N\max}$—The maximum displacement at the Nth cycle;
$y_{N\min}$—Cumulative displacement at the Nth cycle.

2.6.2. Fractal Dimension

In fractal theory, fractal dimension is a quantitative index whose value increases with the increase of complex shapes and forms of graphics [53]. The effective value of fractal dimension depends on the proper calculation method, in which the box-counting method is usually used [68].

In the box counting method, a divided mesh with a grid length of h is used to cover the image whose fractal dimension needs to be calculated. Then, the number of grids overlapped with the graphic characters in the image is counted as "$N(h)$". When h is constantly changing, the relationship between h and $N(h)$ can be obtained and expressed by Equation (4).

$$N(h) = kh^{-D} \quad (4)$$

where D = fractal dimension; h = grid length; $N(h)$ = number of grids; and k is a constant. D can be obtained by removing logarithm from both sides of Equation (4), as shown in Equation (5).

$$\ln N(h) = -D \ln h + \ln k \quad (5)$$

It is easy to know that D is the slope of the logarithmic equation represented by Equation (5). The image processing process is shown in Figure 6. First, the image is binarized, and then meshed. In this paper, the fractal dimension of Figures 7 and 8 is calculated.

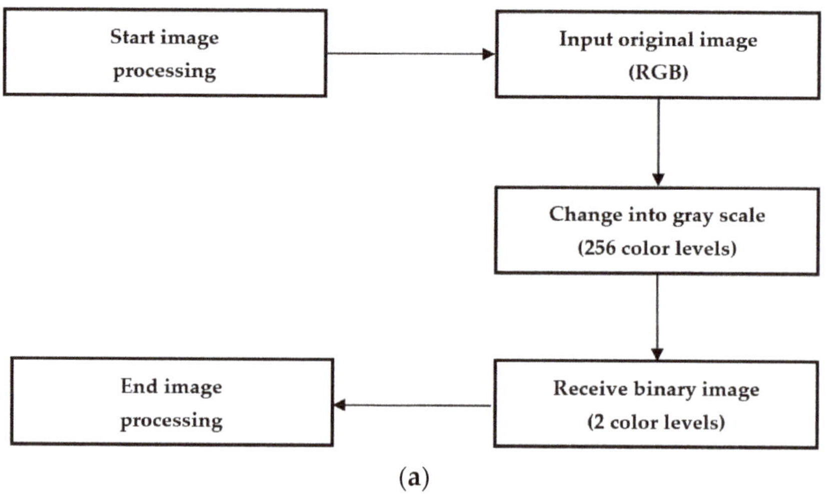

(a)

Figure 6. *Cont.*

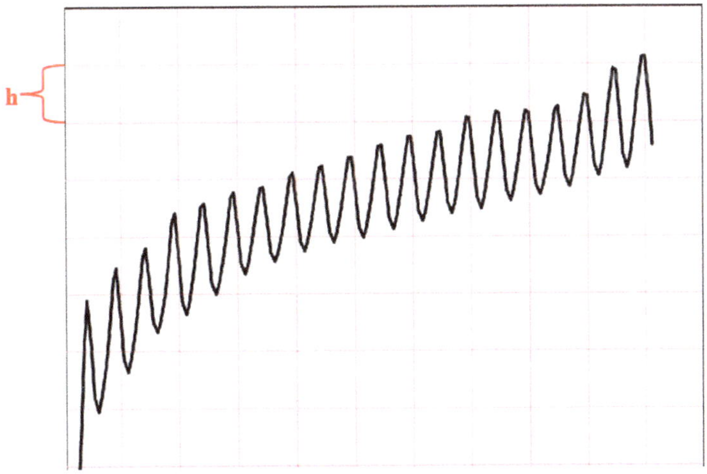

(b)

Figure 6. Image processing procedure. (**a**) flowchart of image processing. (**b**) Meshing (box-counting method).

3. Results and Discussion

3.1. Analysis of Cumulative Displacement of Pile Top Cyclic Loading

The displacement curve of the pile top under each step of cyclic loading was obtained by measuring the displacement of the pile top of the model pile on the soil surface. The test analysis results are shown in Figure 7. Loading each test to the same number of steps, the cumulative pile top displacement with a cyclic loading peak of 3 N is sorted in descending order of group 1, group 2, group 3, group 4, group 5, and the cumulative pile top displacement with a cyclic loading peak value of 4 N is in descending order of group 6, group 7, group 8, group 9, and group 10.

The detailed incremental data of the displacement of each test pile top with the increase of cycle steps are shown in Table 5.

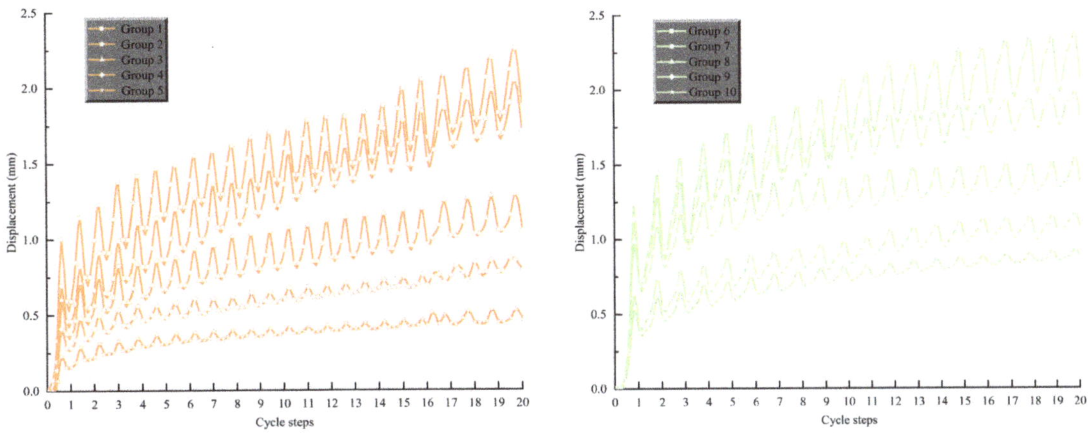

(a) **(b)**

Figure 7. Curve of pile top displacement—cyclic loading steps. (**a**) *H*max = 3 N *H*min = 0 N. (**b**) *H*max = 4 N *H*min = 0 N.

Table 5. Detailed data of pile top accumulated displacement.

Maximum and Minimum Values of Cyclic Loading	Group	y_{1min} (mm)	y_{20min} (mm)	E(δ) (mm)	$S^2(\delta)$	COV
$H_{max} = 3$ N $H_{min} = 0$ N	1	0.595	1.67	0.0566	16.1×10^{-4}	1.29
	2	0.46	1.52	0.0558	5.5×10^{-4}	1.51
	3	0.379	0.961	0.0306	4×10^{-4}	0.754
	4	0.284	0.691	0.0214	2×10^{-4}	0.482
	5	0.159	0.429	0.0142	1.4×10^{-4}	0.313
$H_{max} = 4$ N $H_{min} = 0$ N	6	0.735	2.195	0.0768	55.7×10^{-4}	2.11
	7	0.728	1.87	0.0601	33.7×10^{-4}	1.60
	8	0.694	1.353	0.0347	15.9×10^{-4}	0.904
	9	0.426	1.062	0.0335	8×10^{-4}	0.977
	10	0.378	0.873	0.0261	4.2×10^{-4}	0.753

The detailed data of accumulated pile top displacement are obtained after processing the data in Figure 7. Groups 6 to 10 showed the ratio of the cumulative pile top displacement in the first step to the cumulative pile top displacement in the twentieth step exceeded by 35%. That is to say, the plastic deformation caused by the pile body to the soil around the pile is the largest during the first cyclic loading, and, as the number of cyclic steps increases, the cumulative displacement increment of the pile top of each test pile decreases, which is consistent with the conclusion of previous cyclic loading tests in sand with different densities [67]. The short-term effect of cyclic cumulative displacement of a laterally loaded pile is greater than its long-term effect.

Several more representative data were obtained after processing the cumulative displacement increment of the pile top, as shown in Table 5. y_{Nmin} in the table is the cumulative pile top displacement at the end of the Nth cycle; E(δ) is the average increment of the cumulative pile top displacement between each cycle step after the first cyclic loading; and $S^2(\delta)$ is the variance of the cumulative pile top displacement increment data group between each step. COV is the covariance between the cumulative pile top displacement and the number of cycle steps. COV is a measure of how much the cumulative pile top displacement and the number of cycle steps move together. $S^2(\delta)$ shows the change of pile top displacement better than E(δ). $S^2(\delta)$ shows that, when the laterally loaded pile is subjected to cyclic loading, the degree of dispersion of the cumulative plastic deformation increment data of the pile top is related to the ability of the soil around the pile to limit the lateral displacement of the pile.

It can be seen from Table 5 that the COV of all experimental groups is greater than zero, which means a strong positive relationship between the cumulative displacement of the pile top and the number of cycle steps. Under the lateral cyclic loading with the same peak value, as the water level drops, both E(δ) and $S^2(\delta)$ decreased. This shows that the lowering of the groundwater level can increase the soil mass and limit the displacement of the soil around the pile. At the same time, it can be seen that the E(δ) and $S^2(\delta)$ of dry sand are slightly lower than those of saturated sand. This is because, on the one hand, as the groundwater level is at the soil surface, hydrostatic pressure will reduce the effective stress of the soil particles, thereby, reducing the strength of the soil [24,25]. That is the ability to restrict the movement of the pile is reduced, and finally the displacement of the pile top will increase. On the other hand, as the groundwater level rises, the buoyancy of the groundwater on the pile is also greater, which reduces the effect of soil on the pile and also causes the displacement of the pile top to increase.

3.2. Analysis of Cyclic Stiffness of Sand around Pile

According to Figure 8, the cyclic stiffness of the soil increases gradually with the number of cycles for different cyclic load magnitudes and different groundwater levels. For example (Figure 8e), with cyclic peak load H_{max} of 3 N and H_{min} of 0 N, k_{h1} = 43.82 N/mm and k_{h20} = 91.93 N/mm, the cyclic stiffness at the 20th cyclic loading was 2.1 times higher than that at the 1st. Obviously, this is different from the general conclusion that cyclic loading leads to the weakening of soil stiffness [42]. This is mainly because the self-weight stress provided by the soil in the model test is low, while the deformation of sand under low confining pressure is usually dominated by compression.

It was noteworthy that both the lateral initial and cyclic stiffness decreased with the increase of the cyclic loading amplitude. For example (Figure 8e), at H_{max} = 3 N, k_{h1} = 43.82 N/mm and k_{h20} = 91.93 N/mm. At H_{max} = 4 N, k_{h1} = 24.92 N/mm and k_{h20} = 73.26 N/mm. As the loading amplitude rose, the difference between the maximum and minimum displacements during a single cycle increased. This indicates that the cumulative deformation of the monopile increases with the cyclic loading amplitude. This is similar to the conclusion reached by Malakshah et al. in calcareous sands [47], and it also shows that the response behavior of the pile–soil system under cyclic loading is less affected by the type of soil.

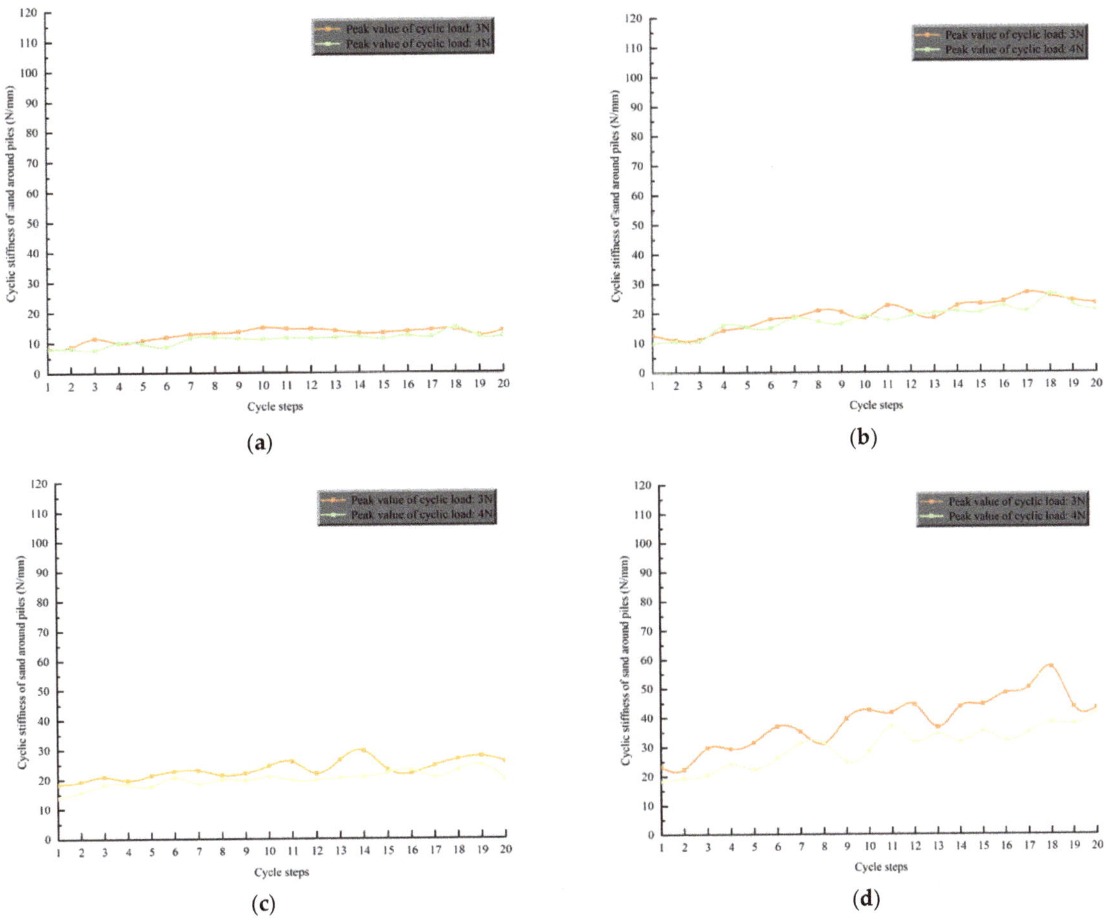

Figure 8. *Cont.*

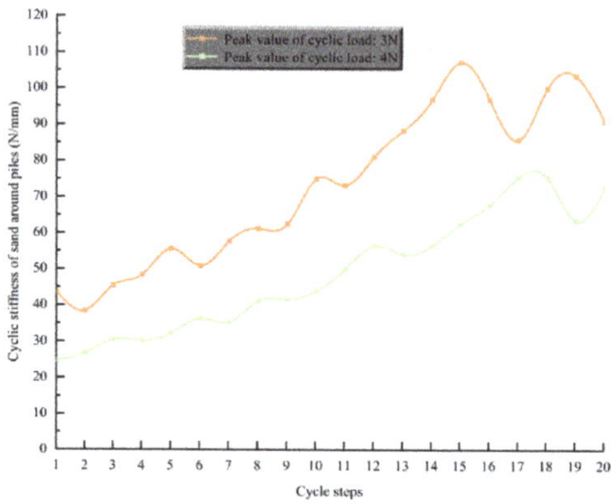

(e)

Figure 8. Curve of cyclic stiffness of sand around piles—cycle steps. (**a**) Group 1 and Group 6, (**b**) Group 2 and Group 7, (**c**) Group 3 and Group 8, (**d**) Group 4 and Group 9, (**e**) Group 5 and Group 10.

Taking the test groups with a loading amplitude of 3 N as an example, when the loading conditions remain unchanged, k_{h20} = 13.82 N/mm in the saturated sand test and k_{h20} = 23.47 N/mm in the dry sand test. Compared with the dry sand test, when the soil was saturated (the groundwater level was at the soil surface), the ability of the sand to limit the displacement of the pile was significantly weakened. The suction stress in saturated soil was the pore water pressure [25]. The grain spacing between soil particles became larger, the effective stress was reduced, and the soil strength decreased. It can be seen that the presence of groundwater can significantly impact the mechanical properties of sand.

When the water level was lower than the soil surface, the constraining capacity of the sand to the laterally loaded piles was obviously enhanced compared to the dry sand test, and the growth rate of the cyclic stiffness of each group was affected by the change in water levels. For example, in the 140 mm groundwater depth tests (Figure 8c), the cyclic stiffness at the 20th cycle increased by 40.3% and 43.2% when compared with the first cycle, the corresponding growth rates in the 90 mm groundwater depth tests (Figure 8d) were 85.9% and 129%, and the corresponding growth rates in the 40 mm groundwater depth tests (Figure 8e) were 109.8% and 194.3%. In Bishop's effective stress, the capillary pressure is regarded as the matrix suction, and this theory is widely adopted in many disciplines [69]. Lowering the water level causes capillary stress or matrix suction to develop above the water level. Capillary water can be considered as a fluid with an isotropic negative pore water pressure in the unsaturated soil. The shrink film produced by the capillary water bears an air pressure greater than the water pressure [23,25], which affects the effective stress of the soil and the strength of the soil is increased. The capillary layer moved down and the growth rate of cyclic stiffness decreased as the groundwater level dropped. This is because the shallow soil has a greater influence on the pile–soil system compared to the deep sand [70].

3.3. *Analysis of Fractal Dimension*

Fractal dimension can describe the fluctuation degree of curve quantitatively. The higher the fractal dimension, the greater the fluctuation of the curve. The curves of the relationship between the pile top displacement and cycle steps are zigzag and has obvious fractal characteristics. The curve of the relationship between the cyclic stiffness of the sand around the pile and cycle steps also has the same fractal characteristics.

Figure 9 shows the fractal dimension of the curve of the relationship between the pile top displacement and the cycle steps. Taking the peak cyclic load 4 N as an example

(Figure 8b), the fractal dimension decreases in the order of saturated sand (group 6), dry sand (group 7), and groundwater depth 140 mm (group 8), 90 mm (group 9) and 40 mm (group 10), which are 1.3728, 1.3107, 1.295, 1.269, and 1.2443, respectively. The lower the fractal dimension, the smoother the displacement curve, which means the difference between y_{max} and y_{min} of the pile in each cycle is small. In other words, the high resistance ability of the soil to deformation represents a small fractal dimension. The displacement of pile top mainly depends on the behavior of shallow sand. Compared with dry sand, the shallower the groundwater table is, the more obvious the matric suction caused by capillary water rising strengthens the shallow sand, which leads to the smaller the fractal dimension.

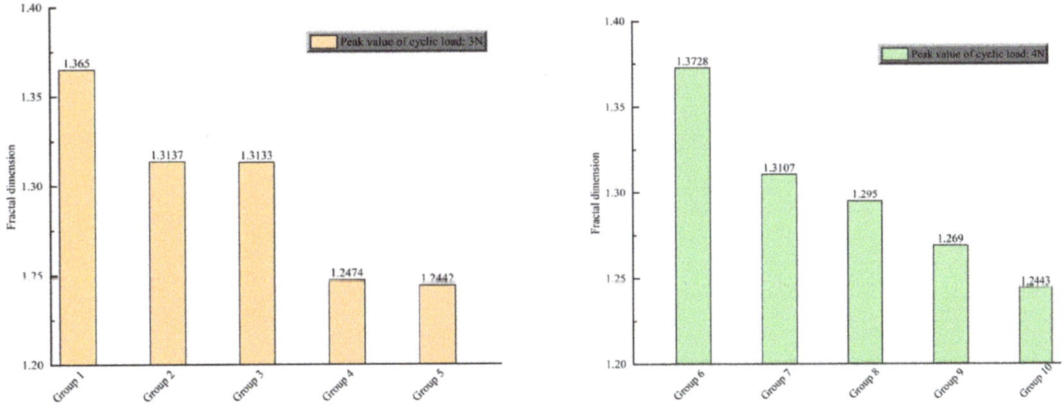

(**a**) (**b**)

Figure 9. Fractal dimension of the curve of the relationship between the pile top displacement and cycle steps. (**a**) Peak value of cyclic load: 3 N. (**b**) Peak value of cyclic load: 4 N.

However, as far as saturated sand is concerned, the shallow sand is saturated and the matric suction has disappeared. The water film between the sand particles in the shallow sand plays a "lubricating role", which makes the sliding between the sand particles easier. Therefore, under the action of cyclic load, the displacement range of sand is naturally the largest, and the fractal dimension is also the largest. When the peak cyclic load is 3 N (Figure 9a), the fractal dimensions are 1.365, 1.3137, 1.3133, 1.2474, and 1.2442 respectively, which is similar to the law when the peak cyclic load is 4 N.

Figure 10 shows the fractal dimension of the curves of the stiffness of sand around piles and cycle steps. It can be seen from Figure 10a that the fractal dimensions of group 1 to group 5 are 1.2289, 1.2144, 1.2065, 1.1709, and 1.1596 respectively, which indicates that under the condition of saturated sand, the change amplitude of sand stiffness is the largest. However, when the buried depth of groundwater level is 40 mm, the change amplitude is the smallest. In Figure 10b, the fractal dimensions of group 6 to group 10 are 1.231, 1.2211, 1.2159, 1.1886, and 1.1502, respectively. By comparing the fractal dimensions of Figure 9a,b, it can be seen that the change amplitude of sand stiffness under the peak cyclic load of 4 N is larger than that under the peak cyclic load of 3 N when the groundwater depth is the same.

In order to explore the sensitivity of pile top displacement to cyclic steps under different groundwater depths, the relationship between displacement fluctuation index and fractal dimension is discussed. Figure 11 shows the linear fitting results of the variance and fractal dimension of the accumulated pile top displacement increment data group between each cycle step. The slope of linear fitting line is 0.04265 when the peak value of cyclic load is 4 N, which is higher than 0.01043 when the peak value of cyclic load is 3 N, indicating that the increase of peak value of cyclic load will increase the sensitivity of pile top displacement to cyclic steps. Obviously, there is a strong positive correlation between them, which means that the larger the fractal dimension is, the stronger the sensitivity of pile top displacement to cyclic steps is. Moreover, the correlation between fractal dimension and the mechanical response of the pile–soil system is also further verified. The value of the fractal dimension can reflect the change of the pile–soil system under various environmental circumstances.

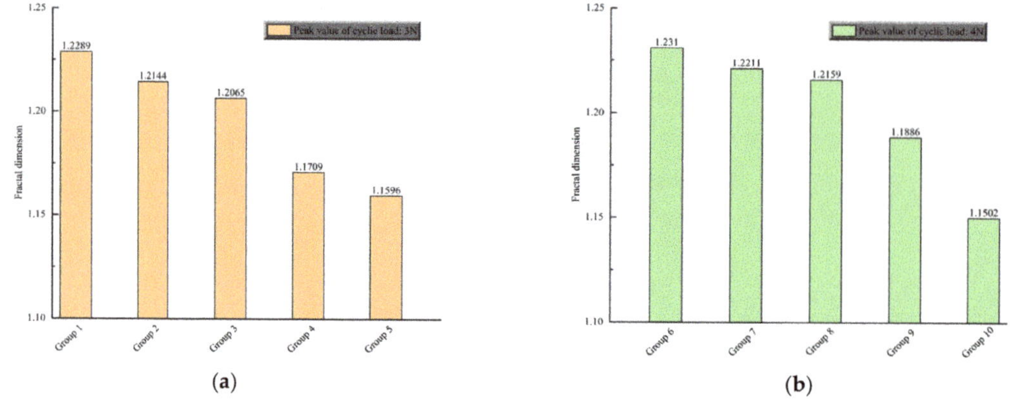

Figure 10. Curve of cyclic stiffness of sand around piles—cycle steps. (**a**) Peak value of cyclic load: 3 N. (**b**) Peak value of cyclic load: 4 N.

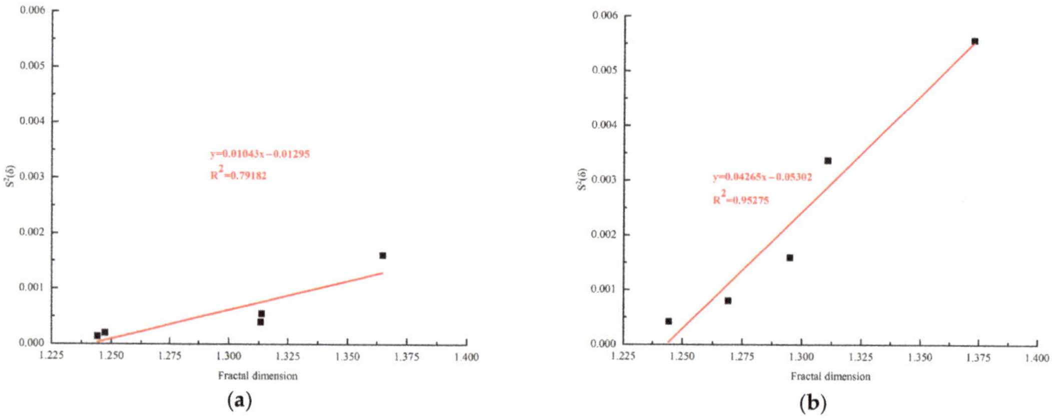

Figure 11. Relationship between fractal dimension and variance of the cumulative pile top displacement increment data group. (**a**) Peak value of cyclic load: 3 N. (**b**) Peak value of cyclic load: 4 N.

In addition, the average increment and fractal dimension of cumulative pile top displacement between each cyclic step after the first cyclic load are fitted and analyzed, and the results are shown in Figure 12. They also showed a good positive correlation.

3.4. Comparative Analysis of the Change Law of Soil Displacement Field

Research by Yuan et al. explained how to obtain the spot pattern needed to calculate the soil displacement field [71]. Using that information, this research process applied PIVview2C software to process the images taken during the cyclic loading peak *H*max = 4 N test process to obtain the soil particle displacement field data between the key frames during each step of the cyclic loading process, and then imported it into the TECPLOT software for displacement field equivalence line drawing [72]. This section analyzes the mechanism of pile–soil interaction in soil under cyclic loading from the macroscopic view of particle movement.

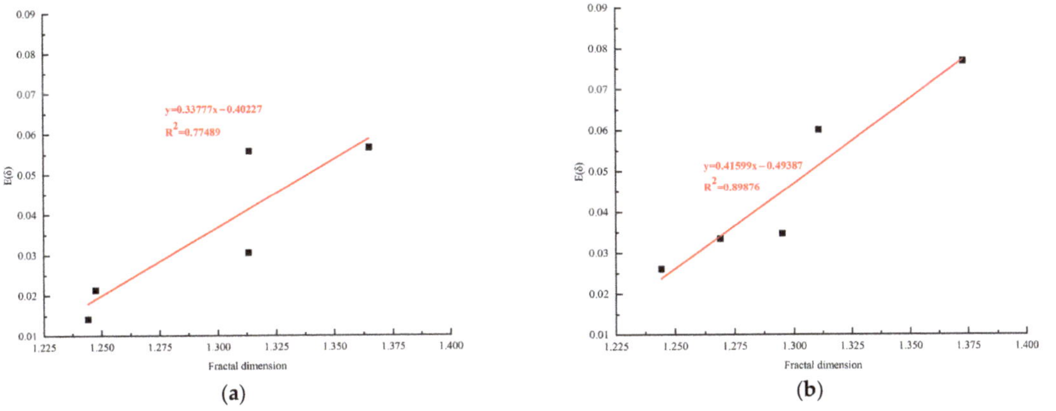

Figure 12. Relationship between fractal dimension and average increment of the cumulative pile top displacement between each cycle step after the first cyclic loading. (**a**) Peak value of cyclic load: 3 N. (**b**) Peak value of cyclic load: 4 N.

There were five sets of experiments in this experiment, and the practice associated with each set of images was the same. Therefore, a set of representative PIV images at a groundwater depth of 90 mm was selected. The PIV images after the first, second, tenth, and twentieth cyclic loadings are as shown in Figure 13. As shown in Figure 13a, the main deformation of the soil around the pile occurred in the first cycle. Initially, the soil displacement was induced within 20 mm of the pile circumference, and the soil displacement was most obvious at a distance of 10 mm around the pile circumference. The image after the second cyclic loading is shown in Figure 8b. Most of the soil displacement was concentrated within 40 mm around the pile circumference, and the soil displacement at a distance of 10 mm around the pile circumference was still the most obvious. The image after the tenth cyclic loading is shown in Figure 13c. Compared with the second loading, the soil displacement range was significantly reduced, and the soil displacement was concentrated within 30 mm of the pile circumference. The image after the twentieth loading is shown in Figure 13d. The soil displacement particles are obviously sparse, and the soil displacement range is reduced to within 20 mm.

The reason for this phenomenon is that the displacement of sand grains in the vertical direction is predisposed to occur, resulting in the volume change of the soil remaining largely compressible and maintaining a lot of room for strength. When the cyclic loading reaches the peak state, the shear zone will form in the core area around the pile. As the deformation further increases, the shear zone begins to develop around the pile, and the sand on both sides of the pile gradually slips and the strength decreases slowly. When the shear band is misaligned to a certain extent, under the action of false cohesion, the soil around the pile reaches the static equilibrium state again, and no new slip occurs. At this point, the soil around the pile reaches a stable residual state, and the displacement of the pile only disturbs the soil around the pile within a certain range.

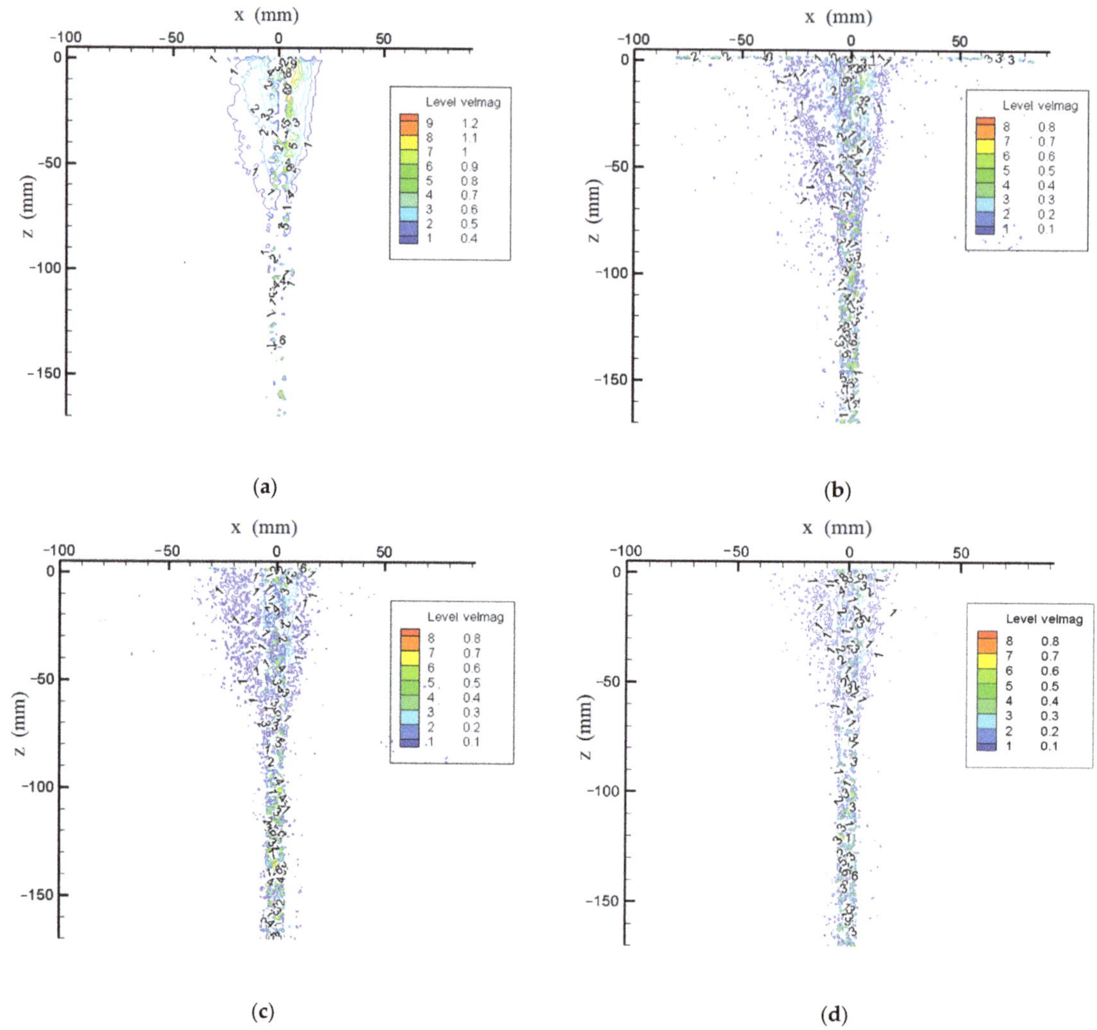

Figure 13. PIV displacement image after cyclic loading of 90 mm groundwater depth test. (**a**) After the first cyclic loading. (**b**) After the second cyclic loading. (**c**) After the tenth cyclic loading. (**d**) After the twentieth cyclic loading.

4. Conclusions

An original pile–soil test platform combined with PIV was designed for conducting horizontal loading tests on a monopile with the groundwater level and loading amplitude as variables under the unidirectional lateral cyclic loading. The pile–soil system was significantly affected by the cyclic loading, and the lateral stiffness of the soil increased as the loading proceeded. It was evident from the PIV images that the short-term effects of cyclic loading on the pile are much greater than the long-term effects. The disturbance range of soil around the pile rose with the increase in the number of cycles in the first ten cycles of loading. The soil structure was reconstructed, and the disturbance range of the soil around the pile gradually shrank and stabilized in the final ten cycles of loading. With a higher loading amplitude, the initial and cyclic stiffness of the soil would be less. The growth rate of cyclic stiffness was reduced as the groundwater level dropped. The stiffness of dry sand was greater than the stiffness of saturated soil, but less than the stiffness of the soil when the water table was below the soil surface, a difference due to matrix suction. Fractal theory was introduced to analyze the curves of the pile top displacement and the curve of the cyclic stiffness of the sand. The larger the fluctuation amplitude of the curve, the larger the fractal dimension. Both the variance and the average increment of the cumulative pile top displacement between each cycle step were positively correlated with fractal dimension. Therefore, a new physical meaning was given to fractal dimension, which meant fractal dimension could reflect the load response of the pile–soil system under cyclic loading.

The study outcome is expected to deepen understanding of the physical process and fundamental principles of the response of laterally loaded piles and sand bodies under different groundwater depths. However, this study was limited to a single soil layer. Further research is needed to explore the interaction mechanism of the pile–soil system during changes in groundwater levels under complex stratified soil in order to provide a better reference for the project monitoring.

Author Contributions: Data curation, Z.L. and W.C.; formal analysis, Z.L. and J.L.; investigation, W.C.; methodology, B.Y.; project administration, B.Y.; software, Z.L. and J.Z.; supervision, X.C. and J.S.; writing—original draft, B.Y. and Z.L.; writing—review and editing, B.Y. and J.L. All authors have read and agreed to the published version of the manuscript.

Funding: The authors would gratefully like to acknowledge the support provided by the National Natural Science Foundation of China (No. 51978177).

Institutional Review Board Statement: Not applicable.

Informed Consent Statement: Not applicable.

Data Availability Statement: The data presented in this study are available on request from the corresponding author. The data are not publicly available due to intellectual property.

Acknowledgments: The authors would like to thank all the referees for their constructive comments and suggestions. The editorial help from Galen Leonhardy of Black Hawk College is also greatly appreciated.

Conflicts of Interest: No conflict of interest exits in the submission of this manuscript, and manuscript is approved by all authors for publication.

References

1. Sunday, K.; Brennan, F. A review of offshore wind monopiles structural design achievements and challenges. *Ocean Eng.* **2021**, *235*, 109409. [CrossRef]
2. van der Male, P.; Vergassola, M.; van Dalen, K.N. Decoupled Modelling Approaches for Environmental Interactions with Monopile-Based Offshore Wind Support Structures. *Energies* **2020**, *13*, 5195. [CrossRef]
3. O'Kelly-Lynch, P.; Long, C.; McAuliffe, F.D.; Murphy, J.; Pakrashi, V. Structural design implications of combining a point absorber with a wind turbine monopile for the east and west coast of Ireland. *Renew. Sustain. Energy Rev.* **2020**, *119*, 109583. [CrossRef]
4. Rathod, D.; Krishnanunni, K.T.; Nigitha, D. A Review on Conventional and Innovative Pile System for Offshore Wind Turbines. *Geotech. Geol. Eng.* **2020**, *38*, 3385–3402. [CrossRef]
5. Vicedo, J.S.; Barba, J.G.; Frades, J.L.; Valdecantos, V.N. Scale Tests to Estimate Penetration Force and Stress State of the Silica Sand in Windfarm Foundations. *Energies* **2021**, *14*, 5904. [CrossRef]
6. Schweizer, J.; Antonini, A.; Govoni, L.; Gottardi, G.; Archetti, R.; Supino, E.; Berretta, C.; Casadei, C.; Ozzi, C. Investigating the potential and feasibility of an offshore wind farm in the Northern Adriatic Sea. *Appl. Energy* **2016**, *177*, 449–463. [CrossRef]
7. Wang, X.F.; Zeng, X.W.; Yang, X.; Li, J.L. Feasibility study of offshore wind turbines with hybrid monopile foundation based on centrifuge modeling. *Appl. Energy* **2018**, *209*, 127–139. [CrossRef]
8. Zou, X.J.; Cao, X.; Zhou, C.L.; Zhou, M.; Zhang, X.H. Experimental study on the bearing capacity of large-diameter monopile in sand under water flow condition. *Ocean Eng.* **2021**, *224*, 108708. [CrossRef]
9. McAdam, R.A.; Byrne, B.W.; Houlsby, G.T.; Beuckelaers, W.J.A.P.; Burd, H.J.; Gavin, K.G.; Igoe, D.J.P.; Jardine, R.J.; Martin, C.M.; Wood, A.M.; et al. Monotonic laterally loaded pile testing in a dense marine sand at Dunkirk. *Geotechnique* **2020**, *70*, 986–998. [CrossRef]
10. Wang, J.Y.; Sun, G.D.; Chen, G.S.; Yang, X. Finite element analyses of improved lateral performance of monopile when combined with bucket foundation for offshore wind turbines. *Appl. Ocean Res.* **2021**, *111*, 102647. [CrossRef]

11. Li, J.L.; Zhang, Y.; Wang, X.F.; Sun, Z.Z. Assessment of offshore wind turbine with an innovative monopile foundation under lateral loading. *Ocean Eng.* **2021**, *237*, 109583. [CrossRef]
12. Li, F.; Zhu, Z.Y.; Tian, P.S.; Hu, D.; Li, Y.F. The behavior of monopile embedded in sand under scour and cyclic loading conditions. *Mar. Georesour. Geotechnol.* **2022**, *40*, 52–63. [CrossRef]
13. Que, X.C.; Zhu, Z.D.; Niu, Z.H.; Lu, W.N. Estimating the strength and deformation of columnar jointed rock mass based on physical model test. *Bull. Eng. Geol. Environ.* **2021**, *80*, 1557–1570. [CrossRef]
14. Shao, J.L.; Zhang, Q.; Wu, X.T.; Lei, Y.; Wu, X.A.; Wang, Z.Y. Investigation on the Water Flow Evolution in a Filled Fracture under Seepage-Induced Erosion. *Water* **2020**, *12*, 3188. [CrossRef]
15. Bai, B.; Zhou, R.; Cai, G.Q.; Hu, W.; Yang, G.C. Coupled thermo-hydro-mechanical mechanism in view of the soil particle rearrangement of granular thermodynamics. *Comput. Geotech.* **2021**, *137*, 104272. [CrossRef]
16. Bai, B.; Nie, Q.K.; Zhang, Y.K.; Wang, X.L.; Hu, W. Cotransport of heavy metals and SiO_2 particles at different temperatures by seepage. *J. Hydrol.* **2021**, *597*, 125771. [CrossRef]
17. Yang, B.B.; Liu, J.W.; Zhao, X.M.; Zheng, S. Evaporation and cracked soda soil improved by fly ash from recycled materials. *Land Degrad. Dev.* **2021**, *32*, 2823–2832. [CrossRef]
18. Yuan, B.X.; Li, Z.H.; Su, Z.L.; Luo, Q.Z.; Chen, M.J.; Zhao, Z.Q. Sensitivity of Multistage Fill Slope Based on Finite Element Model. *Adv. Civ. Eng.* **2021**, *2021*, 6622936. [CrossRef]
19. Wu, Y.; Cui, J.; Huang, J.S.; Zhang, W.; Yoshimoto, N.; Wen, L.W. Correlation of Critical State Strength Properties with Particle Shape and Surface Fractal Dimension of Clinker Ash. *Int. J. Geomech.* **2021**, *21*. [CrossRef]
20. Shen, J.H.; Hu, M.J.; Wang, X.; Zhang, C.Y.; Xu, D.S. SWCC of Calcareous Silty Sand Under Different Fines Contents and dry Densities. *Front. Environ. Sci.* **2021**, *9*, 303. [CrossRef]
21. Liu, M.M.; Yang, M.; Wang, H.J. Bearing behavior of wide-shallow bucket foundation for offshore wind turbines in drained silty sand. *Ocean Eng.* **2014**, *82*, 169–179. [CrossRef]
22. Liu, Z.Y.; Xue, J.F. The deformation behaviour of an anisotropically consolidated kaolin clay under lateral cyclic loading. *Mar. Georesour. Geotechnol.* **2021**. [CrossRef]
23. Maleksaeedi, E.; Nuth, M. Evaluation of capillary water retention effects on the development of the suction stress characteristic curve. *Can. Geotech. J.* **2020**, *57*, 1439–1452. [CrossRef]
24. Yang, B.; Su, Y.; He, N.; Zhang, B.; Zhou, X.S.; Zhang, Y. Experimental study on the stable morphology and self-attraction effect of subaqueous barchan dunes. *Adv. Powder Technol.* **2020**, *31*, 1032–1039. [CrossRef]
25. Zhang, C.; Lu, N. Unified Effective Stress Equation for Soil. *J. Eng. Mech.* **2020**, *146*, 04019135. [CrossRef]
26. Guo, Y.C.; Ye, Y.Y.; Guan-Lin; Lv, J.F.; Bai, Y.L.; Zeng, J.J. Effective usage of high strength steel tubes: Axial compressive behavior of hybrid FRP-concrete-steel solid columns. *Thin Walled Struct.* **2020**, *154*, 106796. [CrossRef]
27. Lozano-Minguez, E.; Kolios, A.J.; Brennan, F.P. Multi-criteria assessment of offshore wind turbine support structures. *Renew. Energy* **2011**, *36*, 2831–2837. [CrossRef]
28. Liu, F.; Chen, G.X.; Li, L.J.; Guo, Y.C. Study of impact performance of rubber reinforced concrete. *Constr. Build. Mater.* **2012**, *36*, 604–616. [CrossRef]
29. Liu, F.; Zheng, W.H.; Li, L.J.; Feng, W.X.; Ning, G.F. Mechanical and fatigue performance of rubber concrete. *Constr. Build. Mater.* **2013**, *47*, 711–719. [CrossRef]
30. Damgaard, M.; Bayat, M.; Andersen, L.V.; Ibsen, L.B. Assessment of the dynamic behaviour of saturated soil subjected to cyclic loading from offshore monopile wind turbine foundations. *Comput. Geotech.* **2014**, *61*, 116–126. [CrossRef]
31. Devolder, B.; Rauwoens, P.; Troch, P. Application of a buoyancy-modified k-omega SST turbulence model to simulate wave run-up around a monopile subjected to regular waves using OpenFOAM (R). *Coast. Eng.* **2017**, *125*, 81–94. [CrossRef]
32. Guo, Y.C.; Xie, J.H.; Zhao, J.B.; Zuo, K.X. Utilization of unprocessed steel slag as fine aggregate in normal- and high-strength concrete. *Constr. Build. Mater.* **2019**, *204*, 41–49. [CrossRef]
33. Zhang, X.Y.; Tang, L.; Ling, X.Z.; Chan, A.H.C.; Lu, J.C. Using peak ground velocity to characterize the response of soil-pile system in liquefying ground. *Eng. Geol.* **2018**, *240*, 62–73. [CrossRef]
34. Cao, X.L.; Dai, G.L.; Gong, W.M.; Zhou, F.X.; Xu, J. Resistance of saturated soil to a laterally vibrating pile. *Soil Dyn. Earthq. Eng.* **2021**, *141*, 106496. [CrossRef]
35. Xie, J. Impedance of Inner Saturated Soil to Horizontally Vibrating Large-Diameter Pipe Piles. *Shock Vib.* **2021**, *2021*, 1845109. [CrossRef]
36. Ding, X.M.; Qu, L.M.; Yang, J.C.; Wang, C.L. Experimental study on the pile group-soil vibration induced by railway traffic under the inclined bedrock condition. *Acta Geotech.* **2020**, *15*, 3613–3620. [CrossRef]
37. Ren, L.W.; Yang, Q.W.; Kong, G.Q.; Dun, Z.L.; Wang, X.Y. Model Tests on Y-Shaped Piles under Compressive and Lateral Loading in Saturated Sand. *Geofluids* **2021**, *2021*, 6978602. [CrossRef]
38. Fattah, M.Y.; Zabar, B.S.; Mustafa, F.S. Effect of saturation on response of a single pile embedded in saturated sandy soil to vertical vibration. *Eur. J. Environ. Civ. Eng.* **2020**, *24*, 381–400. [CrossRef]
39. Li, J.H.; Zhang, B.; Shen, C.; Fu, X.L.; Li, W.C. Experimental Study on Local Scour Depth around Monopile Foundation in Combined Waves and Current. *Sustainability* **2021**, *13*, 13614. [CrossRef]
40. Li, Q.; Askarinejad, A.; Gavin, K. Impact of scour on lateral resistance of wind turbine monopiles: An experimental study. *Can. Geotech. J.* **2021**, *58*, 1770–1782. [CrossRef]

41. Cui, C.Y.; Meng, K.; Xu, C.S.; Liang, Z.M.; Li, H.J.; Pei, H.F. Analytical solution for longitudinal vibration of a floating pile in saturated porous media based on a fictitious saturated soil pile model. *Comput. Geotech.* **2021**, *131*, 103942. [CrossRef]
42. Achmus, M.; Kuo, Y.S.; Abdel-Rahman, K. Behavior of monopile foundations under cyclic lateral load. *Comput. Geotech.* **2009**, *36*, 725–735. [CrossRef]
43. Lombardi, D.; Bhattacharya, S.; Wood, D.M. Dynamic soil-structure interaction of monopile supported wind turbines in cohesive soil. *Soil Dyn. Earthq. Eng.* **2013**, *49*, 165–180. [CrossRef]
44. Arshad, M.; O'Kelly, B.C. Analysis and Design of Monopile Foundations for Offshore Wind-Turbine Structures. *Mar. Georesour. Geotechnol.* **2016**, *34*, 503–525. [CrossRef]
45. Feng, W.H.; Liu, F.; Yang, F.; Li, L.J.; Jing, L. Experimental study on dynamic split tensile properties of rubber concrete. *Constr. Build. Mater.* **2018**, *165*, 675–687. [CrossRef]
46. Luo, R.P.; Yang, M.; Li, W.C. Numerical study of diameter effect on accumulated deformation of laterally loaded monopiles in sand. *Eur. J. Environ. Civ. Eng.* **2020**, *24*, 2440–2452. [CrossRef]
47. Malakshah, R.R.; Moradi, M.; Ghalandarzadeh, A. Centrifuge Modelling of Monopiles in Calcareous Sand Subjected to Cyclic Lateral Loading. *Int. J. Civ. Eng.* **2022**, *20*, 195–206. [CrossRef]
48. Namazi, H. Complexity-based analysis of the correlation between stride interval variability and muscle reaction at different walking speeds. *Biomed. Signal. Process.* **2021**, *69*, 102956. [CrossRef]
49. Phinyomark, A.; Larracy, R.; Scheme, E. Fractal Analysis of Human Gait Variability via Stride Interval Time Series. *Front. Physiol.* **2020**, *11*, 333. [CrossRef]
50. Tabares-Ospina, H.A.; Angulo, F.; Osorio, M. New Method to Calculate the Energy and Fractal Dimension of the Daily Electrical Load. *Fractals* **2020**, *28*, 2050135. [CrossRef]
51. Yan, B.W.; Chan, P.W.; Li, Q.S.; He, Y.C.; Shu, Z.R. Characterising the fractal dimension of wind speed time series under different terrain conditions. *J. Wind Eng. Ind. Aerodyn.* **2020**, *201*, 104165. [CrossRef]
52. Shu, Z.R.; Chan, P.W.; Li, Q.S.; He, Y.C.; Yan, B.W. Quantitative assessment of offshore wind speed variability using fractal analysis. *Wind Struct.* **2020**, *31*, 363–371. [CrossRef]
53. Yang, B.B.; Liu, Y. Application of Fractals to Evaluate Fractures of Rock Due to Mining. *Fractal Fract.* **2022**, *6*, 96. [CrossRef]
54. Yang, B.B.; Du, S.T.; Zhao, X.M.; Tang, D.Q.; Yang, C.D. Decision Making of Curriculum Attainment Degree for Engineering Geology Based on Fuzzy Set Theory. *Adv. Civ. Eng.* **2021**, *2021*, 1743778. [CrossRef]
55. Wang, L.; Luo, R.Y.; Zhang, W.; Jin, M.M.; Tang, S.W. Effects of Fineness and Content of Phosphorus Slag on Cement Hydration, Permeability, Pore Structure and Fractal Dimension of Concrete. *Fractals* **2021**, *29*, 2140004. [CrossRef]
56. Xiao, J.; Qu, W.J.; Jiang, H.B.; Li, L.; Huang, J.; Chen, L. Fractal Characterization and Mechanical Behavior of Pile-Soil Interface Subjected to Sulfuric Acid. *Fractals* **2021**, *29*, 2140010. [CrossRef]
57. Li, Y.; Zhang, H.; Huang, M.H.; Yin, H.B.; Jiang, K.; Xiao, K.T.; Tang, S.W. Influence of Different Alkali Sulfates on the Shrinkage, Hydration, Pore Structure, Fractal Dimension and Microstructure of Low-Heat Portland Cement, Medium-Heat Portland Cement and Ordinary Portland Cement. *Fractal Fract.* **2021**, *5*, 79. [CrossRef]
58. Wang, L.; Lu, X.; Liu, L.S.; Xiao, J.; Zhang, G.; Guo, F.X.; Li, L. Influence of MgO on the Hydration and Shrinkage Behavior of Low Heat Portland Cement-Based Materials via Pore Structural and Fractal Analysis. *Fractal Fract.* **2022**, *6*, 40. [CrossRef]
59. Wang, L.; Guo, F.X.; Yang, H.M.; Wang, Y.; Tang, S.W. Comparison of Fly Ash, Pva Fiber, Mgo and Shrinkage-Reducing Admixture on the Frost Resistance of Face Slab Concrete Via Pore Structural and Fractal Analysis. *Fractals* **2021**, *29*, 2140002. [CrossRef]
60. Duan, H.Q.; Zhang, S. Fractal Characteristics of Coal Specimens' Surface Cracks in Triaxial Conventional Compression and Cyclic Loading Tests. *Geotech. Geol. Eng.* **2020**, *38*, 19–29. [CrossRef]
61. Kim, N.S.; Lee, J.H.; Chang, S.P. Equivalent multi-phase similitude law for pseudodynamic test on small scale reinforced concrete models. *Eng. Struct.* **2009**, *31*, 834–846. [CrossRef]
62. Stanier, S.A.; Blaber, J.; Take, W.A.; White, D.J. Improved image-based deformation measurement for geotechnical applications. *Can. Geotech. J.* **2016**, *53*, 727–739. [CrossRef]
63. White, D.J.; Take, W.A.; Bolton, M.D. Soil deformation measurement using particle image velocimetry (PIV) and photogrammetry. *Geotechnique* **2003**, *53*, 619–631. [CrossRef]
64. Gudehus, G.; Nubel, K. Evolution of shear bands in sand. *Geotechnique* **2004**, *54*, 187–201. [CrossRef]
65. Georgescu, M.R.; Meslem, A.; Nastase, I.; Sandu, M. Numerical and experimental study of the International Space Station crew quarters ventilation. *J. Build. Eng.* **2021**, *41*, 102714. [CrossRef]
66. Yuan, B.X.; Sun, M.; Xiong, L.; Luo, Q.Z.; Pradhan, S.P.; Li, H.Z. Investigation of 3D deformation of transparent soil around a laterally loaded pile based on a hydraulic gradient model test. *J. Build. Eng.* **2020**, *28*, 101024. [CrossRef]
67. Zhang, X.; Huang, M.S.; Hu, Z.P. Model tests on cumulative deformation characteristics of a single pile subjected to lateral cyclic loading in sand. *Rock Soil Mech.* **2019**, *40*, 933–941. [CrossRef]
68. Prabakar, J.; Sridhar, R.S. Effect of random inclusion of sisal fibre on strength behaviour of soil. *Constr. Build. Mater.* **2002**, *16*, 123–131. [CrossRef]
69. Lu, N.; Zhang, C. Soil Sorptive Potential: Concept, Theory, and Verification. *J. Geotech. Geoenviron.* **2019**, *145*, 04019006. [CrossRef]
70. Wang, H.; Wang, L.Z.; Hong, Y.; Masin, D.; Li, W.; He, B.; Pan, H.L. Centrifuge testing on monotonic and cyclic lateral behavior of large-diameter slender piles in sand. *Ocean Eng.* **2021**, *226*, 108299. [CrossRef]

71. Yuan, B.X.; Sun, M.; Wang, Y.X.; Zhai, L.H.; Luo, Q.Z.; Zhang, X.Q. Full 3D Displacement Measuring System for 3D Displacement Field of Soil around a Laterally Loaded Pile in Transparent Soil. *Int. J. Geomech.* **2019**, *19*, 04019028. [CrossRef]
72. Yuan, B.X.; Xiong, L.; Zhai, L.H.; Zhou, Y.F.; Chen, G.F.; Gong, X.; Zhang, W. Transparent Synthetic Soil and Its Application in Modeling of Soil-Structure Interaction Using Optical System. *Front. Earth Sci.* **2019**, *7*, 276. [CrossRef]

fractal and fractional

Article

Fractal Analysis of Particle Distribution and Scale Effect in a Soil–Rock Mixture

Xiaodong Fu [1,2], Haifeng Ding [1,2], Qian Sheng [1,2], Zhenping Zhang [1,3,*], Dawei Yin [4] and Fei Chen [5]

1 State Key Laboratory of Geomechanics and Geotechnical Engineering, Institute of Rock and Soil Mechanics, Chinese Academy of Sciences, Wuhan 430071, China; xdfu@whrsm.ac.cn (X.F.); dinghaifeng21@mails.ucas.ac.cn (H.D.); shengqian@whrsm.ac.cn (Q.S.)
2 School of Engineering Science, University of Chinese Academy of Sciences, Beijing 100049, China
3 School of Architecture and Civil Engineering, Shenyang University of Technology, Shenyang 110870, China
4 China State Construction International Investments (Hubei) Limited, Wuhan 430000, China; yindawei@cohl.com
5 School of Architecture and Civil Engineering, Chengdu University, Chengdu 610106, China; chenfei@cdu.edu.cn
* Correspondence: zhangzp21@sut.edu.cn

Abstract: A soil–rock mixture (SRM) is a type of heterogeneous geomaterial, and the particle distribution of SRM can be described by fractal theory. At present, it is difficult to quantify the fractal dimension of a particle size distribution and understand the scale effect in SRMs. In this study, the fractal theory and discrete element method (DEM) were introduced to solve this problem. First, the particle gradation of SRM was dealt with by using fractal theory. The fractal structure of particle distribution was studied, and a method of calculation of the fractal dimension is presented in this paper. Second, based on the fractal dimension and relative threshold, the particle gradations of SRMs at different scales were predicted. Third, numerical direct shear tests of SRM at different scales were simulated by using the DEM. The scale effects of shear displacement, shear zone, and shear strength parameters were revealed. Last, taking the maximum particle size of 60 mm as the standard value, the piece-wise functional relationship between shear strength parameters and particle size was established. The results are as follows: for SRM in a representative engineering area, by plotting the relationship between particle cumulative mass percentage and particle size, we can judge whether the SRM has a fractal structure; in Southwest China, the frequency of the fractal dimension of the SRM is in the normal distribution, and the median fractal dimension is 2.62; the particle gradations of SRMs at different scales calculated by fractal dimension and relative threshold can expand the study scope of particle size analysis; when the particle size is less than 70 mm, the strength parameters show a parabolic trend with the particle size increases, and if not, a nearly linear trend is found. The proposed method can describe the fractal characteristics of SRM in a representative engineering area and provides a quantitative estimation of shear strength parameters of SRM at different scales.

Keywords: soil–rock mixture; fractal theory; particle distribution; scale effect; discrete element model

Citation: Fu, X.; Ding, H.; Sheng, Q.; Zhang, Z.; Yin, D.; Chen, F. Fractal Analysis of Particle Distribution and Scale Effect in a Soil–Rock Mixture. *Fractal Fract.* **2022**, *6*, 120. https://doi.org/10.3390/fractalfract6020120

Academic Editor: Zine El Abiddine Fellah

Received: 18 January 2022
Accepted: 17 February 2022
Published: 19 February 2022

1. Introduction

A soil–rock mixture (SRM) is a heterogeneous geomaterial that is composed of rock blocks with different particle sizes and a soil matrix. In 1994, Medley and Goodman [1] innovated the term "bimrocks" (block-in-matrix rocks) and pointed out that the particle sizes in bimrocks range from centimeters to hundreds of meters [2]. Lindquist and Goodman [3] called the mixed rock and soil embedded with fragments and rock blocks from itself or outside a "mélange" [4]. You [5] proposed the concept of a SRM for Quaternary loose accumulations, which considered the particularity of its material composition between rock and soil. Xu et al. [6,7] defined the SRM as an extremely heterogeneous loose geomaterial formed since the Quaternary that is composed of rock blocks, fine-grained soil, and pores. According to these definitions, the particle size distribution of soil and rock blocks in the SRM varies greatly. In the traditional soil classification system, the division of coarse-grained and fine-grained soil can be determined according to the limit values in specifications; however, there is no unified standard for the limiting values of particle size divided by particle group, and different countries have different regulations. For example, the Chinese Handbook of Engineering Geology [8] defined the mixtures of fine-grained soil and coarse-grained soil at sizes of less than 10 mm and lacking intermediate grain sizes as gravelly (stony) soils. Thus, to understand the particle distribution of SRM, it is essential to find a method with which to divide the soil and rock block in SRM and determine the threshold.

More importantly, the particle distribution of SRM is closely related to its mechanical properties. When obtaining the mechanical parameters of SRM by mechanical tests, the particle size of some natural SRM is much larger than the maximum size of rock block allowed by test equipment, such as 60 mm [9,10]. Thus, for the large-size rock block in SRM at the engineering scale, before mechanical testing, to make the scaled particle gradation maintain similar mechanical properties to the natural particle gradation as far as possible, the natural particle gradation needs to be treated by the scaling method. There are four common methods of coarse aggregate scaling, namely the removal method, equivalent substitution method, similar gradation method, and mixing method [11]. The variations in strength parameters of SRMs obtained by different scaling methods are different [12–14]. The reason is that the change of average particle gradation and fine particle caused by scaling methods directly affects the structural characteristics of SRM and then leads to the change of mechanical properties. The mechanical properties of a scaled SRM are different from those of natural SRM (the so-called scale effect).

As fractal theory is now widely used [15–23], many researchers [24–31] have found that geomaterials exhibit fractal characteristics, and the rock joints, particle distribution, and pores at different scales have self-similarity, which is also consistent with the evolution of geomaterials under natural conditions. Generally, geomaterials can be regarded as an open non-linear self-organizing system. When there are no external conditions to interfere with its self-organizing characteristics, the evolution of internal spatial structure exhibits certain orderliness [32–35]. The fractal structure is closely related to its composition and natural conditions: the fractal dimension represents the spatial structure characteristics formed by the long-term evolution of geomaterials. Sui et al. [36] discussed the development in rock fracture network descriptions by using traditional geometry and fractal methods and presented an analytical method for the quantitative evaluation of the irregularity and complexity of fracture networks based on the self-similarity of fractals. Xiao et al. [37] investigated the crushing characteristics of single particles and assemblies of rockfill materials with different nominal diameters and discussed the effects of the particle size on the Weibull modulus, compressibility index, and ultimate fractal dimension. He et al. [38] obtained the pore size distribution of the calcareous sand, quartz sand, and glass beads by using nuclear magnetic resonance tests, and fractal theory was introduced to describe the fractal properties of the pore size distribution. These studies show that the fractal theory can describe both structural characteristics and mechanical properties of geomaterials at different scales.

Previous studies [39–41] have shown that natural SRM exhibits a fractal structure. Commonly, the fractal curve of particle gradation of SRM is a broken line with an obvious turning point. According to the turning point, the particle size can be divided into fine aggregate and coarse aggregate. For example, Xu et al. [39] pointed out that SRM had a visible turning point at the particle size of 20 mm in its fractal dimension curve. However, at present, it is difficult to quantify the fractal dimension of the particle distribution of SRM, as well as the scale effect. In this study, the fractal theory was invoked to analyze the fractal characteristics of the particle distribution of SRM, numerical simulations of direct shear tests of SRM samples at different scales were conducted by using the discrete element method (DEM), and the variations in shear strength parameters of SRM at different scales were studied.

2. Fractal Model

2.1. Fractal Mathematical Model

Given the self-similarity of the particle distribution of SRM, the fractal characteristics under two-dimensional conditions can be quantitatively expressed by using the formula proposed by Mandelbrot [42]:

$$A(r > R) = C_a\left[1 - \left(\frac{R}{\lambda_a}\right)^{2-D}\right] \tag{1}$$

where R is any given particle size, r is the random particle size, A is the total area of particles satisfying $r > R$, Ca represents the area shape factor, λa is the maximum particle size under two-dimensional conditions, and D denotes the fractal dimension of particle size.

Under three-dimensional conditions, the formula proposed by Tyler and Wheacraft [43] is introduced:

$$V(r > R) = C_m\left[1 - \left(\frac{R}{\lambda_m}\right)^{3-D}\right] \tag{2}$$

where V is the total volume of particles satisfying $r > R$, C_m is the volume shape factor, and λ_m refers to the maximum particle size under three-dimensional conditions. Without considering the variation of particle density with particle size, in Formula (2), we multiply the density, ρ, on both sides at the same time, and then the total mass of particles satisfies $r > R$:

$$M(r > R) = \rho C_m\left[1 - \left(\frac{R}{\lambda_m}\right)^{3-D}\right] \tag{3}$$

Furthermore, the total mass of all particles comprising an SRM sample can be calculated as follows:

$$M_T = M(r > 0) = \rho C_m\left[1 - \left(\frac{0}{\lambda_m}\right)^{3-D}\right] = \rho C_m \tag{4}$$

The cumulative mass ratio of the SRM sample can be obtained as follows:

$$\frac{M(r > R)}{M_T} = 1 - \left(\frac{R}{\lambda_m}\right)^{3-D} \tag{5}$$

where M_T is the total mass of all particles, and d_{max} is defined as the maximum particle size of an SRM sample. According to Formula (5), when $R = d_{max}$, the cumulative mass ratio is zero, and $\lambda_m = d_{max}$. By combining the commonly used particle gradation curves of SRM samples, Formula (5) can be converted into the following formula:

$$\frac{M(r < R)}{M_T} = \left(\frac{R}{d_{\max}}\right)^{3-D} \tag{6}$$

In Formula (6), for a given particle size, R, the mass percentage is defined as the ratio of the mass of particles satisfying $r < R$ to the total mass; the particle size ratio is defined as the ratio of R to d_{max}; there is a power function relationship between the mass percentage and the particle size ratio; and the index is $3 - D$.

Using the logarithmic transformation of Formula (6), in the double-logarithmic coordinate system, we can plot the relationship curve between lg $(M(r < R)/M_T$ and lg R, and whether the SRM meets the fractal structure can be ascertained. If the relationship is approximately linear, it indicates that the particle gradation of SRM within the range of interest has strict statistical self-similarity, the particle distribution demonstrates a fractal structure, and the slope is $a = 3 - D$. Otherwise, it means that the particle gradation is poor, the pore filling is poor, and natural SRM is loose. In addition, if the relationship shows the characteristics of a broken line, it indicates that the particle gradation has good multifractal characteristics, and the particle distribution has self-similarity in the corresponding linear interval. Based on the above fractal mathematical model, the particle gradation of natural SRM can be processed to judge whether it has a fractal structure, and the corresponding fractal dimension can be obtained.

2.2. Fractal Analysis of SRM Samples

Taking the SRM from Zhaizi village, Lijiang City, Yunnan Province, China, as an example [44], field investigation and sampling were undertaken (Figure 1). Through laboratory particle-size analysis of a typical SRM sample with the maximum particle size of 60 mm, the particle size distribution was obtained (Figure 2a), and its equivalent histogram was plotted (Figure 2b). The statistical results show that the particle-size curves of SRM in this area are bimodal, indicating that the material is well sorted. The particle size corresponding to the trough is about 2 mm, and the threshold between the soil and the rock block of the SRM in this area is determined to be 2 mm.

(a) (b)

Figure 1. SRM in Zhaizi village, Lijiang City, Yunnan Province, China. (**a**) Site investigation; (**b**) SRM sample.

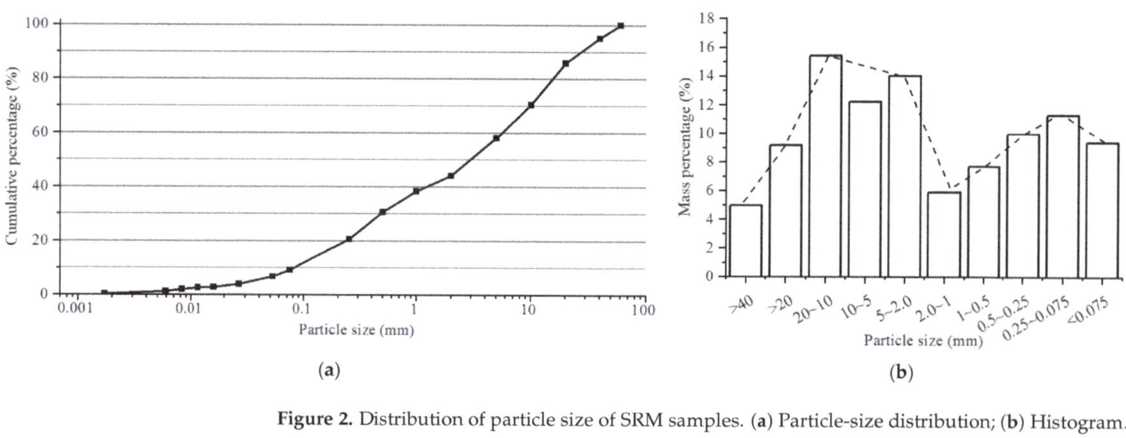

(a) (b)

Figure 2. Distribution of particle size of SRM samples. (**a**) Particle-size distribution; (**b**) Histogram.

Formula (6) was used to process the particle-size-distribution data. The screening results were brought into the fractal mathematical model, and the relationship between lg $(M(r < R)/M_T$ and lg R was plotted in the double-logarithmic coordinate system (Figure 3). The results show that the relationship curve is approximately linear, indicating that the particle size of SRM in this area has a fractal structure. The fractal-curve-fitting formula of the overall grading is $y = 0.33x + 1.50$, the slope is 0.327, and the fractal dimension D is 2.67. In addition, it should be noted that the fractal curve does not meet the strict linear relationship within the full particle-size range. A dividing point is found at the particle size $r = 2$ mm; there is a scale-free interval on both sides, where the fitting formulae are $y = 0.48x + 1.57$ and $y = 0.24x + 1.59$ respectively; and the corresponding fractal dimensions are $D1 = 2.52$ and $D2 = 2.76$ respectively. These results imply that there is a double-fractal structure in the SRM, with a maximum particle size of 60 mm, meaning that the particle size and pore structure change at 2 mm. Thus, through fractal analysis of particle distribution of this SRM, the threshold between soil and rock blocks in the SRM in this area is found to be 2 mm, which is consistent with the results of laboratory grading.

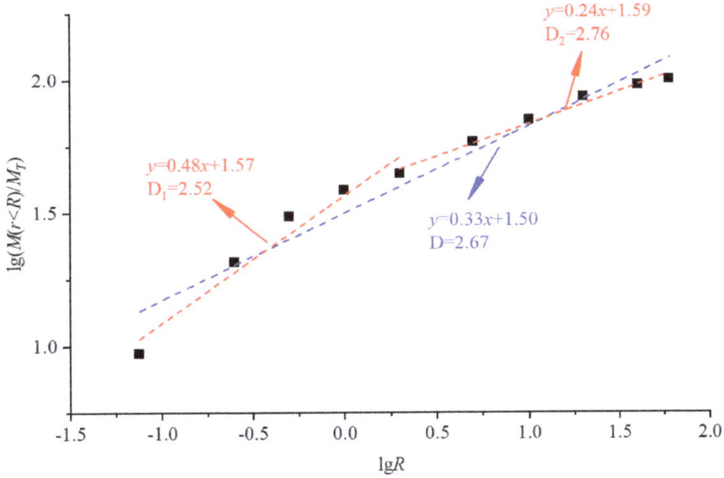

Figure 3. Fractal analysis of SRM sample.

3. Predictive Theory

3.1. Relative Threshold between Soil and Rock Blocks

After analyzing the particle distribution characteristics of an SRM, Medley [2] and Lindquist [4] found that the particle distribution of SRM exhibits proportion-independence. As shown in Figure 4, even if the study area changes, when the particle size of SRM is 0.05 times the study area, the fractal dimension of particle size will always contain a turning point. According to this understanding, it is considered that the threshold between soil and rock blocks of SRM at the engineering scale should satisfy Formula (7):

$$d_{S/RT} = 0.05L_c \tag{7}$$

where $d_{S/RT}$ denotes the threshold between soil and rock blocks; and L_c is the engineering scale that, being the square root of the area for the plane study area, is the slope height for a slope, and it is the height of a single shear box for the direct shear sample.

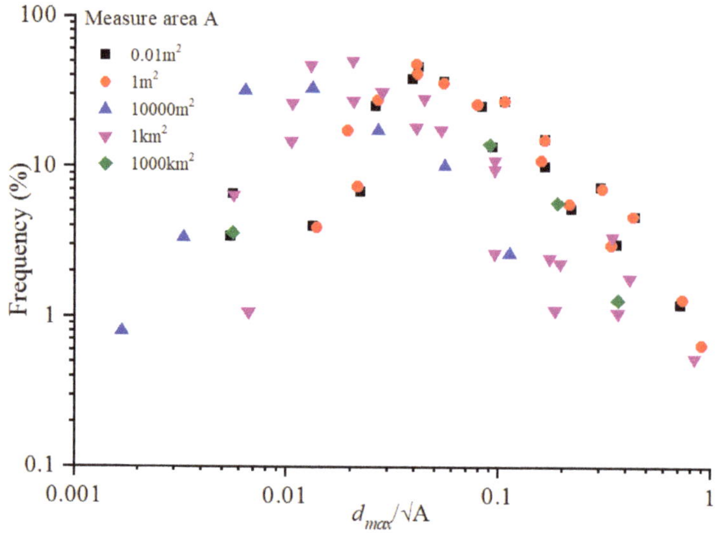

Figure 4. Particle-size-distribution characteristics of SRM after normalization.

This threshold definition method sets the soil and rock blocks in an SRM as a relative concept; across different research scales, the SRM from the same source may have different thresholds of coarse and fine-particle size. Therefore, this definition is called the relative threshold between soil and rock blocks, which is different from the limiting values of particle size divided by particle group defined in prevailing national standards [8]. The relative threshold links the particle gradation of SRM at different research scales by the proportion of rock blocks, which can directly expand the mechanical properties of SRM at test scale to engineering at full-scale.

3.2. Prediction of Particle Gradation Curves at Different Scales

Since the particle gradation of SRM has a fractal structure, using the concepts of fractal dimension and the relative diameter threshold between soil and rock blocks, we can calculate the distribution of particle groups after the maximum particle size increases according to the existing gradation: the particle gradations of SRMs at different scales are thus obtained. Taking SRM in Zhaizi village as an example (Figure 3), the fractal dimension is 2.67. According to geological drilling data, the range of rock block sizes in the SRM is between 30 and 90 mm. For a given maximum particle diameter (d_{max}), substituting the fractal dimension 2.67 into Formula (6), the particle gradation of the SRM sample can be predicted. Figure 5 shows the predicted particle gradation of SRM samples with different d_{max}, including 100, 80, 70, 60, and 50 mm (labeled S1 to S5, respectively). Since tested particle gradation with the maximum particle size of 60 mm is obtained as shown in Figure 2a, the comparison between the tested particle gradation and predicted particle gradation can be performed. Taking 2 mm as the threshold between soil and rock blocks, we see that the result shows that the rock-block content of the predicted particle grading is about 62.46%, while the rock-block content of the tested particle gradation is 55.8%, so both are close to 60%. Therefore, the accuracy of the predicted particle grading curve is acceptable.

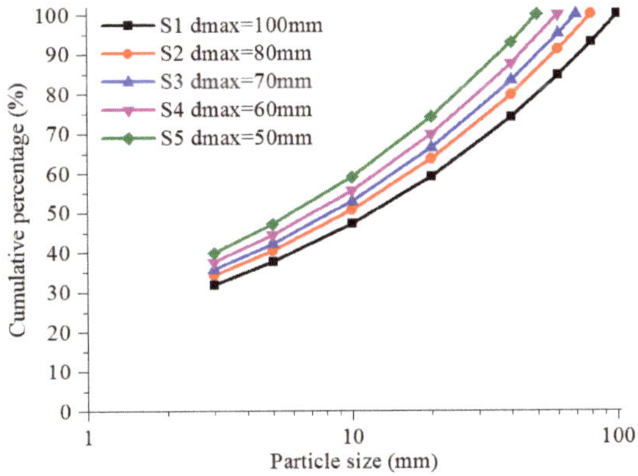

Figure 5. Predicted particle size distributions of SRM samples with different maximum particle diameters.

Then, according to the research scale, based on the particle size distribution in Figure 5, the gradation is modified by using the relative threshold between soil and rock blocks. In the following section of this study, we recount the numerical simulation of direct shear testing of SRM samples at different scales that were conducted, making it necessary to establish numerical models with different maximum particle diameters. Taking this scale as an example, we see that the size of the shear box used in the test is 500 mm (length) × 200 mm (height). According to Formula (7), the research scale L_c is the height of the shear box (200 mm), and the limit particle size of particles should be set to 10 mm. In addition, for a numerical model, the influences of the minimum particle size on the test results should also be considered [45]. Commonly, the particle size range of soil particles is 2 to 3 mm in numerically simulated direct shear tests. If the ratio of the maximum particle size to the minimum particle size remains unchanged at 1.5, the range of soil-particle size after modifying the threshold should be 6.66 to 10 mm. The particle gradation of rock-block particles larger than 10 mm remains the same (Figure 5), and the size of particles less than 10 mm should be corrected. Figure 6 shows the modified particle distributions of SRM samples with different maximum particle diameters.

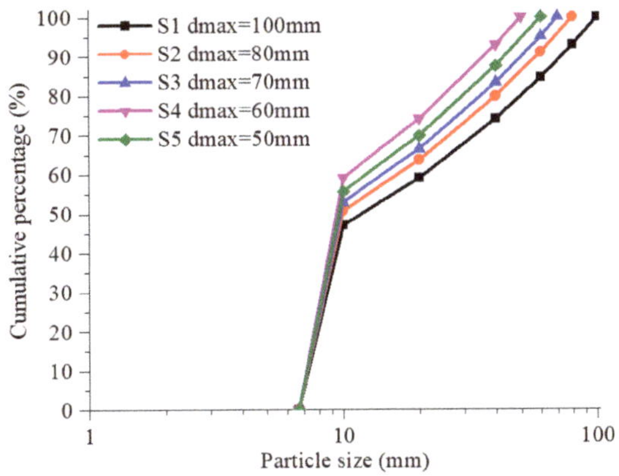

Figure 6. Modified particle distributions of SRM samples with different maximum particle diameters.

4. Simulation Method

4.1. Simulation of Large-Scale Direct Shear Tests of SRM

During laboratory large-scale direct shear testing, the SRM sample needs to be encapsulated in an opaque shear box, so the movement of the particles in the sample cannot be seen [46]. The numerical simulation using particle flow code (PFC) based on DEM can overcome this problem [47]. The particle DEM proposed by Cundall [48] uses the rigid particle as the basic calculation element. In the calculation process, particle DEM continuously updates the contact characteristics between particle and particle or particle and boundary; based on Newton's second law, it uses the explicit difference algorithm for iteratively solving the motion parameters of particle elements, and finally makes each particle element reach an equilibrium state. In this study, PFC2D [49] is used to simulate direct shear tests.

Before the simulation, it is necessary to build the particle flow model of SRM by wall element and particle element in PFC2D. As shown in Figure 7, keeping the same size as the shear box as used in the laboratory, we used a simulated shear box representing 500 mm (length) × 200 mm (height). The numerical model is divided into upper and lower halves: the upper half is composed of Walls 4, 5, and 6, and the lower half of Walls 1, 2, and 8. To prevent particles from escaping during shearing, Walls 3 and 7 are set on both sides of the shear box. During simulated shear, normal stress is applied to Wall 5 and the vertical load is kept stable, the upper shear box is fixed, and the same horizontal velocity is applied to the lower shear box wall; in addition, Wall 7 and the lower shear box move together, and Wall 3 remains fixed to avoid a loss of particles during shearing.

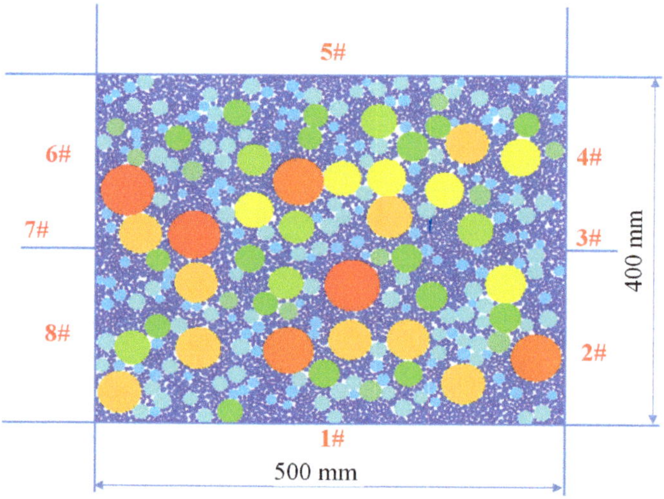

Figure 7. Numerical model of the direct shear test of an SRM sample.

When building the particle model of SRM, due to the limited shape indices pertaining to the rock block, both rock and soil particles are simulated by traditional disk particles in PFC2D. According to particle size distributions of SRM samples in Figure 6, the random generation algorithm provided in the PFC2D is used for producing disk particles, and the target number of elements is generated by setting the maximum and minimum particle sizes and the target void ratio. To make the size of each particle group more consistent with the fractal law of particle gradation of SRM, the volume content of the rock block of each particle group is determined by using Formula (8), and then the volume content provided in PFC2D is employed to produce each particle group.

$$M(d_1 - d_2)/M_T = \left(\frac{d_1}{R_{\max}}\right)^{3-D} - \left(\frac{d_2}{R_{\max}}\right)^{3-D} \quad (8)$$

Figure 8 shows the particle models of SRM samples with different maximum particle diameters. For each type of SRM sample, numerical tests under four different normal stresses of 200, 400, 600, and 800 kPa are simulated. When the shear displacement of the SRM sample reaches 75 mm, the test is deemed to be over. The simulations are summarized in Table 1.

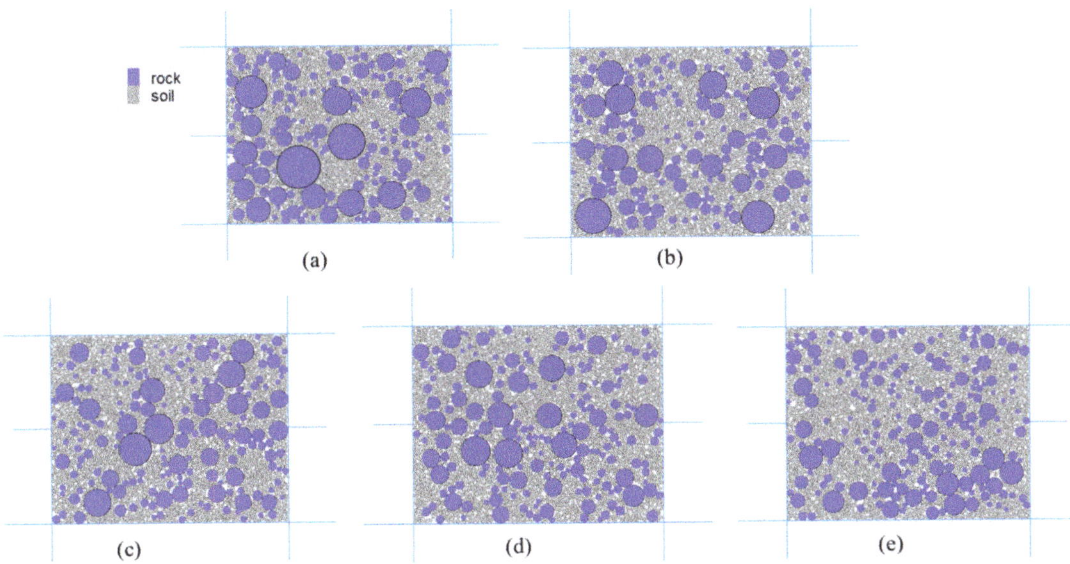

Figure 8. Particle models of SRM samples with different maximum particle diameters. (**a**) S1, (**b**) S2, (**c**) S3, (**d**) S4, and (**e**) S5.

Table 1. Numerical test scheme.

Sample Number	Rock Proportion/%	Fine Particle/%	Maximum Particle Diameter/mm	Normal Stress/kPa
S1	52.9	47.1	100	200, 400, 600, 800
S2	49.3	50.7	80	
S3	47.1	52.9	70	
S4	44.3	55.7	60	
S5	40.9	59.1	50	

4.2. Scale Effect Analysis

4.2.1. Shear Displacement

Figure 9a displays the relationship curve between shear stress and shear displacement as simulated. Under normal stress of 800 kPa, the shear stress increases gradually with the increase of the maximum particle size of SRM: for sample S1 (d_{max} = 100 mm), the peak shear strength is 492 kPa, while for sample S5 (d_{max} = 50 mm), the peak shear strength is reduced to 403 kPa. Under normal stress of 400 or 200 kPa, the change of maximum particle size has relatively little effect on the shear strength of SRM, thus indicating that, with the decrease of normal stress, the relationship is less affected by the maximum particle size in the SRM sample.

Figure 9b illustrates the relationship between vertical displacement and shear displacement as simulated: under normal stress of 200 kPa, the deformation of the SRM samples is similar, and they all show the characteristics of initial shear shrinkage and dilation; with the increase of d_{max}, the dilatancy at the end of the test increases. Under normal stress of 800 kPa, the shear shrinkage of sample S1 is obvious and close to that of sample S5, while shear expansion of sample S3 occurs.

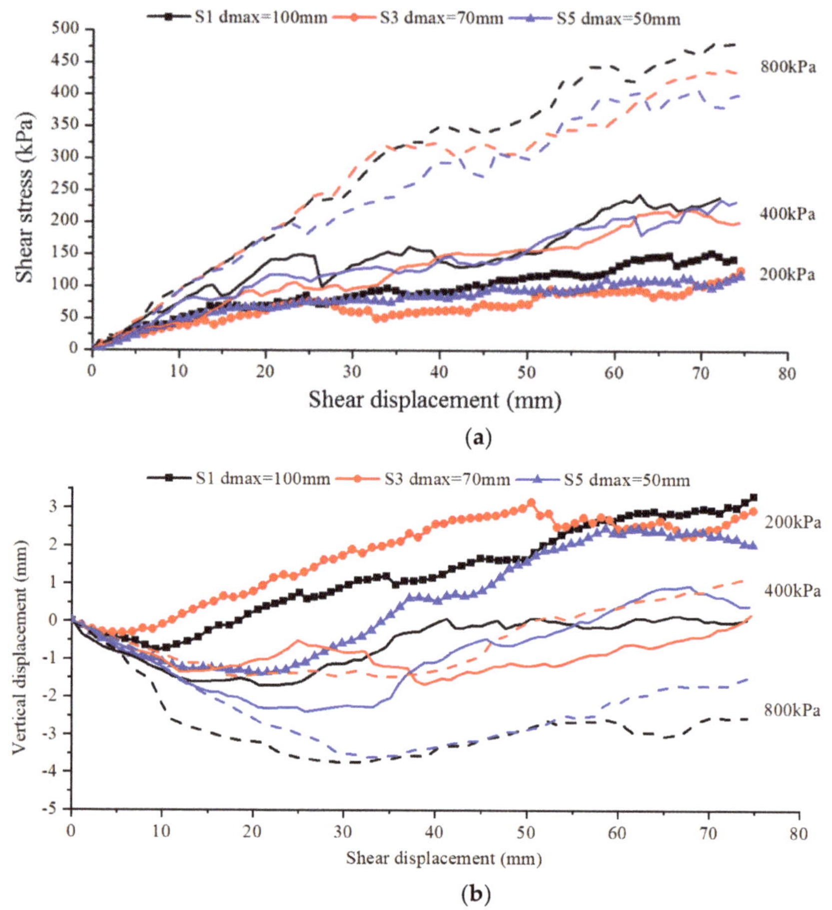

Figure 9. Shear displacements from numerical simulation of SRM samples. (**a**) Relationship between shear stress and shear displacement; (**b**) Relationship between vertical displacement and shear displacement.

4.2.2. Shear Zone

Figure 10 shows the horizontal displacements of SRM samples at the end of the simulations and marks the horizontal displacement of 0 mm as the boundary of the upper shear zone (drawn with a red dotted line). Under different normal stresses, the shape of the shear zone in different SRM samples fluctuates, and the shear zones are mostly distributed in the upper shear box.

As shown in Figure 10a–c, due to the large difference in particle size of sample S1, the shapes of shear zone change little under different normal stresses, and this is mainly controlled by particles with large size; in addition, compared with the case at 200 kPa, the irregularity of shear zone is reduced under normal stresses of 400 and 800 kPa. However, compared with sample S1, the shear zone shapes of sample S5 (Figure 10g–i) are significantly different. Under normal stress of 200 kPa, the fluctuation of the shear zone is more obvious, and the number of soil particles therein is smaller; under normal stress of 800 kPa, the shape of the shear zone is more regular, and the number of soil particles therein increases.

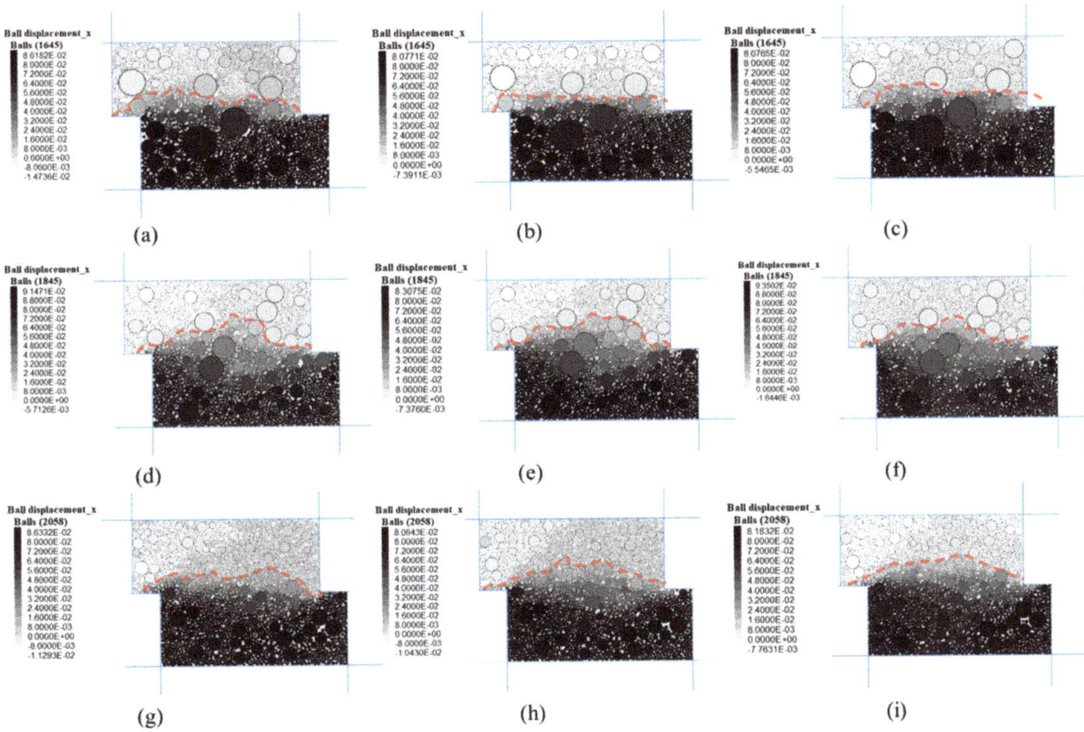

Figure 10. Horizontal displacements of SRM samples at the end of simulations. (**a**) S1-2 $\times$ 10^2 kPa, (**b**) S1-4 $\times$ 10^2 kPa, (**c**) S1-8 $\times$ 10^2 kPa, (**d**) S3-2 $\times$ 10^2 kPa, (**e**) S3-4 $\times$ 10^2 kPa, (**f**) S3-8 $\times$ 10^2 kPa, (**g**) S5-2 $\times$ 10^2 kPa, (**h**) S5-4 $\times$ 10^2 kPa, and (**i**) S5-8 $\times$ 10^2 kPa.

4.2.3. Shear Strength

Figure 11 shows the relationship between the shear strength and d_{max} of SRM samples. As shown in Figure 11a, when d_{max} increases from 50 mm to 100 mm, the corresponding peak strength of the SRM samples under each normal stress increases. Taking the shear strength of sample S5 (d_{max} = 50 mm) as the standard, under normal stress of 800 kPa, the peak strengths of samples S1, S2, S3, and S4 are increased by 21.97%, 15.18%, 9.87%, and 4.31%, respectively.

As shown in Figure 11b, the shear strength indices of the SRM sample, including the cohesion and friction angle, fluctuate with the changes in the d_{max}. The cohesion of SRM samples decreased to 17 kPa at d_{max} = 70 mm, and then it increased to 34 kPa with the increase of d_{max}. The friction angle tended to increase with the d_{max}, from 29.5° to 32.7°.

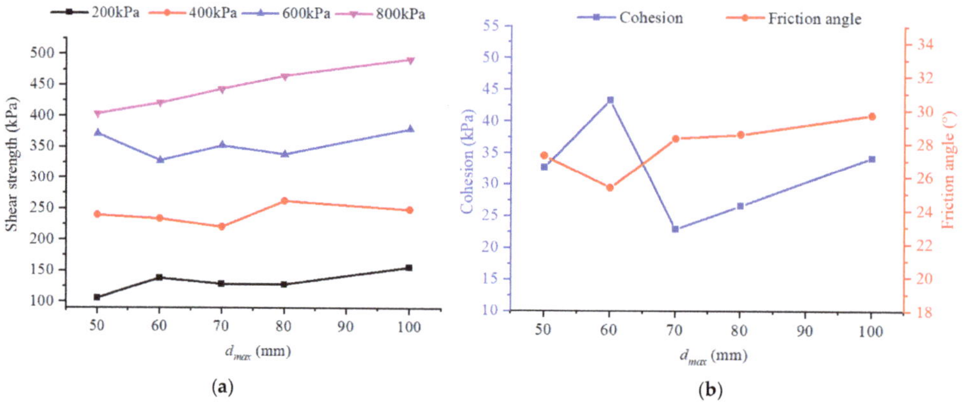

Figure 11. Relationship between the shear strength and the maximum particle size of SRM samples. (**a**) Peak shear stress; (**b**) Strength parameters.

5. Estimation of Shear Strength Parameters of SRM

In fact, apart from the fractal dimension calculated based on the grading curve of the SRM from Zhaizi village in this study, the grading data of the SRM located in Southwest China were collected from the literature [50–54]. Using these grading data, we obtained over 100 fractal dimensions based on the fractal theory. Table 2 lists a part of representative fractal dimensions of the SRM located in Sichuan and Yunnan province, China. Figure 12 presented the frequency histogram of the whole collected fractal dimensions. The average fractal dimension of the SRM is 2.60, and the maximum and minimum values are 2.90 and 2.23, respectively. The results show that the largest frequency of the fractal dimension is in the range of 2.60–2.70, and then it is the range of 2.50–2.60. The frequency of the fractal dimension is in the normal distribution, and the median fractal dimension is 2.62. The fractal dimension utilized in this study for the numerical simulation is 2.67, which is near the most frequency range and the median value according to the discussions above. Therefore, the relationships between the d_{max} and the shear-strength parameters could be considered as the representations of the material distributed in the southwest of China.

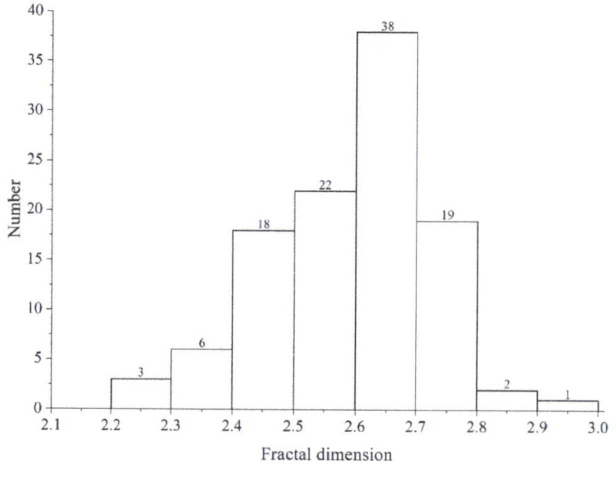

Figure 12. Frequency histogram of the fractal dimensions of the SRM in Southwest China.

Table 2. Representative fractal dimensions of the SRM in China.

Number	Fractal Dimension	Location of the SRM	Resource
1	2.67	A deposit slope in Zhaizi village	This study
2	2.52	Soil–rock mixtures in the Hutiao Gorge area	Xu et al. (2007) [6]
3	2.55	Gravelly soil in Jiangjiagou Ravine in Yunnan	Wei et al. (2008) [50]
4	2.62 2.64 2.59 2.64 2.66	Fujiapingzi, Xiluodu Reservoir Ganhaizi, Xiluodu Reservoir NiuGudang, Xiluodu Reservoir Shuanglongba, Xiluodu Reservoir Shaniwan, Xiluodu Reservoir	Hu (2014) [51]
5	2.68 2.65 2.66 2.59 2.69 2.65 2.64	Zhangmu soil–rock mixture deposit in Southwestern China	Gao et al. (2014) [52]
6	2.85	Soil–rock mixture of Nuozhadu hydropower station in Yunnan	Zhang et al. (2016) [46]
7	2.66	An open-pit limestone mine in Esheng, Sichuan Province	Ma et al. (2019) [53]
8	2.75 2.75 2.74 2.75 2.45	Accumulated crushed stone soil in Deqin County, Yunnan	Tu et al. (2019) [54]

To establish the relationship between the d_{max} and the shear strength indices of SRM, the particle size of sample S4 (d_{max} = 60 mm) is taken as the reference particle size. The particle size ratios of SRM samples can be calculated by using their corresponding d_{max} divided by 60 mm. The particle-size ratios of samples S1, S2, S3, S4, and S5 are 1.67, 1.33, 1.17, 1, and 0.83, respectively. Then, the cohesion and friction angle of sample S4 is selected as a reference, the functional relationship between strength parameter ratio and particle size ratio is established. As shown in Figure 13, both the cohesion and friction angle ratio present the complex variation law as the particle-size ratio, x, increases. When $x \leq 1.17$ ($d_{max} \leq 70$ mm), the strength parameters show a parabolic trend with the x increases with an obvious variation. After d_{max} larger than 70 mm, the responses of the parameters keep increasing in a nearly linear trend. In this study, the relationships between the strength parameters and particle size ratio were expressed by using the piece-wise functions:

$$c/c_0 = \begin{cases} -12.9x^2 + 25.13x - 11.23 & 0.83 \leq x \leq 1.17 \\ 0.52x - 0.08 & 1.17 < x \leq 1.67 \end{cases} \quad (9)$$

$$\varphi/\varphi_0 = \begin{cases} 3.45x^2 - 6.78x + 4.32 & 0.83 \leq x \leq 1.17 \\ 0.045e^{0.81x} + 1 & 1.17 < x \leq 1.67 \end{cases} \quad (10)$$

Using Formulas (9) and (10), we can estimate the cohesion and friction angle of SRM at different scales on the basis of obtaining the shear strength parameters of SRM with the reference particle size by large-scale direct shear test. The result provides a reference for the quantitative evaluation of mechanical properties of geomaterials at the engineering scale.

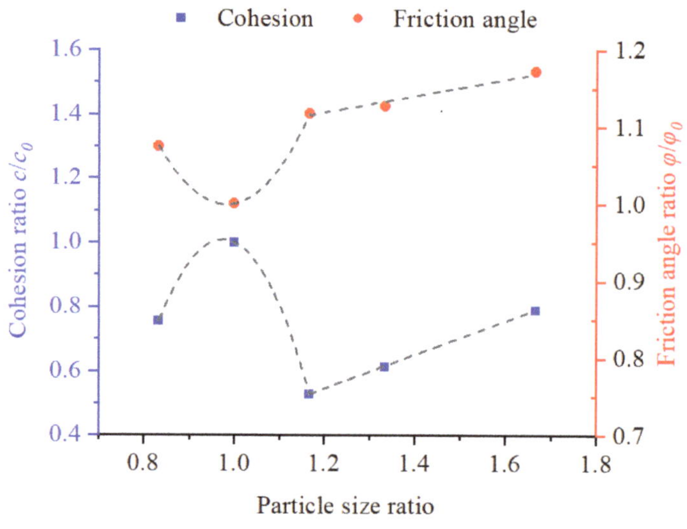

Figure 13. Relationship between strength-parameters ratio and particle-size ratio.

6. Conclusions

In this study, the fractal theory was invoked to explore the fractal characteristics of the particle distribution of SRM; the particle gradation curves of SRM at different scales were predicted by using the fractal dimension and relative threshold; the numerical large-scale direct shear tests were simulated by using PFC2D, and the variations in shear strength parameters of SRM with different particle sizes were studied. The main conclusions could be drawn as follows:

(1) Based on the particle gradation data of SRM, the relationship curve between particle cumulative mass percentage lg $(M(r < R)/MT$ and particle size lg R in the double logarithmic coordinate system can be drawn, whether the SRM has a fractal structure can be judged and the corresponding fractal dimension can be obtained. The case study shows that the SRM in Zhaizi village has a fractal structure, and the fractal dimension is 2.67. In addition, there is an obvious double-fractal structure in the SRM, with a maximum particle size of 60 mm, and the threshold between soil and rock blocks of the SRM in this area was determined to be 2 mm.

(2) Based on the self-similarity and fractal structure of the natural SRM, the grading curve and the particle group mass under the condition of different maximum particle diameters could be obtained. This method not only can expand the study range of particle size of SRM, but also overcome the deficiency of conventional direct shear tests, which remain limited by sample size.

(3) The increase of the maximum particle size leads to a more obvious positive influence on the shear stress. The shear contraction at the beginning of shearing then becomes dilated under the low normal stress with the fluctuation in the shear zone. With the increase of the particle size ratio, the friction-angle ratio increases. Meanwhile, the cohesion ratio increases up to a particle-size ratio of 1.3 and then decreases.

(4) In Southwest China, the frequency of the fractal dimension of the SRM is in the normal distribution, and the median fractal dimension is 2.62. In the range of this fractal dimension, by taking the particle diameter of 60 mm as a standard value, the piece-wise functional relationship between the strength-parameter ratio and particle-size ratio is established. When particle size is less than 70 mm, the strength parameters show a parabolic trend and the particle size increases. When the particle size is larger

than 70 mm, the responses of the strength parameters keep increasing in a nearly linear trend.

Author Contributions: Conceptualization, Z.Z.; methodology, Q.S.; validation, X.F. and D.Y.; formal analysis, H.D. and F.C.; writing—original draft preparation, X.F.; supervision, X.F. and Q.S.; funding acquisition, X.F. and D.Y. All authors have read and agreed to the published version of the manuscript.

Funding: This research was funded by the National Natural Science Foundation of China (No. 52179117) and the Youth Innovation Promotion Association CAS (No. 2021325).

Data Availability Statement: The data are available from the corresponding author upon request.

Conflicts of Interest: The authors declare no conflict of interest.

References

1. Medley, E.; Goodman, E. Estimating the block volumetric proportions of melanges and similar block-in-matrix rocks (Bimrocks). In Proceedings of the 1st North American Rock Mechanics Symposium, Austin, TX, USA, 1–3 June 1994.
2. Medley, E. The Engineering Characterization of Melanges and Similar Block-In-Matrix Rocks (Bimrocks). PhD Thesis, University of California, Berkeley, CA, USA, 1994.
3. Lindquist, E.S.; Goodman, E. Strength and deformation properties of a physical model melange. In Proceedings of the 1st North American Rock Mechanics Symposium, Austin, TX, USA, 1–3 June 1994.
4. Lindquist, E.S. The Strength and Deformation Properties of Melange. PhD Thesis, University of California, Berkeley, CA, USA, 1994.
5. You, X. *Stochastic Structural Model of the Earth-Rock Aggregate and Its Application*; Northern Jiaotong University: Beijing, China, 2001.
6. Xu, W.J.; Hu, L.; Tan, J. Some geomechanical properties of soil–rock mixtures in the Hutiao Gorge area, China. *Geotechnique* **2007**, *57*, 255–264. [CrossRef]
7. Xu, W.-J.; Xu, Q.; Hu, R.-L. Study on the shear strength of soil–rock mixture by large scale direct shear test. *Int. J. Rock. Mech. Min.* **2011**, *48*, 1235–1247.
8. Editorial Committee of Handbook of Engineering Geology. *Handbook of Engineering Geology*; China Building Industry Press: Beijing, China, 1982.
9. Wang, S.; Li, Y.; Gao, X.; Xue, Q.; Zhang, P.; Wu, Z. Influence of volumetric block proportion on mechanical properties of virtual soil-rock mixtures. *Eng. Geol.* **2020**, *278*, 105850. [CrossRef]
10. Zhang, Z.; Sheng, Q.; Fu, X.; Zhou, Y.; Huang, J.; Du, Y. An approach to predicting the shear strength of soil-rock mixture based on rock block proportion. *Bull. Eng. Geol. Environ.* **2020**, *79*, 2423–2437. [CrossRef]
11. Marachi, N.D. Strength and Deformation Characteristics of Rockfill Materials. PhD Thesis, University of California, Berkeley, CA, USA, 1969.
12. Afifipour, M.; Moarefvand, P. Mechanical behavior of bimrocks having high rock block proportion. *Int. J. Rock. Mech. Min.* **2014**, *65*, 40–48. [CrossRef]
13. Kalender, A.; Sonmez, H.; Medley, E.; Tunusluoglu, C.; Kasapoglu, K.E. An approach to predicting the overall strengths of unwelded bimrocks and bimsoils. *Eng. Geol.* **2014**, *183*, 65–79. [CrossRef]
14. Wang, Y.; Li, X. Experimental study on cracking damage characteristics of a soil and rock mixture by UPV testing. *Bull. Eng. Geol. Environ.* **2015**, *74*, 775–788. [CrossRef]
15. Avnir, D.; Farin, D.; Pfeifer, P. Surface geometric irregularity of particulate materials: The fractal approach. *J. Colloid Interface Sci.* **1985**, *103*, 112–123. [CrossRef]
16. Turcotte, D.L. Fractals and fragmentation. *J. Geophys. Res. Atmos.* **1986**, *91*, 1921. [CrossRef]
17. Goldstein, G.; Benaroya, H. Fractal modelling of one-dimensional layered medium. *Appl. Math. Model* **1989**, *13*, 555–558. [CrossRef]
18. Jafari, F.; Asgari, M.S.; Pishkoo, A. The fractal calculus for fractal materials. *Fractal Fract.* **2019**, *3*, 8. [CrossRef]
19. Sun, Y.F.; Sumelka, W. Fractional viscoplastic model for soils under compression. *Acta Mechanica* **2019**, *230*, 3365–3377. [CrossRef]
20. Sarkar, D.; Knig, D.; Goudarzy, M. The influence of particle characteristics on the index void ratios in granular materials. *Particuology* **2019**, *46*, 1–13. [CrossRef]
21. Sun, Y.F.; Sumelka, W.; Gao, Y.F. Reformulated fractional plasticity for soil-structure interface. *Mech. Res. Commun.* **2020**, *108*, 103580. [CrossRef]
22. Bohaienko, V.; Bulavatsky, V. Fractional-fractal modeling of filtration-consolidation processes in saline saturated soils. *Fractal Fract.* **2020**, *4*, 59. [CrossRef]

23. Sun, Y.F.; Chen, C. Fractional order creep model for coral sand. *Mech. Time Depend. Mater.* **2019**, *23*, 465–476. [CrossRef]
24. McDowell, G.R.; Bolton, M.D.; Robertson, D. The fractal crushing of granular materials. *J. Mech. Phys. Solids* **1996**, *44*, 2079–2101. [CrossRef]
25. Yang, Z.Y.; Taghichian, A.; Huang, G.D. On the applicability of self-affinity concept in scale of three-dimensional rock joints. *Int. J. Rock. Mech. Min.* **2011**, *48*, 1173–1187. [CrossRef]
26. Wu, L.; Zhu, S.; Wang, Y.; Wei, K.; Lu, C. A modified scale method based on fractal theory for rockfill materials. *Eur. J. Environ. Civ. Eng.* **2014**, *18*, 106–127. [CrossRef]
27. Luo, M.; Glover, P.W.J.; Zhao, P.; Li, D. 3D digital rock modeling of the fractal properties of pore structures. *Mar. Pet. Geol.* **2020**, *122*, 104706. [CrossRef]
28. Li, X.; Luo, M.; Liu, J. Fractal characteristics based on different statistical objects of process-based digital rock models. *J. Pet. Sci. Eng.* **2019**, *179*, 19–30. [CrossRef]
29. Sun, Y.F.; Sumelka, W. State-dependent fractional plasticity model for the true triaxial behaviour of granular soil. *Arch. Mech.* **2019**, *71*, 23–47.
30. Basirat, R.; Goshtasbi, K.; Ahmadi, M. Determination of the fractal dimension of the fracture network system using image processing technique. *Fractal Fract.* **2019**, *3*, 17. [CrossRef]
31. Zhou, Y.Q.; Sheng, Q.; Li, N.N.; Fu, X.D. The dynamic mechanical properties of a hard rock under true triaxial damage-controlled dynamic cyclic loading with different loading rates: A case study. *Rock Mech. Rock Eng.* **2022**, *55*, 1–22. [CrossRef]
32. Sun, Y.F.; Gao, Y.F.; Chen, C. Critical-state fractional model and its numerical scheme for isotropic granular soil considering state dependence. *Int. J. Geomech.* **2019**, *19*, 4019001. [CrossRef]
33. Tao, G.; Wu, Z.; Li, W.; Li, Y.; Dong, H. Simplified relation model of soil saturation permeability coefficient and air-entry value and its application. *Fractal Fract.* **2021**, *5*, 180. [CrossRef]
34. Sun, Y.F.; Sumelka, W.; Gao, Y.F.; Nimbalkar, S. Phenomenological fractional stress-dilatancy model for granular soil and soil-structure interface under monotonic and cyclic loads. *Acta Geotechnica* **2021**, *16*, 3115–3132. [CrossRef]
35. Zhao, Z.; Ni, X.; Cao, Y.; Shi, Y. Application of fractal theory to predict the coal permeability of multi-scale pores and fractures. *Energy Rep.* **2021**, *7*, 10–18. [CrossRef]
36. Sui, L.; Yu, J.; Cang, D.; Miao, W.; Wang, H.; Zhang, J.; Yin, S.; Chang, K. The fractal description model of rock fracture networks characterization. *Chaos Solitons Fractals* **2019**, *129*, 71–76. [CrossRef]
37. Xiao, Y.; Meng, M.; Daouadji, A.; Chen, Q.; Wu, Z.; Jiang, X. Effects of particle size on crushing and deformation behaviors of rockfill materials. *Geosci. Front.* **2020**, *11*, 375–388. [CrossRef]
38. He, S.H.; Ding, Z.; Hu, H.-B.; Gao, M. Effect of grain size on microscopic pore structure and fractal characteristics of carbonate-based sand and silicate-based sand. *Fractal Fract.* **2021**, *5*, 152. [CrossRef]
39. Xu, W.J.; Yue, Z.Q.; Hu, L. Study on the mesostructure and mesomechanical characteristics of the soil-rock mixture using digital image processing based finite element method. *Int. J. Rock. Mech. Min.* **2008**, *45*, 749–762. [CrossRef]
40. Wu, L.Q.; Zhu, S.; Wei, K.M.; Lu, Y. The fractal properties of scale effect on the density of rock-fill materials. *Adv. Mat. Res.* **2013**, *706–708*, 520–525. [CrossRef]
41. Pi, Z.; Zhou, Z.; Li, X.; Wang, S. Digital image processing method for characterization of fractures, fragments, and particles of soil/rock-like materials. *Mathematics* **2021**, *9*, 815. [CrossRef]
42. Mandelbrot, B.B. The fractal geometry of nature. *Am. J. Phys.* **1983**, *51*, 286–287. [CrossRef]
43. Tyler, S.W.; Wheatcraft, S.W. Fractal scaling of soil particle-size distributions: Analysis and limitations. *Soil Sci. Soc. Am. J.* **1992**, *56*, 362–369. [CrossRef]
44. Fu, X.; Zhang, Z.; Sheng, Q.; Zhou, Y.; Huang, J.; Wu, Z.; Liu, M. Applications of an innovative strength parameter estimation method of the soilrock mixture in evaluating the deposit slope stability under rainfall. *Front. Earth Sci.* **2021**, *9*, 868. [CrossRef]
45. Zhou, W.; Wang, D.; Ma, G.; Cao, X.; Hu, C.; Wu, W. Discrete element modeling of particle breakage considering different fragment replacement modes. *Powder Technol.* **2020**, *360*, 312–323. [CrossRef]
46. Zhang, Z.L.; Xu, W.J.; Xia, W.; Zhang, H.Y. Large-scale in-situ test for mechanical characterization of soil–rock mixture used in an embankment dam. *Int. J. Rock. Mech. Min.* **2016**, *86*, 317–322. [CrossRef]
47. Shi, D.D.; Cao, D.; Deng, Y.B.; Xue, J.F. DEM investigations of the effects of intermediate principal stress ratio and particle breakage on the critical state behaviors of granular soils. *Powder Technol.* **2021**, *379*, 547–559. [CrossRef]
48. Cundall, P.A.; Strack, O.D.L. A discrete numerical model for granular assemblies. *Geotechnique* **1979**, *29*, 47–65. [CrossRef]
49. Itasca Consulting Group. *PFC2D (Particle Flow Code in 2 Dimensions) Use's Guide: Verification Problems and Example Applications;* Itasca Consulting Group: Minneapolis, MN, USA, 2008.
50. Wei, H.Z.; Wang, R.; Hu, M.J.; Zhao, H.Y.; Xu, X.Y. Strength behavior of gravelly soil with different coarse-grained contents in Jiangjiagou. *Rock Soil Mech.* **2008**, *29*, 48–51.
51. Hu, W. *Experimental Study on Shear Strength of Soil-Rock-Mixture in Xiluodu Reservoir;* Institute of Rock and Soil Mechanics, Chinese Academy of Sciences: Wuhan, China, 2014.
52. Gao, W.; Hu, R.; Oyediran, I.A.; Li, Z.Q.; Zhang, X.Y. Geomechanical characterization of Zhangmu soil-rock mixture deposit. *Geotech. Geol. Eng.* **2014**, *32*, 1329–1338. [CrossRef]

53. Ma, C.; Zhan, H.B.; Zhang, T.; Yao, W.M. Investigation on shear behavior of soft interlayers by ring shear tests. *Eng. Geol.* **2019**, *254*, 34–42. [CrossRef]
54. Tu, G.X.; Huang, D.; Huang, Q.; Deng, H. Effect of locally accumulated crushed stone soil on the infiltration of intense rainfall: A case study on the reactivation of an old deep landslide deposit. *Bull. Eng. Geol. Environ.* **2019**, *78*, 4833–4849. [CrossRef]

MDPI
St. Alban-Anlage 66
4052 Basel
Switzerland
Tel. +41 61 683 77 34
Fax +41 61 302 89 18
www.mdpi.com

Fractal and Fractional Editorial Office
E-mail: fractalfract@mdpi.com
www.mdpi.com/journal/fractalfract

www.ingramcontent.com/pod-product-compliance
Lightning Source LLC
Chambersburg PA
CBHW061846210926
51850CB00006B/984
* 9 7 8 3 0 3 6 5 7 0 5 4 9 *